The Vanishing Mew Gull

THE VANISHING
MEW GULL

A Guide to the Bird Names of the Western Palearctic

RAY REEDMAN

PELAGIC PUBLISHING

First published in 2024 by
Pelagic Publishing
20–22 Wenlock Road
London N1 7GU, UK

www.pelagicpublishing.com

*The Vanishing Mew Gull: A Guide to the Bird
Names of the Western Palearctic*

https://doi.org/10.53061/XNWK2436

British Library Cataloguing in Publication Data
A catalogue record for this book is available from the British Library

ISBN 978-1-78427-462-7 Hbk
ISBN 978-1-78427-463-4 ePub
ISBN 978-1-78427-464-1 PDF

Typeset in Minion Pro by BBR Design

Cover photograph: Mew (Common) Gulls *Larus canus*
© Erlend Haarberg/naturepl.com

For Paul, Lynne-Marie and Justin

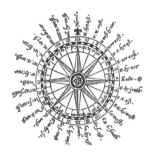

Bag, Barrel Tit, Bum Barrel, Bottle Tit, Bottle Jug, Bush Oven,
Bum Towel, Feather Poke, Poke Bag, Poke Pudding…

Just some of the now-disused local English names of the
Long-tailed Tit recorded by Lockwood, 1984.

Nomenclature is proverbially a vexed subject, but there is one
necessity, which all, however diverse their views, must agree
is of the first importance – the necessity for uniformity.

– Hartert, Jourdain, Ticehurst and Witherby, 1912.

From somewhere deeper in the heart of the wood
came the droning of ring-doves.

– George Orwell, *Nineteen Eighty-Four*

Contents

MOCKINGBIRDS TO CHATS ETC.

DIPPERS TO PIPITS

FINCHES TO CARDINALS

PREFACE

Some nineteenth-century explorers were collecting specimens of natural history in the forests of the north. A sharp-eyed young midshipman raised his gun and brought down a bird that nobody had ever seen before. It had purple-tinged grey plumage and a short bill.

A few months later, a debate raged in the back rooms of a national museum. In a welter of passion, politics and infighting, the new bird was discussed and analysed to the last tip of its tiny feathers. A weary chairman finally announced with a flourish that this newly discovered species now truly existed in the world of science because at last it had – A NAME! A clerk solemnly attached a neatly penned label to the dead bird's leg.

Meanwhile, back in the forest, as summer drew to an end, a large gathering of small, short-billed, blue-grey birds prepared to migrate in a sociable flock. Two old acquaintances shared a branch while they waited.

'Good to see you again, Jim. I see Cousin Fred didn't make it this year. Shame!'

All this is a facetious way of saying that the process of naming birds may matter a great deal to us humans, but that our perspectives on the topic do not matter one jot to the birds themselves. It also underlines the fact that the world of science has taken control of bird names, replacing their historical spontaneity and colour with a form of labelling that is often quite utilitarian. As David Attenborough put it, 'the first task of ornithology was to name birds'. Cousin Fred probably finished up with a boringly descriptive English name, and with a scientific name that has already been reviewed and changed three times in accordance with taxonomic advances.

What started me off on the current analysis was just that fact: the scientists analyse the DNA and juggle speciation, change names and leave the rest of us feeling a bit puzzled. In my previous book, *Lapwings, Loons and Lousy Jacks*, I took a broad look at how names came about and what they meant, particularly in my own birding life, but my framework there was deliberately not a taxonomical one. In writing this new book, I felt that it was time to look at names from that more scientific perspective, with reference to the birds that occur in the biogeographic zone most relevant to British birdwatchers.

In *Lapwings, Loons and Lousy Jacks*, I looked at the processes that cause bird names to be formed and to change. One section concluded with the comment that 'it still seems that the maintenance of bird names is an ongoing process: indeed, the IOC (International Ornithological Congress) *World Bird List* website speaks of "the active industry of taxonomic revision", while its list of principles governing English-language names hints at areas of future revision. Maybe we are closer to a definitive set of names, but I am willing to wager that a hundred years from now, they will probably look different again.'

When I revisited that comment, just five years after publication, I saw that the IOC had already noted many changes in the classification and naming of species: many scientific names had changed and a number of English-language names too. Among those modifications, I was relieved to see that the name Common Gull was to be formally reinstated on the IOC World List; that Mew Gull had been dropped; and that the name Short-billed Gull had been taken up in America. I had always felt that the previous imposition of Mew Gull as the world name had not been a wise decision: it was, after all, the name of the American subspecies, and it was tautological, so in a way good riddance!

But then I looked a bit more closely at those three names...

• Common Gull is a dreadfully dull name, of course, with all the charm of a well-worn boot sock, but we are used to it and put up with it, in spite of the fact that it is generally inaccurate. Perhaps the attraction lies in the irony of the fact that the bird is not that common.

• Short-billed Gull is not the new name that it appears to be: it resurrects one created by the Scottish explorer and naturalist John Richardson (1787–1865), who described that bird in 1831 under the name Short-billed Mew Gull.

• The use of the simpler Mew Gull evolved from that over a period of time. As for that name, W.B. Lockwood pointed out in the 1984 *Oxford Book of British Bird Names* that '[Mew] is in fact the native English name for a Sea Gull, going back via Middle English *mēw* to Old English *mēw*, *mǽw*. It remained in ordinary literary use until the seventeenth century, when it was largely ousted by the newcomer GULL.'

Seen in the torchlight of that statement, the last name has a completely different value. On the one hand, it is apparently as old as the Lindisfarne Gospels: it was used and understood by migrating Anglo-Saxons and by marauding Vikings. On the other hand, it is a historical symbol of a much later migration, one in which colonists took with them aspects of the English language that would fall into disuse in the 'old country'. About 150 years after those first American settlers sailed off, *Mew* was formally dropped in Britain and was replaced by the Celtic word *Gull*. If the word were an artefact it would be treasured in a museum for either of two substantial historical reasons, but now, if not in the actual waste-bin, it has yet again been quietly pushed to the back of a cupboard.

The Vanishing Mew Gull as a title thus symbolises the many changes of recent years. The action of 'splitting' former subspecies to create full species has been a fairly frequent occurrence. In the same timescale, a smaller number of changes have reversed earlier 'splits': we have, for example, acquired Eastern and Western Bonelli's Warblers, while Eastern and Western Ospreys are once more embraced (or 'lumped' in birder parlance) by the single word Osprey, having been denied their rather brief separate species status. Genetic analysis has also resulted in some substantial changes to genera: among European species, for instance, only Blackcap and Garden Warbler have survived the demolition of the genus *Sylvia*, while the Lesser Whitethroat, long treated as the type-species for the *Sylvia* warblers, has now not only donated its specific name to a whole new genus of *Curruca* warblers, but also is apparently poised to gather to its bosom the former splits of Hume's and Desert Whitethroats.

So what is going on?

The website of the British Trust for Ornithology (BTO) underlines the fact that nothing is as simple as it seems; science is necessarily fluid:

• Science uncovers new relationships between taxa, or disproves others.

• New species are discovered, even among birds, in the field or in museum collections.

• Consensus changes on where the boundaries between species should be drawn.

• Expert opinion changes on how data should be interpreted and uncertainties managed.

Again, the BTO explains that 'in a "taxonomic" or "systematic" list of birds, species are placed in a linear sequence according to how long ago they are thought to have evolved, oldest first. Related taxa are grouped together' (www.bto.org, 2022).

This has long been the convention; the developing knowledge-base of a given era partly accounts for the fact that reference books of different vintages show different sequencing. However, field guides need to be arranged somewhat artificially so that similar but unrelated birds can be compared: it makes eminent sense to place the 'crests' next to the quite similar leaf warblers, even if they are no longer considered particularly close relatives. In recent decades, changes to taxonomy have been more radical and more frenetic than previously because the tools of DNA analysis have allowed scientists to focus on speciation in a new way. In 2013, Nigel Collar wrote a thought-provoking article in *British Birds* entitled 'A species is whatever I say it is'. In that, he underlined the fact that the science does not have hard edges: there has to be room for differing opinions and for the use of different tools.

In short, taxonomic sequencing has been revised radically and some surprising results have emerged, especially in the fact that falcons are not at all closely related to the other 'birds of prey'. Several new species have been proposed and new names created: even the familiar Robin seems set to branch out, while a 'brand new' wheatear bears the name of a librarian at the British Museum of Natural History.

In addition to all of this, there have been some radical changes at organisational levels. In 2016, the British Ornithologists' Union (BOU) announced its decision to align with the taxonomic decisions taken by the IOC, rather than pursue its own programme. It still maintains the British list, of course, which has its own versions of some English names in parallel to the IOC version. This decision removed doubts about the status of a few disputed species splits. Also in 2016, the American Ornithologists' Union (AOU), long the accepted authority on American nomenclature, amalgamated with the California-based Cooper Ornithological Society to become the American Ornithological Society (AOS).

As I focused on these issues, the case for an up-to-date comprehensive reference grew in my mind and this compilation began to evolve. If the Mew Gull could tell a story, so could a good many other species…

Ray Reedman

ACKNOWLEDGEMENTS

After the publication in 2016 of *Lapwings, Loons and Lousy Jacks*, I was genuinely surprised by the number and the geographical spread of the reviews and critiques. Their diversity was particularly interesting, and some useful lessons were gleaned from the experience. The positive responses were encouraging, but a number were good for my humility.

The concept of this book had been around for a while, but I probably would never have started had it not been for the Covid lockdowns and unrelated matters that combined to curtail other activities, including my work with the Berkshire Ornithological Club. In short, I soon needed a new and rather more static focus.

It cannot be too strongly stated that I could not have sustained the effort and concentration needed without the support and feedback of my wife, Mary. She has been alongside me ever since our school and university days and somehow puts up with my spells of intensity and focus.

At the same time, the dedication to our son, Paul, and to our daughter and son-in-law, Lynne-Marie and Justin, is in appreciation of the support they have given us, particularly in recent years through the difficult times in which this work began to develop. Our grand-children, Robert, Jade and Amber, have been there for us too.

Both Jade and Lynne-Marie contributed some useful perspectives, and the former did a bit of critical reading, as did my friend Gray Burfoot. Gray also ensured that I finally caught up with the elusive Squacco Heron.

I must now turn to those commentators, past and present, whose works have provided the material that I needed. The most accessible works on bird names in general are those of William Lockwood, James Jobling and Ernest Choate, and I acknowledge the various permissions to use these sources. Together they illustrate very different perspectives on the examination and recording of bird names. I have referred to all three regularly, and with great respect for their individual skills and authority. A great many other authors, past and present, have contributed information, as have the many authoritative websites available online. There is a list of resources at the end that give details, though I fear that the list is far from exhaustive.

Finally, I appreciate the faith of Nigel Massen in taking on the publication. It is a nightmare of detail for the author and must be even more so for the proof-reader and fact-checker. My thanks to the Pelagic team and in particular to David Hawkins who has steered this project through the processes of editing and production.

ABOUT THIS BOOK

RATIONALE

The birds themselves are central to the concept: each bird is depicted via its names and other supporting material. In that respect this guide serves as an inverted dictionary, the organisation of which keeps all aspects of the names with the bird, rather than spread across the alphabetical sequencing of the traditional dictionary. The bird appears next to its closest relatives and in the taxonomic sequencing currently agreed by the IOC.

Bearing in mind that there are well over 10,000 species of bird in the world, it seemed wise to limit the scope of this project to the Western Palearctic (see definition below). Even that limited target yields almost 1,100 species in just over 100 families. That is in part because a much wider geographical area is represented, since many vagrant birds stray in from outside the region, while many otherwise extralimital species occur as the result of deliberate introduction or of accidental escape. The list therefore becomes a much wider cross-section of the birds of the world: Asia, Africa, the Americas, and even Australia and the Southern Ocean are represented to a greater or lesser degree.

The structural starting point is a list of the birds of the Western Palearctic based on version 13.2 of the IOC list. This is used as a reference base for both content and for taxonomic sequencing. However, that list is treated flexibly, since some of the introduced birds that have been included in this text do not appear on it. What counts in this interpretation is that the birds do occur as free-flying individuals within the region in some form or another: they may be breeding native species; seasonal migrants; escapes with a permanent or an ephemeral population; introduced gamebirds; irregular visitors; or the scarcest of vagrants. A brief note indicates the bird's general status within the region.

STRUCTURE

The bird families and species at the centre of the structure are arranged according to the IOC's current taxonomic sequencing, which is in keeping with the 2016 decision made by the BOU to harmonise with that particular international body on taxonomy. The rationale behind the taxonomic approach is that it places birds in an evolutionary order (as currently understood) and treats individual species in conjunction with their closest relatives. That also permits some general discussion of each family.

Conventional dictionaries are wordcentric, and therefore disperse the component parts of current and alternative names across the alphabet. This guide places the birds themselves at centre stage. That allows an integrated discussion of the names, and offers scope for some supplementary material too.

Each family may be introduced in a short note and then each species is outlined as follows:

- The first comment briefly outlines the bird's relevance to Britain, to Europe or to the wider Western Palearctic.

- The second element discusses aspects of both vernacular and scientific names. Names that have widespread use, such as woodpecker, may be discussed in a note at the start of the family section or in the first relevant example in a sequence. Unnecessary repetition is avoided.

- Eponyms (names based on people) offer minimum data on the person concerned, each noting dates, nationality and occupation or status. These are collected in a simple biographical index that brings together examples of multiple dedications. That list also includes a brief context for others mentioned in the accounts.

- Further notes may include elements of supplementary interest, including some synonyms and some archaic forms, though those elements are not treated exhaustively. Clearly not all names justify a great deal of comment or explanation, so some entries are more succinct than others.

A typical species entry works as follows:

Cinereous Vulture *Aegypius monachus*

This bird breeds in Spain in small numbers and has been recorded as far north as Norway, though not in Britain.

Cinereous literally means 'ash-grey', though this seems to apply to the hood and neck of the adult, since the rest of the bird is in fact far darker (see *Black* below).

Vulture evolved via Old French from the Latin *vultur, voltur*.

Aegypius appears as a character in the same Greek legend as Neophron (see Egyptian Vulture).

The specific name *monachus* means 'monk', an image created by the hooded appearance of the adult.

* This bird was formerly known as the *Black Vulture* (sometimes with *Eurasian* attached), but the current name is now preferred in order to avoid confusion with the American species, which now formally owns that simple form.

* The name *Monk Vulture* appeared for a while, obviously based on *monachus*, but with the Hooded Vulture also carrying the same specific name (though in a different genus) it had potential to confuse.

USING THIS GUIDE

It is fairly obvious that about 1,100 names produce an amorphous mass of data. In any case, the disadvantage of the birdcentric approach is that any one name will have to be sought via the index. To help the user find a species entry more readily, the contents list at the beginning has a series of divisions and the families are numbered, so it serves as a quick locator. Each division is marked by a relevant illustration to support a quick search.

A DEFINITION OF THE WESTERN PALEARCTIC

Scientists divide the world's surface into eight biogeographic regions. These do not correlate exactly to geopolitical boundaries. The region of the Western Palearctic is generally accepted to include the whole of Europe, parts of the Middle East, much of North Africa and a large part of the North Atlantic, particularly those archipelagos that are governed by, or affiliated to, European countries.

The map here shows that the region's eastern boundary is approximated by the Ural Mountains, the Caspian Sea and the border between Turkey and Iran down to the Persian Gulf. A line drawn across North Africa includes the western Sahara and the whole of each country with a Mediterranean coast, but also takes in northern parts of Mauretania, Mali, Niger and Chad, with a rather more northerly line drawn across Saudi Arabia: the Sahara and Arabian deserts form a more general barrier than do any political boundaries.

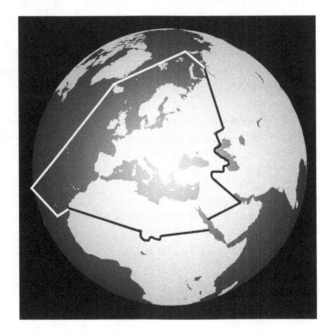

For their inclusion into the formal list for this region, birds must have been recorded in the wild at least once, though the exact status of some species may differ locally or be in a state of flux. The reasons for the birds being included here are complex, and that complexity will emerge from the notes for each species. It would be foolish to think that the lists are definitive: even without the catalyst of taxonomic revision, factors as diverse as climate change and sheer chance make it inevitable that new species will be added in future.

(I am very conscious of the fact that, in the following section, there is some reiteration of material covered in *Lapwings, Loons and Lousy Jacks*. Obviously I could not assume that everyone had read that material. Any repetition is therefore as concise as possible.)

THE LINGUISTIC BACKGROUND

In considering the names of birds, in either the English language or in their scientific forms, it is useful to have some linguistic reference points. English is a complex hybrid language: it blends elements of the Germanic languages with a lesser portion of Romance (Latin-based) elements, both of which groups have their own origins and, perhaps surprisingly, a common root.

Languages are not static: they evolve over many centuries, and with them the names that we use for our birds. It helps to realise that there are links between many of the language elements discussed, so the figure below follows the threads of relationship between the languages of Europe.

The evolution of the name of the Common Crane, *Grus grus*, offers an example of how the elements of a name may evolve. At face value, Crane and *Grus* are unalike, but they do have common roots.

The ancestral word *geranós* is found in Classical Greek. That word compressed with usage into the Latin *grus*, which eventually, via the influence of the Roman Empire, gave rise to the very similar forms *grua*, *grue*, *grulla* and *grou* in Italian, French, Spanish and Portuguese respectively.

The Greek word *geranós* had a very close relative, *kranô*, which was at the root of the Germanic languages. The latter evolved into names such as *Kranich*, *Kranvogel* and *crane* (German, Dutch and English), in the Western Germanic evolution, and to *Trana* and *Trane* (Swedish, Danish/Norwegian) in the Northern Germanic forms.

The relationships of the main European languages

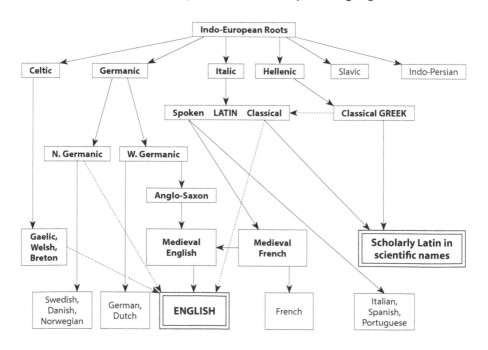

That sort of evolution of language takes place over many centuries and results largely from the usage of the spoken language: one only has to look at Shakespeare to realise just how much English has changed over a few hundred years. The example given here represents well over 2,000 years of linguistic evolution.

THE EVOLUTION OF ENGLISH

The evolution of the English language is a complex story, and the evidence of its various component elements can be found in many bird names.

Pre-Roman Celtic languages survived on the fringes of Roman Britain after the conquest in 43 CE, but the influence of 400 years of Roman occupation of Britannia was virtually erased from the linguistic record by 400 years more of invasion and colonisation by the Angles, Saxons and Jutes from mainland Europe. These spoke variants of what are known as Western Germanic languages. Anglo-Saxon eventually metamorphosed into the language of England. That was eventually heavily modified, particularly in the northern and eastern parts of the country, by a further 200 years of Nordic Viking invasion and settlement, which introduced an element of Northern Germanic into the mix.

By the eleventh century, Old English was an evolved Germanic language. Meanwhile, the Viking settlement in Normandy had espoused the Latin-based French language. After the Norman Conquest of 1066, the English language became heavily infused with the French of the ruling classes, whose culture was constantly reinforced by French alliances and marriages over the next few centuries.

By the end of the Middle Ages, English had become a unique hybrid of Germanic and Romance languages. In the fifteenth century, the Renaissance (the rebirth of Classical languages and learning) swept through Europe. Over the next three centuries, that culture enriched the English language with elements of Greek and of Latin. Meanwhile the politics and culture of England evolved, as Wales, Scotland and Ireland became part of a political union. By the end of the eighteenth century, English had been subtly infused with elements of language originating from those areas too.

The background to these sweeping movements was one in which the ordinary residents of these isles were largely untutored. Language was transmitted predominantly in spoken forms, which might vary from village to village, so local and regional dialects had a lot of influence on the naming of wild creatures, including birds. The eighteenth-century Age of Enlightenment brought a new generation of scientists prepared to study and classify birds, but it also gave rise to Dr Johnson's influential dictionary and to the standardisation of the English language. The eventual formalisation of English bird names is discussed later.

Inevitably, English-language bird names tend to reflect the complex history of the language itself. The pre-Roman Celts are represented by names such as the Cornish gull; the Romans (eventually and indirectly) by pheasant; the Anglo-Saxons by hawk; the Vikings by fulmar; the Norman French by falcon; and the post-Renaissance scholars by gallinule.

SCIENTIFIC BIRD NAMES

The crane illustration given earlier results in a rather disparate list of vernacular names: *grua, grue, grulla grou, Kranich, Kranvogel,* crane, *Trana* and *Trane* – and, of course, the crane had names in many other languages too. That was also the problem with finches, woodpeckers, herons, and a whole range of other birds.

This matter was addressed during the eighteenth century by the Swede, Carl Linnaeus, who rationalised the process by concocting a system that eventually led to the delightfully simple two-word method (the binomial system) of naming all species of living things in a Latinate structure, which used mainly Latin and Greek roots. This was part of a hierarchical framework, which Linnaeus called the *Systema Naturae*. It separated plants from animals (Kingdom); vertebrates from invertebrates (Phylum); mammals from birds (Class); duck-like birds from heron-like birds (Order); herons from spoonbills (Family); herons from bitterns (Genus); and Grey Heron from Purple Heron (Species).

From now on, the familiar European heron would be called *Ardea cineria* and the similar American species *Ardea herodias*. The first name showed that they were close relatives and the second that they were different in some way. What they were called in the different languages of the world was local business, but when a new generation of ornithologists shared notes and learned papers across national boundaries they now had a common currency: scientists of the future could be confident that a discussion of the breeding biology or of the gut parasites of *Ardea cineria* would not be readily confused with those of *Ardea herodias*.

Many scientists adopted the system voluntarily, and it was embedded firmly during the nineteenth century, during which time a number of scientists promoted a further division of species into subspecies, a term that now crops up frequently in the naming of birds. This involves the use of a third name to label the subspecies, such as *Branta bernicla hrota*. That is described as a trinomial system. A full examination of the topic would expand this work considerably, but it is illustrated in the discussion of some species, such as Brent Goose or White Wagtail, where named subspecies may be familiar. It is also increasingly relevant to the discussion of recent or potential 'splits', where subspecies have been upgraded to full species.

The Linnaean system still pertains and has since been formalised. In 1901, biological scientists from all over the world met to establish the International Code of Zoological Nomenclature (the ICZN), which formally adopted the Linnaean System as a structure for the names of all living things. That discipline was not universally adopted, so in 1912 Hartert and his colleagues produced a checklist of British birds to remind their contemporaries that a standard had been agreed. Curiously, that agreement was not reinforced with any real authority until as late as 1961. This is not the place to discuss the full set of rules for the use of the system, but some general observations are appropriate here.

- It seems obvious than no two names should be exactly the same, though some overlap is allowed between the Kingdoms: the word *calendula* is one example that occurs in both plants and birds.

- The rule of precedence crops up a number of times. Under that convention, older names prevail over newer ones, even when there is an attempt to correct an earlier error. This was described on an American website as the 'oldest fool wins'. In short, some names remain plain wrong, some nonsensical, some banal, others inaccurate. The advantage of the rule is one of continuity, but it does mean that the Laughing Gull still retains an erroneous black tail, the result of a small slip of the pen by Linnaeus himself.

- The system tends to use Latin and Greek very loosely, though in a Latinate structure. In any case there are now many examples of non-Latin words, ranging from Turkish, through Russian, Hindi, Japanese, colonial-French patois and much more. Somehow these get woven into the system and mercilessly Latinised to fit the need. Eponyms are little better: some people just do not have the sort of name that is suited to a Latinate form. The advantage of such loose adaptability is that changes or additions to scientific names may be provoked by any scientist, working in any language or any culture.

In summary, the framework of the *Systema* might be considered to be a very workable mechanism, even though it is now clothed by an expedient Babel of languages. It suits its purpose very well. Its great drawback in the modern world is that times have moved on, because other languages now tend to dominate the world of science, and international scholarship is no longer published in Latin. The Latin and Greek forms are not so readily understood by the average bird enthusiast and are probably not a lot clearer to the average scientist. Fortunately, James Jobling's *Dictionary of Scientific Bird Names* makes those names much more accessible.

THE MODERN DEVELOPMENT OF ENGLISH BIRD NAMES

Development of English-language bird names was mainly in the hands of ornithologists from the seventeenth century onwards. Prior to that, names had evolved locally and spontaneously, giving rise to a confusing variety of mostly regional forms. During the seventeenth and eighteenth centuries, a number of important pioneers in British ornithology gathered and collated data on birds and some of their preferred bird names survived them. Figures such as Christopher Merrett, Walter Charleton, Francis Willughby and John Ray laid the foundations of the milestone work of Thomas Pennant in the late eighteenth century. The influence of such successors as John Latham, William Yarrell, George Montagu and several others was also considerable, but it was not until the 1858 foundation of the BOU that some 'authority' was imposed on the English names of birds, under the stern eye of Alfred Newton.

However, a series of English-speaking subcultures in the former colonies of America, Australia, Africa and Asia had begun to influence the naming of birds in different ways. These often duplicated English names when applied to local forms that were similar but unrelated: for example, the Australian Magpie and the Baltimore Oriole belong to different families from their apparent European counterparts. Conversely, Common Loon and Great Northern Diver are the same species, but wearing labels that depend on different cultural heritages.

Since the mid-nineteenth century, the BOU has been the principal domestic authority on the nomenclature of birds in their English forms. The AOU developed an equivalent role and continues that in its new guise as the AOS. The former Royal Australasian Ornithologists Union (now part of BirdLife Australia) and other English speaking areas have their own voices. Apart from the major agencies mentioned earlier, many other non-English-speaking nations also have an interest in bird names in the English language simply because that is often now a common language of the scientific world. Discussions have not always been smooth and it seems likely that elements of macro- and micro-politics make harmonisation more difficult. Towards the end of the twentieth century, some consensus was sought, and the English-speaking nations agreed a forum, within the IOC, which would provide a common set of names to remove any confusion.

THE IOC MASTER LIST

The IOC first met in 1884 as a forum for bird biologists working everywhere. Just over 100 years later, in 1994, the organisation started to work on standardised sets of international names in English, French and Spanish. The IOC Master List now tracks and records agreed changes to scientific names and discusses amendments to English-language names.

Their website's home page explains that: 'Our primary goal is to facilitate worldwide communication in ornithology and conservation based on an up-to-date evolutionary classification of world birds and a set of English names.' This list is updated twice a year, and 2024 saw the launch of version 14.1.

CLASSIFICATION

Consensus is not always easy, and the IOC does not find itself in total harmony with other major agencies in matters of classification or of resultant nomenclature. It is clear that there is sometimes a discrepancy between the decisions of various agencies on acceptance or rejection of recommendations made by scientists working on speciation. This may be down to timing as much as to disagreement. In 2016, the BOU chose to harmonise with the IOC on taxonomic decisions, though they still maintain their own version of some names. To simplify matters for the purposes of this work, the IOC's current decisions are treated as authoritative.

IOC ENGLISH NOMENCLATURE

As emphasised in *Lapwings, Loons and Lousy Jacks*, the various English-speaking nations of the world have their own cultural forces, and there are inevitable differences that need to be reconciled. From a British perspective, we do not always appreciate that Shakespeare was still alive when the English language was transplanted into America. Five centuries of a separate culture and tradition are bound to have fostered differences in usage, and those differences have their own cultural validity. The Mew Gull is a symbol of that aspect. The time span for divergence in other areas, such as Australasia, has generally been shorter, but local cultures, and sometimes a very different avifauna, have factored their own colour into local English-language bird names.

The IOC has established ten principles for acceptable international English names. These all seem perfectly reasonable, but the one that they clearly find hardest to apply is that 'simplicity and brevity are virtues'. There have been attempts in the past to make all names follow the pattern of the scientific names and that sort of succinctness would perhaps be desirable. In reality, most English bird names started life as single words, such as Bunting, Robin and Wren, because the names were applied to a single representative local species long before there was an awareness of its relatives elsewhere in the world. Miraculously, a few single-word traditional names still survive and suffice, such as Dunnock and Dunlin. However, many simple names have now become generic because those relatives are now recognised and need distinctive names. The dynamics of usage are such that some local names die out or become curios, while some enter into wider use and are qualified by a second word. The name Wheatear, for instance, originated as an Anglo-Saxon word for the one species commonly seen in Britain, but today it is a generic name for several dozen related species: Desert Wheatear, Isabelline Wheatear and so on. Those structures are as neat as the Linnaean forms and work well.

We do, of course, use a number of three-word names such as Great Crested Grebe and Lesser Spotted Woodpecker, which could arguably be shortened if we had to take a strictly binomial approach. One of the traps for many users of those names is the temptation to slip in a meaningless hyphen. One of the wheatears, the Black-eared Wheatear, demonstrates the accurate use of the hyphenated qualifier. That device has long been accepted in English bird names.

The problem now is that the Black-eared Wheatear was split in 2020, with inelegant results: two former subspecies have morphed into the Eastern Black-eared Wheatear and the Western Black-eared Wheatear. These names are not new: they appeared as subspecies names in T.A. Coward's 1926 list in Volume 3 of the Birds of the British Isles and their Eggs. Even so, a more concise solution would have been desirable: one strange feature of a lot of recent taxonomic changes is that names often get longer as a result of splitting, and one wonders whether that is really necessary. In any case, a closer look at 'Eastern' and 'Western' in those names reveals another problem: it is somehow assumed that the user knows that the terms are relative to the Mediterranean and do not have a universal value.

In short, many modern names in English appear to be clumsy and untidy, but one imagines that the back-room team working on the Master List at the IOC has enough of a task for now in just keeping up with many changes wrought by the 'taxonomic industry'. In any case, locally used names are often jealously guarded and solutions have to be negotiated rather than imposed: the Australians, for example, eventually reclaimed such local names as Swamp Harrier and Pied Stilt that were, for some time, overridden by IOC decisions.

With all that in mind, it follows that some species entries may be more complicated than others.

Wildfowl and Gamebirds

1. ANATIDAE: Ducks, Geese and Swans

The family name, **Anatidae**, derives from the Latin *anas* (duck). It embraces all ducks, geese and swans, with just one exception: the Australian species Magpie Goose is the unique member of the family Anseranatidae, which evolved from a common ancestry with the Anatidae.

The traditional popular division of this family into swans, geese and ducks is somewhat over-simplified, even in Britain, where the taxonomic picture is less complicated than in some areas. Shelducks, for example, do not fall neatly into one of those three major categories. The taxonomy of the family is complex and seems to be under constant revision, so the current sequencing may well be modified.

WHISTLING DUCKS

These are a somewhat odd subset of wildfowl that are found mainly in the tropics, but occur as escapes from collections. While none appear as wild birds on the British list, odd individuals have been recorded within the region covered.

Fulvous Whistling Duck *Dendrocygna bicolor*

This widely distributed species is not normally found north of the Sahara, but it has been recorded in Morocco. Escaped birds have been noted in Britain.

Fulvous describes a tawny-coloured bird, though the specific name *bicolor* implies that it is two-toned.

Whistling ducks do generally whistle or make other high-pitched calls.

Dendrocygna points to two features of these odd ducks, first that they are tree-ducks (Greek *dendron*, tree) and secondly that they are generally shaped like small swans (Latin *cygnus*, swan).

Lesser Whistling Duck *Dendrocygna javanica*

This Asian species has been recorded in Israel.

Lesser describes the smallest species, which is also known as the *Javan*, given the specific name *javanica*, or as the *Indian* Whistling Duck.

White-faced Whistling Duck *Dendrocygna viduata*

This is a South American and southern African species, recorded historically in Spain and the Canary Islands.

White-faced is self-evident.

The specific name *viduata* means 'bereaved' or 'mourning' and refers to the black 'hooded and cloaked' appearance.

GEESE

Goose is rooted in the Germanic word *gans*, which in turn has much older Indo-European roots that also give rise to the Latin *anser*.

Other related words have similar roots: the male goose is the gander and the young goose is the gosling. The Anglo-Saxon diminutive suffix *-ling* is in evidence in the latter word, as it is in 'duckling'.

'Black' Geese and 'Grey' Geese

In Britain, geese have traditionally been divided in popular usage into two main groups.

* Black geese are those of the genus *Branta*.

* Grey geese are those of the genus *Anser*.

Brant Goose *Branta bernicla* (INTERNATIONAL NAME)

Dark-bellied Brent Goose *Branta bernicla bernicla* (BRITISH NAME)
Light-bellied Brent Goose *Branta bernicla hrota* (BRITISH NAME)
Black Brant *Branta bernicla nigricans* (AMERICAN NAME, ALSO USED IN BRITAIN)

This species offers a few complications, largely because three very recognisable subspecies occur in Britain, two as regular wintering species and one as a scarce vagrant.

Dark- and **Light-bellied** both winter in large numbers in Britain, though in different regions. Vagrant **Black Brants** may be 'carried' by normal flocks of either form.

The words **Brant** and **Brent** are both rooted in Germanic forms (e.g. the Norse name *brandgas*, burnt goose), which are clearly based on their blackness. The Americans have traditionally favoured the form *Brant*, while *Brent* has been the normal usage in Britain since Pennant preferred it over Ray's earlier use of Brant.

The generic name *Branta* is a Latinised version of those words.

The qualifying names **Dark-bellied, Light-bellied** and **Black** all refer to distinguishing features that can be noted in the field.

The specific name tells of past confusion of the two species of 'black geese' that winter in Britain, because *bernicla* refers to 'barnacles'. That is explained by reference to the next species.

The subspecific name *hrota* is based on the Icelandic name *hrot goes* (snoring goose), referring to the guttural call.

For the Black Brant, the subspecific name is *nigricans*, which simply underlines the blackness.

* This species is circumpolar in the breeding season. Light bellied (*hrota*) breed in the Canadian Arctic; Dark-bellied (*bernicla*) breed in Siberia; Black Brant (*nigricans*) breeds in areas adjacent to both, so rare strays may be 'carried' with either group.

Barnacle Goose *Branta leucopsis*

This species also winters in Britain in good numbers, particularly in the North-West, though small numbers of feral birds, originating as collection escapes, are seen all year round in other areas.

The **Barnacle Goose** is associated with a legend that was first recorded by the monk Giraldus (aka Gerald of Wales) in 1185. The phenomenon of migration was not then understood: these birds appeared in winter but disappeared in the breeding season. Folklore explained their annual appearance by claiming that they hatched from goose barnacles. Mark Cocker has pointed out that some people found it convenient to perpetuate the legend well into modern times: by treating the birds as a form of fish, they could avoid the Catholic Church's Friday embargo on meat.

The specific name *leucopsis* refers to the distinctive white face.

* The legend of the goose barnacle seems to have applied fairly equally to both species of black geese; this has resulted in the fact that one species has the legend in the vernacular name and the other in its scientific name. This is confusing, but less confused than the mindset that claimed a bird could be born from shellfish.

Red-breasted Goose *Branta ruficollis*

This is a species that breeds in the Siberian Arctic and winters near the Black Sea. It is a scarce vagrant to Britain, where wild birds are sometimes 'carried' by Brent Geese. Escapes from collections may confuse the picture.

Red-breasted is echoed in the specific name *ruficollis* (red-necked). Both are valid, since the rufous-red extends over the throat, breast and cheeks of an otherwise extraordinarily patterned black and white bird. In the field, the total effect breaks up the silhouette of the bird.

Canada Goose *Branta canadensis*

The Canada Goose was introduced into Britain in the seventeenth century by Charles II, so it is well established. A good many are relatively tame, and it is arguably the most familiar of all goose species for many Britons. These birds are descended from the large, and mainly sedentary, nominate form.

In its native North America, there are seven identifiable subspecies, some of which have occurred as vagrants to Britain. Though the BOU is usually reluctant to commit to field identification at subspecies level, such birds are not of the nominate form.

Canada Goose and *canadensis* are self-evident, although some forms breed in the United States and many winter there.

Cackling Goose *Branta hutchinsii*

The former Canada Goose complex originally embraced a dozen subspecies. Four of the smallest forms, which average about the same size as the Barnacle Goose, now constitute a full species, which has been represented by a few identified vagrants to Britain.

Cackling is self-evident: the smaller birds make a higher-pitched call than the larger Canada Geese, more akin to a yelp or a laugh.

The name *hutchinsii* was dedicated to Thomas Hutchins (1730–1790) of the Hudson Bay Company.

The nominate form is often known as **Richardson's Goose**. It was first identified and named by the Scottish explorer and naturalist John Richardson, who explored Canada with Franklin in 1819–22 and 1825–7 (see also Mew Gull). However the AOS announced in 2023 that all eponyms in vernacular names of American species will eventually be replaced.

Greylag Goose *Anser anser*

This is the 'default' species of goose, which is represented in Britain both by a large feral reintroduced population and by winter-visiting wild flocks from Iceland.

Greylag is thought to be rooted in an Old English word *lag* (goose) and therefore simply means 'grey goose'. In that case, Greylag Goose is a tautology. The word was historically used by goose-herds to call their flocks. It has also been suggested that the origin of the name may have been that this species lagged behind the other migrating geese at the end of winter.

Anser is the Latin for 'goose' and the genus of the 'grey geese'.

* This is the species that furnished the base stock for the farmyard goose in the Western world. It was reputedly first domesticated in Ancient Egypt. The Romans later favoured the white morph in breeding. The 'farmyard goose' is still sometimes in evidence in feral flocks, where the odd white or patchy form is not uncommon.

Pink-footed Goose *Anser brachyrhynchus*

This species is a winter visitor to Britain in large numbers from the Arctic north.

Pink-footed is self-evident, but the colour is not always obvious at a distance.

The short bill of this species gives it the name **brachyrhynchus** (Greek *brakhus*, short, and *rhunkhos*, bill).

Lesser White-fronted Goose *Anser erythropus*

This is a rare vagrant to Britain, which is usually 'carried' with flocks of other wild geese. The nearest breeding population is in northern Scandinavia; these generally winter in the Balkans.

Lesser means that this is the smaller of two 'white-fronted' species. (It is in fact generally smaller than any other grey geese.)

White-fronted refers to the patch of white on the forehead (*frons*). This is generally more extensive in this species than in the Greater White-fronted, extending further back on the crown.

The specific name **erythropus** refers to the bird's 'red legs' (Greek *eruthros*, red, and *pous*, foot).

Greater White-fronted Goose *Anser albifrons*

Eurasian White-fronted Goose *Anser albifrons albifrons*
Greenland White-fronted Goose *Anser albifrons flavirostris*

The formal name '**Greater**' is rarely used in practice. This is a fairly common winter species in Britain, where the two populations may overlap.

Eurasian is the formal name of the species.

Two forms of Greater White-fronted occur annually in Britain in winter. These are currently treated as subspecies from two different populations and are designated accordingly:

* The Eastern subspecies is often colloquially called the Russian Whitefront.

* **Greenland** is self-evident.

White-fronted and *albifrons* mean the same (Latin *albus*, white, and *frons*, forehead). White-fronted Goose is often abbreviated to *White-front.*

The Greenland subspecies has an orange-yellow bill, hence *flavirostris* (Latin *flavus*, yellow, and *rostris*, billed).

* In good views the two forms can be distinguished by the practised eye.

> The traditional name **Bean Goose** originally covered two forms, which were long considered conspecific and are now treated as full species (see below), though some consider the spilt to be premature. Both forms occur in only small numbers in Britain in winter.

Taiga Bean Goose *Anser fabalis*

This is a winter visitor in small numbers originating in the northern Eurasian forest region.

Taiga comes from its forested sub-Arctic breeding habitat.

Bean appears to derive from its preference for feeding on field beans, though Pennant originally related the name to the bean-shaped 'nail' at the tip of the bill.

The specific form *fabalis* derives from the Latin *faba* (broad bean, known in America as the fava bean). This species was formerly treated as the nominate form of the original Bean Goose complex, so retains the scientific name, *fabalis*.

Tundra Bean Goose *Anser serrirostris rossicus*

This also occurs in small numbers in Britain in winter.

Tundra comes from the open Arctic breeding habitat.

The specific name *serrirostris* is Latin for 'saw-billed', referencing the rough edge evolved for grazing grass. (Compare the 'saw-billed' ducks, which have a similar device for holding fish.)

The subspecies that visits Britain is the form *rossicus* (Russian), while the nominate form has a more easterly bias.

Bar-headed Goose *Anser indicus*

In Britain at least, this Asian species is treated as an escape, though there is some evidence of breeding, and feral populations do exist on the continent.

Bar-headed refers to the black lines extending from the stripe at the back of the head.

The specific name *indicus* relates to the fact that part of the wild population winters in the Indian subcontinent.

* These birds are renowned for the feat of migrating at high altitude over the Himalayas, though research now shows that they use mountain passes to wend their way between the highest peaks.

Snow Goose *Anser caerulescens*

This North American species occurs occasionally in Britain as a genuine winter vagrant, often 'carried' with other regular species arriving from North America, but there are also some feral flocks and occasional escapes to be found all year round.

Snow is an apt name for a species that is mainly white and breeds in the Arctic.

The specific name *caerulescens* indicates a more complicated picture, since a proportion of the population is 'bluish'. These purple-grey morphs are known popularly as *Blue Geese*.

* In America, the terms *Greater* and *Lesser* Snow Goose are used. These are subspecies, the Greater being the Eastern form and less numerous.

* The mixing of white and blue morphs is apparently a relatively recent phenomenon, as the two appear to have resulted from the mingling of two formerly separated populations.

* The genus *Chen* (from the Greek *khēn*, goose) is no longer used. For some years, all the white geese (Snow, Ross's and Emperor) were classified in this genus, but have now been placed back in *Anser*.

Ross's Goose *Anser rossii*

This small American relative of Snow Goose is recorded rarely as a vagrant to Britain, but also occurs as an escape.

The Hudson Bay Company's Bernard **Ross** (1827–1874) was the subject of this dedication. When the explorer Robert Kennicott died in 1866, Ross saw to the safe transportation of his specimens, thus earning the gratitude of ornithologist John Cassin, who named the goose after him. This is certainly one of the more unusual reasons behind the attachment of a person's name to a species. However, the AOS announced in 2023 that all eponyms in vernacular names of American species will eventually be replaced.

Emperor Goose *Anser canagicus*

This Alaskan species is present as an escape in some European countries, including Britain.

Emperor: Explanations and alternative names vary, but many suggest that the white-hooded appearance suggested an imperial robe. However, Rockwell pointed out that the name arose when early Russian explorers confused a local Aleut name for the goose with *tsarskia* (imperial). The 'robed' appearance therefore seems to be a happy coincidence.

The specific name *canagicus* refers to Kanaga, an island in the Aleutian chain where these birds breed.

SWANS

Swan is based on an Anglo-Saxon word with much older Germanic roots that seem to mean 'singer', but probably originally simply meant 'sound'. In spite of legends that tell of 'swan song', no swan species could be described as a 'singer': the names Whooper, Whistling and Trumpeter tell part of the story of the more vocal species, but perhaps the 'sound' element, and its links to folklore and legend, could be explained by the slightly unearthly whistle of the beating wings of the Mute Swan, the most familiar species through much of Europe.

Cob, Pen and Cygnet

It is unsurprising that such large and obvious birds should have their own subset of specialist terms for parents and young.

Cob is the male swan, which in other contexts is a 'big man' or leader.

Pen for the female is more obscure, but may be related to the Latin *penna* (feather). It is probably not a coincidence that swans' feathers made excellent quills (pens), though it has been suggested that the females keep their wings flat (pinioned), while the males raise theirs in display.

Cygnet is a diminutive form of the Latin *Cygnus* (swan).

Tame Swan and Wild Swan

Historically in Britain, swans were divided into two types: 'tame' swans, which were the relatively sedentary Mute Swans, and 'wild' swans, which were the migratory species. That seems to have been a fairly universal distinction, since Linnaeus recorded the Latin and French versions of the 'wild' form as *Cygnus ferus* and *Cygne sauvage*. All 'wild' swans were long seen as a single species, but that changed in the early nineteenth century, when they were shown to be two species, today known as Whooper and Bewick's swans.

Whooper Swan *Cygnus cygnus*

This is a regular winter visitor to Britain from Iceland, though a few breed in the north of Scotland, while there are small pockets (at Welney, Norfolk, for instance) of now relatively sedentary birds.

Whooper is based on the loud bugling call. It was first seen as *Hooper* in 1566 and was noted by Ray in 1678. Pennant had made the odd choice of the Northumberland form, Whistling Swan, for the wild swan, but that did not catch on, because Yarrell's 1843 alternative *Hooper* was preferred. Later in the nineteenth century, the spelling was modified to Whooper Swan and standardised by the BOU.

Cygnus cygnus is a Latinised form of the Greek (*kuknos*) for 'swan'. When Linnaeus prioritised this as the 'type species' over the Mute Swan, it was possibly better known in Sweden, since Mute Swans had been greatly reduced by over-hunting in mainland Europe and were probably far less familiar to him.

Mute Swan *Cygnus olor*

The species is resident in Britain, but is partly migratory in colder parts of Europe.

Mute suggests that the bird is silent but it is not wholly mute. Its vocalisation is limited to a small range of grunts, croaks and hisses, but the contrast with the vocal "wild" swans is evident. It does, however, make a noise of a different sort in flight, since stiff flight feathers produce a recurring whistling sound.

The specific name in this case is ***olor***, the Latin for swan.

* The static and confiding behaviour in Britain was long considered to be evidence of introduction and semi-domesticity, but archaeological and fossil evidence now shows that to be erroneous and that the Mute Swan is a full native species. On the other hand it has always been very biddable, especially where food is its reward. All unmarked Swans were long considered royal property, a fact that saved them from being exterminated by hunting.

* The term '*Polish*' Swan was coined by Yarrell to describe a rare colour morph that can be identified at the cygnet stage by its pure white (not grey) down and at the adult stage by its pinkish-grey (not black) legs. *Polish* in this case was used in a loose sense to indicate that the observed bird had been imported from the Baltic region as a potential food item.

Tundra Swan *Cygnus columbianus* (INTERNATIONAL NAME)
Bewick's Swan *Cygnus columbianus bewickii* (EURASIAN SUBSPECIES)
(Whistling Swan *Cygnus columbianus columbianus*) (AMERICAN NOMINATE FORM)

The **Bewick's Swan** breeds in Siberia and winters in Western Europe, including Britain.

The cluster of names here is complicated and confusing. The current consensus is that all the names represent a single species, although that has not always been the case and may yet change again.

Tundra is a formal umbrella name for the species, because the birds breed in Arctic tundra zones. Unusually, that name is used less than are the vernacular names of the two subspecies.

The Eurasian form, **Bewick's Swan** *C. c. bewickii*, winters in Western Europe, including Britain.

When Yarrell identified the smaller 'wild swan' as a species separate from Whooper Swan in 1830, he named it after the recently deceased engraver Thomas Bewick. (Both species have black and yellow bills, but Bewick's Swan is a smaller bird.)

The **Whistling Swan** carries a Northumbrian name that first gained currency in Britain when it was adopted in 1785 by Pennant for the Wild Swan (see earlier note). It was discarded by later British ornithologists, but had meanwhile travelled with colonists to the Americas, where it was retained popularly for what was to became the nominate form of Tundra Swan. That form was named as a species by the American, George Ord, in 1815. Since this action preceded Yarrell's by 15 years, the American form is now the nominate one of the unified species. It is on average a bit larger than Bewick's and has little or no yellow on the bill. The specific name *columbianus* is for the Columbia River, Oregon.

Black Swan *Cygnus atratus*

This Australian species has escaped from collections. Small numbers survive in Britain, with occasional reports of breeding. There is a feral population in the Netherlands, though this may be declining.

Black is self-evident, though there is a surprising amount of white when the bird is on the wing.

The specific name *atratus* simply means 'black'.

The next two species do not fit neatly into the usual categories.

Spur-winged Goose *Plectopterus gambensis*

This sub-Saharan species has been reported in Morocco.

It is a large, unique species and is considered more closely related to the shelducks than to the true geese.

Spur-winged describes the primitive bony wing projections used by the aggressive males in fights during the breeding season.

Plectopterus echoes the vernacular name exactly, with the Greek words *plēktron* (spur) and *optatus* (winged).

The specific name *gambensis* relates to the Gambia in West Africa.

Knob-billed Duck *Sarkidiornis melanotos*

This sub-Saharan African and Asian species was split from the South American Comb Duck in 2020. It has been recorded in Mauretania, north-west Africa.

Knob-billed describes the curious projection on the upper mandible of the male.

Sarkidiornis derives from the Greek for a small bit of flesh (the 'knob' or 'comb') plus *ornis* (bird).

The specific name *melanotos* means 'black-backed'.

SHELDUCKS

Several species are intermediate between goose and duck, sharing with geese and swans the fact that males and females are generally similar in appearance. In spite of its name, the Egyptian Goose is more closely allied to these and sometimes hybridises with Ruddy Shelduck.

Egyptian Goose *Alopochen aegyptiaca*

This species was introduced into Britain in the seventeenth century and established in the wild. Numbers remained relatively low until the late twentieth century, when they proliferated and spread.

Egyptian and *aegyptiaca* remind us that the species originates in Africa.

It is not a 'true **goose**', but is more close-related to the shelducks.

Alopochen derives from the Greek *alōpēx* (fox) and *khēn* (goose) and is rooted in a word shared with the Ruddy Shelduck for its foxy-red colour.

Common Shelduck *Tadorna tadorna*

This is a common breeding species in Britain, but most adults disappear in late summer to join moulting flocks at traditional sites, particularly in Germany's Waddensee.

Common is used in an international context because there are several other forms of shelduck around the world.

Shelduck literally means 'variegated duck', since *sheld* was an interpretation of the boldly patched plumage. The name has nothing to do with 'shields' or 'shells', in spite of some misleading historical forms and interpretations. It was first in evidence in the form Sheldrake in the twelfth century.

Tadorna was adapted from a French form of an original Italian name.

* Lockwood records a fairly widespread alternative historical name, *Bergander*, which he feels was even older than Sheldrake and which appears to be rooted around the word

gander. He hypothesises that the *Ber-* element is for the berry-shaped red knob on the bill of the male.

Ruddy Shelduck *Tadorna ferruginea*

Not to be confused with the Ruddy Duck (see below), this bird has a complicated status in Britain. Most sightings are treated as escapes, though it is fairly certain that some birds stray from sustained feral populations on the Continent, while others may well be genuine vagrants from their natural range further east.

Ruddy and *ferruginea* (rusty, from Latin *ferrum*, iron) both relate to the colour (see also Ferruginous Duck below).

South African Shelduck *Tadorna cana*

This sometimes occurs as an escape and has bred in Sweden.

South African is self-evident. It was previously known as the Cape Shelduck.

The specific name *cana* is Latin for 'grey', for the head of a bird that otherwise resembles Ruddy Shelduck.

DUCKS

Unlike swans, geese and shelducks, male ducks generally have a more showy plumage than the females. The two are so unalike that the Mallard and the Wild Duck were named separately until the twentieth century (see below). Curiously, however, such dimorphism is not so pronounced in some close relatives of the Mallard, such as Mottled and Black Ducks.

Duck originates from an Old English word meaning 'to dive'. Although not all adult ducks dive, their young often can and do. Drake distinguishes the male duck and first appeared in the twelfth century. Lockwood suggests that the term has ancient Germanic roots and that it is an echoic rendition of the male's quieter voice. Duckling, with its diminutive *-ling* ending, denotes the young duck and is in keeping with *Gosling*.

Ducks attract a number of informal terms to describe their differences:

* **'Dabblers'** is a term used for ducks that feed by skimming from the surface; by 'upending'; or by sifting from muddy margins (e.g. Mallard, Gadwall, Common Teal).
* **'Diving ducks'** dive below the surface to feed. This involves both freshwater and sea-water species (e.g. pochards, scoters).
* **'Sea-ducks'** are species that are primarily maritime or estuarine in habit (e.g. Common Eider, Red-breasted Merganser).
* **'Sawbills'** are those diving species that have longer, serrated bills to grip slippery fish (e.g. Smew, Red-breasted Merganser).
* **'Stifftails'** are diving ducks whose tails form short fans of hardened feathers that act as aquaplanes when diving (e.g. Ruddy Duck).

The next four species are all exotics that do not link closely to other species native to the Western Palearctic.

Muscovy Duck *Cairina moschata*

This Central/South American species has been kept in a domesticated form in Europe almost since the first Europeans set foot in the Americas, so most free-flying birds on this side of the Atlantic descend from the domesticated form rather than the truly wild race. The BTO's 2007–11 Bird Atlas reported a number of small, self-sustaining populations in Britain. It is in any case worth inclusion here for its odd names, which, implausibly, link both Moscow and Cairo to a bird from the New World.

Muscovy is an enigmatic name that seems to have been linked to a sixteenth-century trading company, which took its name from the former Duchy of Moscow. One theory is that the root name originated from one of two native-American peoples – the *Muiscas* of Colombia or the *Miskitos* of Honduras and Nicaragua, who domesticated the ducks – and that a form such as *Muiscas duck* became corrupted into a more familiar word.

That seems quite feasible when we consider that an alternative name, *Cairo Duck* gives rise to the genus *Cairina*. The example of the name *Turkey* (see later) for another American species, alerts us to the fact that such names were conveniences which simply implied an exotic origin.

However, the specific name *moschata* offers a different explanation for the source, since several other European languages have names equivalent to *musk duck* (viz. the French *canard musqué*). In the late seventeenth century, John Ray commented that *moschata* was rooted in the word *musk*. The head wattles give off a slight musky odour. A word such as *musky* would convert quite readily into *Muscovy*.

Wood Duck *Aix sponsa*

This American species is not yet on the British list, but may well reach Britain as a vagrant, since it has been recorded in Iceland. It also exists as an escape that has sometimes formed temporary breeding populations, though it seems not to be self-sustaining.

Wood Duck underlines the fact that this species is found where there is woodland and water: it perches readily and it nests in tree-holes. An alternative local name is *Carolina Duck* (see Green-winged Teal for the use of *Carolina*).

Aix originates in Greek and was used by Aristotle to describe a waterbird.

The specific name *sponsa* likens its sumptuous plumage to that of a 'bride', even though it is the male that has the showy garb.

Mandarin Duck *Aix galericulata*

This is an introduced Asian species, which has been established in the wild in Britain for over 100 years, though it did not appear in the 1923 BOU list. It seems that, with the Far-Eastern population at endangered levels, the feral populations here and elsewhere are now vital to the bird's existence.

Mandarin originated from the fact that the plumage seemed to resemble the elaborate garb of Chinese Imperial officials. The male has the most elaborate plumage of all duck species.

The drake's superb mane earns the specific form *galericulata*, since it means 'bewigged'.

Cotton Pygmy Goose *Nettapus coromandelianus*

This African species has been recorded in Jordan. It also breeds in Asia and Australia.

The **Cotton** element of its name is explained by an alternative Australian use of *White-quilled*, which describes a broad white wing-band set against a green-black ground.

Pygmy is for the tiny size: this species has some of the smallest individuals among wildfowl.

Goose is a misleading and perverse element of the name: it is in fact a true duck, also referred to as the *Cotton Teal*, but to some it resembled a tiny goose in the shape of the neck and bill.

Nettapus means 'duck-footed' (see the genus *Netta* later).

The specific name *coromandelianus* refers to the Coromandel Coast of eastern India, a term created by the Dutch and used widely for some centuries by mariners of all nations. Misleadingly, that is now the formal name of a part of New Zealand and should not be confused

Dabblers

A 2009 genetic study split the birds formerly classified in the genus *Anas* into four genera (*Anas*, *Spatula*, *Sibirionetta* and *Mareca*) in a move that resurrected three defunct generic names.

> The genus *Sibirionetta* is a resurrected genus. It was originally created by the German naturalist Johann G. Georgi in 1775, uniquely for the species below.

Baikal Teal *Sibirionetta formosa*

This Asian species has been recorded rarely as a vagrant to Britain, but now seems to be more frequent in the Western Palearctic.

Baikal relates to Lake Baikal, a vastly reduced inland sea in Siberia.

Teal: see Eurasian Teal below.

Sibirionetta literally means Siberian duck: *netta* is a Greek word for 'duck'.

The specific name *formosa* means 'beautiful' – the drake is very colourful.

> The genus *Spatula* was also resurrected. It was first used by Friedrich Boie in 1822 and eventually set aside for some years. It includes the ten 'blue-winged' species of dabbling ducks, so-called for the blue or blue-grey panel on the upper forewing. Two of these are native to Britain, one is an irregular visitor and another is sometimes seen as an escape.

Northern Shoveler *Spatula clypeata*

This is a resident species that is boosted in winter by large numbers of migrants.

Northern pertains because there are other related species on the World List.

Shoveler clearly derives from the shape of the bill rather than from its manner of use. The name was originally used for the Spoonbill, but Ray adopted it for this duck in 1678. Note that there is only one 'l' in the name.

Spatula is, of course, from the flat spreading tool used in cooking and art, for example.

The specific name *clypeata* means 'shield-bearing'. Anyone who has watched a Shoveler preening its breast feathers will see the point immediately.

* This imagery overload all emphasises the fact that the bird has a rather dispropor-tionate flattened bill that maximises the area of filters needed to extract very small food items from the water. While they sometimes up-end, one of their most common feeding methods is to swim in tight circles, in pairs or in small flocks, to create a vortex to bring such items to the surface.

Garganey *Spatula querquedula*

This species migrates up from Africa to breed in small numbers in Britain and in larger numbers in Europe.

Garganey evolved from an echoic Italian word, *garganello*, and travelled via French into English.

The specific name **querquedula** is also an echoic word that was used by the Roman writers Varro and Columella. It is therefore similar in origins to the word Garganey.

* An alternative archaic Norfolk name, *Summer Teal*, underlines the fact that this is the one duck that is generally not seen in winter, but travels north from Africa to breed in Eurasia. Here the *teal* emphasises the word's wide use as a smaller sort of duck. Another local East Anglian name *Cricket Teal* focused on its voice.

Blue-winged Teal *Spatula discors*

This American species is a scarce vagrant to Britain.

It is **Blue-winged** because it has a large blue area on the inner forewing. It is distinguished in America by that name from the Green-winged Teal.

Teal: see Eurasian Teal below.

The specific name *discors* (different) is enigmatic, but there seems to be a clue in Mark Catesby's original use of *variegata*: it may therefore mean 'of different colours'.

Cinnamon Teal *Spatula cyanoptera*

This American species is seen occasionally in Britain and Europe, but such birds are generally treated as escapes.

Cinnamon refers to the body colour, which is orange-brown.

Teal: see Eurasian Teal below.

The specific name *cyanoptera* refers to the 'blue wings'.

Another recent resurrection is the genus *Mareca*, created by James F. Stephens in 1824. It contains just five extant species, four of which are recorded in Britain.

Gadwall *Mareca strepera*

There were few Gadwall in Britain before 1850, but releases in that year boosted the population to make it now one of the most successful species, with winter visitors adding to the numbers.

Gadwall seems to be an echoic idea, as in 'gabble', as evidenced by the old form, *Gaddel*.

Mareca originates from a Portuguese word *Marrèco*, for a small duck, a name taken from a Roman water nymph, *Marica*.

The specific name *strepera* echoes the concept of the vernacular name, since it means 'noisy'.

Eurasian Wigeon *Mareca penelope*

This is probably the most numerous of the winter-visiting species to Britain.

Eurasian and American (below) are needed to distinguish two closely related species, though both qualifiers are dropped in their local usages.

Wigeon is an enigmatic word that seems to be rooted in the whistling call 'wee-oo' (its French name, *canard siffleur*, literally means 'whistling duck'). It was first noted in 1513 as *wegyons* and evolved through a series of variants to the current form. The call is certainly present in the old alternative names *Whistler* and *Whewer*.

In Ancient Greek, *pēnelops* was a sort of duck. Legend tells that a girl child, cast into the water by her Spartan father, was saved by the ducks and named *Penelope* as a result.

American Wigeon *Mareca americana*

This widespread **American** species is a scarce vagrant to Britain.

Its names need no further explanation, though the use of *americana* is surprisingly scarce among American species.

Falcated Duck *Mareca falcata*

This East Asian species is widely kept in collections, so sightings in Britain are traditionally treated as probable escapes. In 2019 the BOURC agreed that at least one and probably more historical records are valid as vagrant wild birds.

Falcated and *falcata* are words related to 'falcon', since they also derive from the Latin *falx*, *falcis* (sickle). In the case of the falcon it is the sickle-shaped wings that count, but in this case the male has sickle-shaped plumes that curve down from his back.

The genus *Anas* is at the root of the word Anatidae, and was long considered to be the default genus for most dabbling species.

Mallard *Anas platyrhynchos*

This widespread species provided the root stock for many domestic species, but a mix of truly wild, released and feral birds now populates Britain, with an influx of winter migrants.

The name **Mallard** was once used for male ducks in general, which supports the view that it was derived from Norman French *masle* (male) with the addition of the suffix *-ard*.

Anas is the Latin word for 'duck'.

The specific name, *platyrhynchos*, derives from two Greek elements meaning 'flat bill'.

* The female was known throughout most of history by the name *Wild Duck*. In fact males and females were recorded separately by Linnaeus (the latter as *A. boschas*) with the result that the original scientific name was changed to match his entry of the male duck,

which was recorded a few pages before the female. Coward still listed them as *Mallard or Wild Duck* in 1926; *Wild Duck* was dropped as a formal name in the following years.

American Black Duck *Anas rubripes*

This species occurs in Britain as a scarce vagrant.

The word **American** is required in a world context to separate this species from the related *Pacific* Black Duck, for instance.

The birds may be perceived from a distance as **black**, but they are in fact a sooty brown, like a very dark female Mallard, but with a greyish-tan face. The lack of distinct white borders to the speculum is diagnostic and makes the upper wings look darker when in flight. Most unusually, the sexes are very similar, with the male's yellow-green bill showing the only significant difference.

The specific name *rubripes* refers to the orange-red feet (which are similar to those of a Mallard).

* Black Duck × Mallard hybrids are not uncommon in parts of their range.

Northern Pintail *Anas acuta*

Northern Pintail is a relatively common winter visitor to Britain.

Northern is not normally needed unless the bird is discussed in the context of the three other 'pintail' species (all living outside the Western Palearctic).

Pintail refers to the elegant, elongated tail feathers of the male.

That idea is echoed in *acuta* (sharp).

Eurasian Teal *Anas crecca*

This is a common species with many winter visitors boosting the moderate resident British population.

Eurasian is needed when other 'teal' species are involved.

Teal was first recorded as *teles* in 1314 and is probably based on a Germanic word of an echoic nature. It eventually came to be used as a term for several other small dabbling duck species, as in the next three entries.

The specific name *crecca* is a Latinised version of the Swedish name *kricka*.

Green-winged Teal *Anas carolinensis*

This American species is a scarce vagrant to Britain. It was long considered to be a subspecies of *Anas crecca* and was listed as such by Coward in 1926.

Green-winged is a description that applies equally to the previous species, but it is an important feature in America, where there is also a *Blue*-winged Teal.

The specific form *carolinensis* has a wider scope than the modern Carolinas, since it referred to the original Carolina Colonies, named after a 1663 charter issued by King Charles II.

Cape Teal *Anas capensis*

This sub-Saharan African species has been recorded as far north as Lebanon and Israel.

Cape and *capensis* both refer to the Cape of Good Hope in South Africa.

Red-billed Teal *Anas erythrorhyncha*

This African species has been recorded in Israel.

Red-billed is self-evident and is echoed exactly in the Greek of the specific name *erythrorhyncha*.

Marbled Duck *Marmaronetta angustirostris*

This species occurs in some Mediterranean areas and has been recorded as a very rare vagrant to Britain.

Marbled refers to the pattern of the plumage.

That idea is echoed in the first part of *Marmaronetta* (Greek *marmaros*, marble, while *netta* is the Greek for duck).

The specific name *angustirostris* is also composed of two elements, the first part deriving from the Latin *angustus* (narrow) and the second from *rostrum* (bill).

> The following two species are unusual in being both 'dabblers' and 'divers', roughly in equal parts. They share their genus with just one other species.

Red-crested Pochard *Netta rufina*

This European species has established a sturdy feral population in Britain, though some Continental birds may also appear, since it is fairly widely distributed in Europe.

For **Pochard** see Common Pochard below.

Netta is the Greek for 'duck'.

The specific name *rufina* refers the drake's spectacular orange-red crown.

Southern Pochard *Netta erythropthalma*

This African species has been noted in Israel.

As the name **Southern** implies, it has a more southerly range than its congener above. Two races are found, one in Africa and the other in South America.

The specific name *erythropthalma* describes the male's red eye (Greek *eruthros*, red, and *opthalmus*, eye.)

Diving Ducks

Birds of the genus *Aythya* are diving ducks and are often referred to as 'the pochards' after the name of one of the most common Eurasian species. Aythya hybrids occur fairly regularly.

Common Pochard *Aythya ferina*

This species breeds widely in Britain, but numbers are increased by passage and wintering birds.

The word **Common** is used internationally because other species have Pochard in their name.

Pochard is rooted in two dialect forms, *Poacher* and *Poker*, both of which reflected the idea of 'poking about'. Pennant underwrote the current form in 1768. Today the word is sometimes used as a collective to describe birds of the genus *Aythya* and their closest allies.

Aythya is a name used by both Aristotle and Hesychius for an undefined waterbird.

The specific name *ferina*, is related to the word feral in suggesting 'wild' (though not to be confused with *Wild Duck*, as in Mallard).

Canvasback *Aythya valisineria*

This American species has been recorded as a very rare vagrant to Britain. Like the Redhead, its similarity to Common Pochard may make it hard to distinguish at a distance.

Canvasback describes the finely vermiculated pale grey of the mantle by likening it to raw canvas, the coarse material used, for example, in sails.

The specific name *valisineria* is unusual in coming from the name of one of the bird's food plants, a form of aquatic celery. This in turn commemorates the Italian botanist Antonio Vallisnieri (1661–1730). Unfortunately for the bird, its flesh is pleasantly flavoured by its diet, making it a favourite with hunters.

Redhead *Aythya americana*

This American species has been recorded as a very rare vagrant to Britain.

The name **Redhead** underlines the fact that this species has a brighter head-colour than the similar Canvasback. That feature only applies to the male, the female being a uniform tawny-brown.

The specific name *americana* is self-evident.

Ferruginous Duck *Aythya nyroca*

This is a species of Eastern Europe and the Middle East that is seen fairly regularly in Britain. While some birds seem to originate from their native region, others may originate as escapes or as feral birds from reintroduction programmes on the Continent.

Ferruginous is a clumsy Latinate name, related to *ferrum* (iron) in its rusty form, and describing the male's colour. The female is similar, but lacks the reflective glow.

The specific name *nyroca* derives from *nyrok*, the Russian for duck.

* The birder's usage, *Fudge Duck*, seems to be a neat, if ironic, short-cut to the vernacular name.

* In the past, the species was also known as the *White-eyed Duck* or *White-eyed Pochard*, for its distinctive eye, but that idea was dropped to avoid confusion with the name of an Australian relative, which has now also formally changed and is known by the enigmatic name *Hardhead*.

Tufted Duck *Aythya fuligula*

This is a common species of inland waters.

Long head-plumes give the **Tufted** Duck its name. Those of the drab-brown female are shorter, but still give the head a distinctive form.

The specific name *fuligula* means 'sooty-throated', though 'sooty' could equally be applied to many other parts of the bird.

* The contrasting black-and-white of the male accounts for an old name *Magpie Duck*, but most birders settle for '*Tufty*'.

Ring-necked Duck *Aythya collaris*

This American species occurs as a regular vagrant to Britain, where it might be overlooked among flocks of very-similar Tufted Ducks.

Both **Ring-necked** and *collaris* imply a feature that is very subdued and, in the field, most often completely hidden by the bird's natural posture. More obvious on the male are the lack of a tuft and a grey flank, both of which help to distinguish it from Tufted Duck. Both males and females are readily distinguished by the bold white band above the black bill-tip.

Greater Scaup *Aythya marila*

This bird is native to both America and Eurasia.

Greater is used mainly in America, where there is also a *Lesser* Scaup. In Britain it is not normally needed.

The word **Scaup** began life as an English dialect word for mussel. From that came two names, the *Mussel Duck* or *Scaup Duck*, feeding on the shellfish. Both Ray and Pennant favoured the latter, and that was formally shortened by the BOU during the early twentieth century to become 'the Scaup'.

The black elements of the bird's plumage are conveyed in the specific name *marila*, which is based on a Greek word meaning 'embers' or 'charcoal dust'.

Lesser Scaup *Aythya affinis*

This American species occurs as a scarce vagrant to Britain.

The name **Lesser** separates this species from the more robust *Greater* Scaup. The bird is slightly smaller and distinctly slimmer.

The 'affinity to' its larger relative is explicit in *affinis*.

* Some male hybrids of Tufted Duck × Pochard can be confusingly similar to this species.

Sea-ducks

Next is a group generally called 'sea-ducks', though some do not spend a great deal of time in coastal waters.

Steller's Eider *Polysticta stelleri*

This very rare vagrant breeds in Siberia and Alaska.

G.W. **Steller** (1709–1746) was a German naturalist and explorer whose name is attached to a number of birds and other animals (seal, sea-cow, jay, eagle, etc.).

Polysticta means 'many-spotted'. The colourful breeding plumage of the male has a few black spots and patches that may account for that.

Common Eider *Somateria mollissima*

This species is found all year round, often in more northerly coastal waters on both sides of the Atlantic, though rarely inland.

It is **Common** because the word Eider, originally coined for the one British species, has grown to be used as a generic name for several related species.

Eider was first used by the Dane, Ole Worm, in 1655. It is based on an Icelandic word rooted in an older Norse word for the female and meaning 'down-bird'. The bird plucks copious amounts of down from its breast when nesting to provide a cosy medium for its eggs. This was long harvested by people as a resource for stuffing clothing and bed coverlets called 'eiderdowns'. The French *duvet* is also based on a word meaning 'down', and has largely replaced the older word in modern usage. The word Eider stands alone (though Pennant confused matters for a while by adding the word *duck*) and now serves as a generic name.

Somateria is a hybrid word linking the Greek for 'body' with the Latin for 'wool'.

The specific name *mollissima* assures us that the substance is very soft.

* The Western Isles of Scotland used the name *Colk*, which has a Gaelic root meaning much the same as Eider.

* The Orkney and Shetland name *Dunter* derived from a Norse word meaning 'to bob up and down in the water'.

* In North-East England, a number of local variants derived from an association with St Cuthbert, the seventh-century Bishop of Lindisfarne. *St Cuthbert's Duck* and corruptions thereof survived for many centuries. The association was first formally recorded in 1165 as the *aves Beati Cuthberti* (the birds of the blessed Cuthbert). Even today, females can be seen nesting in spring close to St Cuthbert's Chapel on Inner Farne Island.

King Eider *Somateria spectabilis*

This is a fairly frequent vagrant to northern British waters from its Arctic breeding areas.

King conveys the majesty of a spectacular male that wears a frontal shield like a gold crown.

A black and white line from the rear of the eye creates the pattern of a pair of spectacles to give the specific name *spectabilis* (which is somewhat confusing, given the vernacular name of the next species).

Spectacled Eider *Somateria fischeri*

This species, normally found in the northern Pacific, has been recorded in Norway.

Spectacled refers to the facial pattern, which has a large area of dark-outlined white around the eye.

J.G. **Fischer** von Waldheim (1848–1886), a German naturalist, is commemorated in the specific name *fischeri*.

Harlequin Duck *Histrionicus histrionicus*

The nearest breeding population is in Iceland, but this bird is rarely recorded in Britain.

Harlequin was a colourfully costumed character in the Italian *Commedia dell'arte* theatrical tradition: this duck has a bright suit of feathers.

Histrionicus relates to the theatre and is therefore linked to the origins of the name Harlequin.

Older references show only three species of scoter, but in recent years genetic studies have brought about several splits, which have, in effect, doubled the number of species. These species can all be separated visually in good viewing conditions.

Common Scoter *Melanitta nigra*

This is the **commonest** of the scoter species normally found in British waters. Small numbers breed in northern Scotland.

Scoter was long undeciphered, declared by the OED as 'of unknown origin'. Lockwood suggested that it was a corruption of a hypothetical name *sooter* (a black duck), but American vernacular usage and the roots of the French name *macreuse* together strongly suggest that the origin is *Sea-Coot* (see *British Birds*, July 2013, and *Lapwings, Loons and Lousy Jacks*, 2016). The word is now applied to a number of closely related species in a generic sense.

The Greek *melas* (black) and *netta* (duck) combine for the generic name *Melanitta*.

The blackness is underlined by *nigra*, the Latin for 'black'.

Black Scoter *Melanitta americana*

This was formerly treated as the American subspecies of Common Scoter, but was split in 2010.

It has been recorded in British waters.

Black is a fairly obvious name to choose, though the Latin equivalent is still attached to Common Scoter.

The specific name *americana* is self-evident.

Velvet Scoter *Melanitta fusca*

This is a regular winter visitor to British waters.

Velvet was first used by Ray in 1678 because the feathers were 'soft and delicate', though his was the *Velvet Duck*. The current form was first used by John Fleming in 1828.

The Latin *fusca* conveys a sooty blackness.

White-winged Scoter *Melanitta deglandi*

This American species was formerly considered to be a subspecies of Velvet Scoter, but was split in 2010. It has been recorded in British waters.

The name **White-winged** is not that helpful, since the feature is shared with Velvet and Stejneger's Scoters. The white is in fact limited to the secondaries, but shows clearly in flight in contrast to the all-black wings of Common, Black and Surf scoters.

The specific name *deglandi* is an 1850 dedication by Charles-Lucien Bonaparte to his contemporary, the French naturalist, C.D. **Degland** (1787–1856).

Stejneger's Scoter *Melanitta stejnegeri*

This is the former Eastern subspecies of Velvet Scoter, also known as the *Siberian* Scoter. It has been reported in European waters.

It is named after the Norwegian-American biologist L.H. **Stejneger** (1851–1943).

Surf Scoter *Melanitta perspicillata*

This American species has long been noted as a rare winter vagrant to Britain.

The name **Surf** comes from the bird's propensity to forage on the surf line.

The specific name *perspicillata* echoes that of Steller's Eider, as another way of saying 'spectacled'. The male has a white eye but no eye-ring, so a large white oval at the base of the bill is the more likely reason for the image. The most distinguishing features are a frontal white patch and another on the nape, which give the head a broken profile.

Long-tailed Duck *Clangula hyemalis*

This widespread duck occurs on both sides of the Atlantic, mainly in colder waters.

Long-tailed is self-evident, where the male wears the longest tail feathers of all ducks. That name was first formally used by George Edwards in 1750.

Clangula is unique to this species and is rooted in the Latin *clangere* (to resound). That seems to be rooted in its far-carrying yodelling call, which also dominated in a number of local British names (see note below).

The specific form *hyemalis* is linked to the winter, possibly because of the drake's whiteness, though probably because this very northerly breeder is seen by more people in winter.

* In the past the bird carried a good many colourful local names, particularly in northern Britain. Lockwood lists *Calloo, Coldie* and *Darcall* as names based on the call. *Coal-and-Candlelight* seems to be a fanciful extension of *Coldie*. *Sea Pheasant* was also used for the Pintail and is self-evident in its analogy.

* The Americans, on the other hand, formalised a settlers' name, *Oldsquaw*. E.A. Choate explains the linguistic and social origins of this. The second element of the word was a native Massachusetts word for a woman, which was often used disparagingly or in a racist manner. On a lighter level, Choate cites A.C. Bent who provided a long list of mischievous names (including the delightful *My Aunt Huddy*), but generally on the *Noisy Duck* theme. The formal name became increasingly embarrassing for its political incorrectness. In the early twenty-first century, the need to harmonise names internationally coincided more or less with a breeding study requiring the cooperation of local native peoples. The formal adoption of the British name was both practical and diplomatic.

Common Goldeneye *Bucephala clangula*

This Eurasian species breeds in Scotland and occurs more widely in winter in both coastal and inland waters.

Common is only needed to separate it from Barrow's Goldeneye, so it is rarely used.

Goldeneye is self-evident, though some other ducks do have golden eyes, notably the Tufted Duck.

Bucephala derives from the Greek *bous* (ox) and *cephalē* (head) and is thus related to the word buffalo. It is an apt image for a bird with a large, round head.

The specific form *clangula* may point more to its relationship to the Long-tailed Duck than to either the sound of its voice of or of its wings.

Barrow's Goldeneye *Bucephala islandica*

This is a North American species, which also breeds in Iceland.

It is a very rare vagrant to Britain.

Sir John **Barrow** (1764–1848) was second Secretary to the British Admiralty and founded the Royal Geographical Society, where he supported the exploration of Arctic areas. However, the AOS announced in 2023 that all eponyms in vernacular names of American species will eventually be replaced.

The specific name *islandica*, for Iceland, was originally chosen by the German naturalist Johann F. Gmelin.

Bufflehead *Bucephala albeola*

This American species has been recorded occasionally in Britain as a genuine vagrant.

Bufflehead relates directly to the roots of the genus *Bucephala* and is clearly very close to *buffalo*.

The specific name *albeola* is a diminutive form of *alba* (white), which seems appropriate for a small duck with a lot of white.

∗ Coward had it as *Buffel-headed Duck* in his 1926 list.

Sawbills

Several species of duck are known colloquially as 'sawbills'. They have in common long, narrow bills with saw-like edges that enable them to grip slippery fish.

'REDHEADS'

The females of both Goldeneye species and of the 'sawbills' differ considerably from their respective males, but share certain characteristics with each-other, in that they all have a grey body and a rufous-brown head. First winter males resemble females. For that reason, females and immature birds are known colloquially as 'redheads'. (N.B. This should not be confused with the American *Aythya* species.)

Smew *Mergellus albellus*

This is a scarce winter visitor to Britain from the Eurasian Arctic.

Mergellus is a diminutive of *Mergus* (see below).

In a ploy similar to the one used with Bufflehead above, the specific name *albellus* uses a different diminutive of *alba*.

A traditional and slightly thoughtless name for the drake was *White Nun*.

Hooded Merganser *Lophodytes cucullatus*

This American species is usually treated as an escape in Britain, though one record from the year 2000 was finally accepted as a wild bird.

Hooded refers to one of the most dramatic mobile crests exhibited in the duck family. The male's is white and black, while the female's is buff-orange.

Lophodytes is derived from Greek and means 'crested diver'.

The Latin *culcullus* (hood) is the root of **cucullatus**.

Common Merganser *Mergus merganser* (INTERNATIONAL AND AMERICAN NAME)
Goosander *Mergus merganser* (BRITISH NAME)

This species occurs on both sides of the Atlantic.

Common is needed in North America, where Merganser covers three different species.

Merganser is a late confection of Conrad Gessner in 1555, which was Anglicised by Edwards in 1747 for the Red-breasted Merganser and adopted by Pennant in 1768. It unites *mergus* (see below) with *anser* (goose). The name is the preferred form in North America.

Goosander is the older word in English, first in evidence in 1622 in the form *Gossander*. It appears to combine 'goose' with the German *ende* (duck), but Lockwood was not wholly convinced by that simplicity.

Mergus seems to have been some sort of diving bird: it is recorded by Pliny, Varro and others. (Compare 'submerge'.)

Red-breasted Merganser *Mergus serrator*

This species occurs on both sides of the Atlantic.

Red-breasted is self-evident (though it is more brown than red). The qualifier is generally not needed in Britain, where *Merganser* suffices alongside *Goosander*.

The specific name **serrator** is for the particularly pronounced serrations on the bill.

Stifftails

Birds of this group are not native to Britain, but one species became an important and eventually unwanted guest. Another is native to Spain.

Ruddy Duck *Oxyura jamaicensis*

In the late twentieth century, this American species escaped from Sir Peter Scott's Slimbridge wildfowl collection and established itself as a feral breeding species in Britain. Scott, a dedicated conservationist, did not foresee that some of the birds would make their way to France and begin to threaten the genetics of the endangered wild Spanish population of its close relative, the White-necked Duck. He died before a massive drive was put in motion to eradicate the non-native species. The controversial eradication was an almost total success, but was hugely costly.

It has been argued that individuals may turn up as genuine vagrants from America, but that seems fairly unlikely.

Ruddy Duck refers to the russet-brown body of the bird. (N.B. This name should not be confused with the Ruddy Shelduck, which is a wholly different species.)

The Greek *oxus* and *ura* combine to make *Oxyura* (a sharp-pointed tail).

The specific name *jamaicensis* dates from 1789, when it was described by Gmelin. Jamaica is at the southern extremity of its natural range.

White-headed Duck *Oxyura leucocephala*

The sedentary and isolated Spanish population is the only one in Western Europe. It also occurs in Greece and Near Asia. A great deal of effort has been put into its conservation in Spain, so its protection from the Ruddy Duck was seen as a matter of great urgency. The species has not been recorded as a wild bird in Britain.

White-headed and *leucocephala* mean the same thing.

2. NUMIDIDAE: GUINEAFOWL

The family name is based on *Numida*, see below.

Helmeted Guineafowl *Numida meleagris*

This African species is probably extinct in Morocco, where it was once native, but it is farmed in parts of Europe, especially France, where escapes have sometimes established feral flocks.

Helmeted describes the rather elaborate pattern of facial skin, wattle and plumage that give the head a distinct shape.

Guineafowl implies, erroneously, that these birds were originally only from the West African country of Guinea, but the family is much more widespread than that. *Guinea* has Portuguese roots and originally described the peoples south of the Sahara. The word became a generalisation and was eventually applied to exotics, such as the Guinea Pig (from South America), with little regard for geographical accuracy (see Muscovy Duck and Turkey for similar cases).

Numida refers to an ancient region of North Africa (now in modern Algeria), though there are probably no wild populations north of the Sahara today.

The specific name *meleagris* is the Greek for 'guineafowl'.

3. ODONTOPHORIDAE: NEW WORLD QUAIL

The family name comes from the wood quails of the genus *Odontophorus*, which means 'bearing teeth'. In ornithology this normally suggests a notched or serrated bill. Confusingly, a succulent plant genus bears the same name because of its stubby spikes, though under the rules of the Linnaean system there can be no duplication of specific names.

California Quail *Callipepla californica*

This American species was introduced into France and has a feral population in Corsica.

California is the core native region of this species.

Quail is from the Old French *quaille* (see Common Quail below).

Callipepla means 'beautifully clothed'.

The specific name *californica* is self-evident.

Northern Bobwhite *Colinus virginianus*

This is another American species that has established in parts of France and Italy after introduction.

Northern separates this from three more southerly species in America.

Bobwhite is echoic, based on the call.

Colinus derives from *zolin*, a word for partridge in the Central American Nahuatl language.

The specific name **virginianus** relates to the State of Virginia, though the bird is much more widespread in its natural range. (The first colony of Virginia was chartered by Queen Elizabeth I of England, who was dubbed the Virgin Queen.)

4. PHASIANIDAE: Pheasants, Partridges, Grouse

The family name is based on *Phasianus*.

Many members of this family have traditionally been introduced and/or managed for so-called field sports. As a result, one of the most widely used collective terms is 'Gamebirds'. This expedient brings together very disparate names such as pheasant, partridge, grouse, ptarmigan, snowcock, quail and others in a convenient collective term.

To supply the shooting industry, huge numbers of partridges and pheasants are raised in captivity (many imported) and released every year. In addition to that, several very ornamental species are also kept in a semi-domestic way. As a consequence there are plenty of instances of birds being transplanted into non-native regions. Within this family, extra-limital records are almost all due to the intervention of humans.

Among these species there are several examples of considerable sexual dimorphism, in which showy males compete for females. In some cases the males can be physically aggressive and have developed spurs to damage an opponent, but some have developed 'lekking' behaviour, in which competition takes the form of display in a chosen arena.

Wild Turkey *Meleagris gallopavo*

The well-known farm bird originates from wild birds found in America. Today there are thriving introduced populations of the latter in Germany and the Czech Republic, where they are sustained for commercial hunting.

Use of the adjective **Wild** emphasises the fact that the Turkey exists in both natural and domesticated populations. Its domestication has been traced back almost 2,000 years, but the truly wild populations have long been popular with shooters in North America.

The name **Turkey** originates in a confusion with guineafowl (see above), which were first imported into Europe via Turkey. Importation of the American species began very soon after the first Spanish and Portuguese incursions into the New World. In England the name Turkey was one of several exotic misnomers attached to the species in various European languages: the French *dinde*, for example, began life as *un poulet d'Inde* (an Indian hen). One of the most curious facts is that the English name was well established before it travelled with early settlers to North America at the start of the seventeenth century to become the standard name for the species there. Not surprisingly, given that there remain very few traces of pre-Columbian names among North American birds, it took precedence over a number of Native American names, some of which are still on record.

Meleagris is originally Greek for 'guineafowl' (see above).

To complicate the matter further, the specific name *gallopavo* unites the words gallus (hen) with *pavo* (peafowl).

Hazel Grouse *Tetrastes bonasia*

This is a sedentary species of mixed forests in Northern and Eastern Europe. It does not occur in Britain.

Hazel relates (loosely) to a habitat type.

Grouse replaced the long-preferred form Hazel *Hen*.

Tetrastes appears to be related to *tetrix* (Black Grouse) and *Tetrao* (Capercaillie) as Greek words meaning some sort of gamebird.

The specific name ***bonasia*** seems most likely to be from an Italian name for the Hazel Grouse.

Red Grouse *Lagopus scotica*

This is the British endemic form of the Willow Ptarmigan. The decision as to whether this is a species or a subspecies is a long-running saga: it is treated here as the former, as that is a recommendation pending action by the IOC. The bird is mostly confined to upland moors, where it is often intensively – and controversially – managed by driven shooting interests.

Red is for the rufous coloration of the male in particular. Unlike Willow Ptarmigan it does not have white body plumage at any season.

Grouse is probably related to the Late Latin *gruta* (field hen).

Lagopus likens the well-feathered legs to the fur-covered feet of a hare: cf. Lagomorphs (hares and rabbits).

The specific name ***scotica*** (Scottish) covers birds found in all areas of the British Isles.

* A confirmation of the proposed split would return the bird to the status recognised by Coward in 1926. S. Madge and P. McGowan made a case for the split from Willow Ptarmigan in 2002. Current IOC Recommendations to that effect by George Sangster et al. (2022) suggest that it should be treated separately on the basis of molecular evidence and plumage differences. A secondary aspect of the decision to confirm the split will be whether *scotica* will revert to *scoticus*: there seems to be no consistency at present.

Willow Ptarmigan *Lagopus lagopus*

This is the Continental equivalent of the Red Grouse, formerly called the Willow *Grouse*.

Willow is misleading since the bird is more likely to be found in birch and conifer forests.

Ptarmigan (see below).

Rock Ptarmigan *Lagopus muta*

This species breeds in Britain only at high altitude in Scotland.

Rock indicates that the species favours a high mountain habitat above the tree-line. That qualifier was added in the early twenty-first century when the name Ptarmigan, which had long served for this species alone, was pressed into wider use for other birds of this genus.

Ptarmigan is an echoic name of Gaelic origins, *tàrmachan* (croaker). It was given its current pseudo-Greek spelling by Sir Robert Sibbald in 1684.

The specific name ***muta*** means 'silent' or 'dumb', and applies because of the limited voice of the bird.

(Western) Capercaillie *Tetrao urogallus*

This is a scarce and declining species in the north of Scotland, but is also found in Eurasian forests.

The qualifier **Western** is only used to distinguish this from the Eastern (Black-billed) species.

Capercaillie derives from a Gaelic word meaning 'horse of the woods'.

Tetrao is a Greek word for a gamebird, possibly the Black Grouse.

The specific name *urogallus* appears to mean 'long-tailed fowl'.

* In this case, the males are much larger than the females, and win control of a number of them by establishing their dominance in the lek, where older males dominate the centre-stage.

Black Grouse *Lyrurus tetrix*

This is a scarce, localised breeding species in parts of upland Britain, but it is also found in northern Eurasia.

Black: only the male is black (though with some white patches). The female, like many others in this family, has a cryptic plumage suited to blending with surrounding vegetation when nesting and feeding.

Lyrurus means 'lyre-tailed', relating the curved tail feathers to the ancient musical instrument.

In the writings of Aristotle, *tetrix* was some sort of ground nesting bird.

* Like Capercaillie, the males are much larger than females and win the right to multiple mates in a lek.

Caucasian Grouse *Lyrurus mlokosiewiczi*

This is a localised montane species.

Caucasian is so-named because it is restricted to the Caucasus Mountains and to a similar area of Turkey.

L.F. **Mlokosiewicz** (1831–1901) was a Polish naturalist specialising in the Caucasus region.

* This species is similar to the previous one, and is sometimes called *Caucasian Black Grouse*.

Grey Partridge *Perdix perdix*

This species is native to Britain and to a wider area of Eurasia.

Grey is apposite, but it is a rich purple-grey. It was long also known in Britain as the *Common* Partridge, but has sadly declined generally.

Partridge evolved in the English language from the imported Old French word *pertriz* (Mod. French *perdrix*) from the verb *perdre*, to lose.

That in turn is rooted in the Latin *perdix*, because the cryptic plumage of the bird camouflages (loses) it.

Reeves's Pheasant *Syrmaticus reevesii*

This introduced Chinese species breeds in the wild in France and the Czech Republic.

Reeves and *reevesii* are used because the Rev. John **Reeves** (1774–1856) introduced the species to Britain.

Syrmaticus derives from the Greek *surma, surmatos* (a robe with a long thin train, referring to the tail).

Golden Pheasant *Chrysolophus pictus*

Small self-sustaining, ornamental populations of this introduced Chinese species exist very locally in Britain, though their future currently appears very precarious.

Golden is for the dominant colour of the handsome male.

Chrysolophus refers to the 'golden crest'.

There are several other colours in the plumage too, so *pictus* here means 'painted'.

Lady Amherst's Pheasant *Chrysolophus amherstae*

A small feral, ornamental population of this Southeast Asian species lasted for many years in Bedfordshire, but is almost certainly now extinct.

Sarah, Countess **Amherst** (1762–1838) is commemorated. She was an English naturalist living in India when she collected the first specimen.

* See also Mrs Anna Blackburne of the Blackburnian Warbler for another rare example of a commemorated female naturalist of this era.

Common Pheasant *Phasianus colchicus*

The Common Pheasant was kept in domestication in Britain by the Romans but has been living here in the wild since it was reintroduced in Norman times.

As a result of constant replenishment of the population, it is literally **common**, and has been estimated, in biomass terms, to outstrip the population of native Woodpigeons in Britain.

Pheasant evolves from the Old French *faisan*, which in turn is rooted in the Latin form *phasianus*.

Phasianus relates to the River Phasis (the modern River Rioni in Georgia), where legend tells us that Jason and the Argonauts first found the birds in the wild while seeking the Golden Fleece.

The specific name *colchicus* also relates to that legend, since the River Phasis was in the land of *Colchis* (now part of Georgia).

* The original form lacks the white neck ring, which belongs to birds from a more easterly heritage. The wide range of colour variants seen in this species is due to the fairly indiscriminate introduction of forms from various parts of a wide Asian range. The population is massively reinforced by millions of captive-bred birds released by the shooting industry each year.

Indian Peafowl *Pavo cristatus*

This highly ornamental Asian species has long been kept in semi-domesticity in many parts of Europe since it was introduced into the region by the Phoenicians. It is often self-sustaining in a feral state, including a population that has existed near Oxford since the nineteenth century.

Indian (also sometimes *Common* or *Blue*) distinguishes this from the lesser-known *Green Peacock*.

Peafowl (and **Peahen**) came into English usage after the establishment of the word *Peacock* in Medieval English (Chaucer's fourteenth-century spelling was *pekok*). The *pea* element was a Germanic evolution of the Latin *Pavo*, which provides the modern generic name.

The specific name *cristatus* refers to the crest.

Black Francolin *Francolinus francolinus*

This is a species found in Cyprus and Turkey and eastwards.

Black rather exaggerates the plumage of the dark, mottled male with its prominent white cheek patch.

Francolin (and its Latin form) has roots in an Italian name (*francolini*) for a free-running fowl that was protected by the authorities. It travelled into English usage via French.

Caucasian Snowcock *Tetraogallus caucasicus*

This is a species found in the **Caucasus** region, hence the names.

Snowcock marries the dominant weather conditions of their high-altitude lifestyle with a word of Anglo Saxon roots that was originally used as *fowl* or *bird* might be today.

Tetraogallus marries the name *tetrao* (see above) with *gallus* (hen).

Caspian Snowcock *Tetraogallus caspius*

The mountains of the **Caspian** region and south-eastern Turkey are home to this species.

Sand Partridge *Ammoperdix heyi*

This is a Middle-Eastern species.

Sand Partridge tells of its desert lifestyle.

Partridge: see Grey Partridge above.

Ammoperdix is from the Greek *ammos* (sand) and *perdix* (partridge).

The German collector Michael **Hey** (1798–1832) is commemorated in *heyi*.

See-see Partridge *Ammoperdix griseogularis*

This is another Middle Eastern species.

See-see is based on its whistling call.

Partridge: see Grey Partridge above.

The name *griseogularis* refers to the male, which is in fact 'grey throated'.

Common Quail *Coturnix coturnix*

This migratory species migrates north from Africa to breed in Britain.

The use of **Common** is only needed when other quail species are under discussion. It is unique in Europe.

Quail is rooted in the Old French *quaille*. It seems probable that the root was a three-syllable word, e.g. the Old High German *quahtala* or the Medieval Latin *quaccula*, since it is thought to be echoic. That idea makes sense alongside the common English rendition of the call, which is the three-syllable phrase 'wet-my-lips'.

Coturnix – another three-syllable rendition – is Latin for 'Quail'.

Red-legged Partridge *Alectoris rufa*

This species was originally introduced into Britain by Charles II, when he returned from exile in France in 1660. Red-legged Partridges are normally native to France, Spain and Portugal. For that reason they are still often called *French* Partridges.

Red-legged is a feature common to all *Alectoris* partridges, but its use here originates with the need to distinguish the introduced bird from the native *Common* or *Grey* Partridge.

The shooting industry replenishes the stock by the annual release of literally millions of hand-reared birds, a fact that renders the generic name *Alectoris* even more meaningful, since that is from the Greek for 'farmyard chicken'.

Partridge: see Grey Partridge above.

The specific name, *rufa*, refers to the rufous tones of the body plumage.

Rock Partridge *Alectoris graeca*

This species replaces *rufa* from south-west France through to Greece.

Rock underlines the fact that this species generally breeds in rocky, treeless landscapes.

Partridge: see Grey Partridge above.

The specific name *graeca* simply means 'Greek'.

* In contrast to the image of semi-domesticity in the name *Alectoris*, this species was made famous as 'the king of gamebirds' in Marcel Pagnol's 1957 autobiographical work, *La Gloire de Mon Père*. His father's 'glory' is the local fame of having shot a pair (albeit by a lucky accident). The French name *la bartavelle* links the bird's call to the clicking of a mill's mechanisms.

Barbary Partridge *Alectoris barbara*

This mainly North African species is also found in Sardinia.

Barbary is an archaic English name for the north-west African coast, which derives from the Berber people of the area.

Partridge: see Grey Partridge above.

The specific name *barbara* is the Latin version of 'Barbary'.

Chukar *Alectoris chukar*

This species is found from the Eastern Mediterranean and eastwards.

Chukar is the Hindi name for the species.

* For a while, birds were bred and released into the English countryside to give variety to the shooting industry, but the practice was outlawed in 1992 to protect the genetic integrity of Red-legged Partridges, which were deemed to have some historical right to their status.

Double-spurred Spurfowl *Pternistis bicalcaratus*

This is a species of western Morocco.

Double-spurred describes the two rear-facing spurs on the foot.

Spurfowl has recently replaced the inaccurate use of Francolin for a genus that is generally similar, but not closely related.

Pternistis tells why the spurs exist since it means 'one who strikes with his heel'. Males of several species in this family use such weapons to compete with rivals.

The double spurs of the name are underlined in the word ***bicalcaratus***.

Erckel's Spurfowl *Pternistis erckelii*

This North-East African species has a local feral population in Italy.

This species, formerly classed in the genus *Francolinus*, was reclassified in 2020.

Theodor **Erckel** (1811–1897) was an assistant to Eduard Rüppell, the German naturalist, who described this species.

Nightjars to Pigeons

5. CAPRIMULGIDAE: Nightjars

The family name is based on *Caprimulgus*.

Nightjars were once considered to be related to owls, a fact that is reflected in several alternative folk names for the European form. They are, in fact, closely related to potoos, frogmouths and the Oilbird, and are not too far removed from the swifts.

Nightjar is one of several names that emerged in England. It was first recorded in 1630 and was formalised after Yarrell used it in 1843. In the Americas, the name Nighthawk, an old traditional English name, which travelled with the settlers, is used alongside several other names of a more local origin.

Common Nighthawk *Chordeiles minor*

This American species is a rare vagrant to Britain.

It is **Common** in the context of a number of other species of nighthawk.

Nighthawk, as an alternative to Nightjar, was used in seventeenth-century Britain and was eventually preferred in the Americas. In flight, the birds are shaped somewhat like a small falcon, hence the comparison with a bird of prey.

Chordeiles is an ambiguous name, based on two Greek words: it appears to combine the idea of 'dancing' or 'moving about' with that of 'early evening' to convey the bird's crepuscular habits.

The specific name *minor* means 'lesser'.

Red-necked Nightjar *Caprimulgus ruficollis*

There appears to be only one nineteenth-century British record of this species, which normally breeds in Spain and North Africa.

Red-necked and *ruficollis* describe a rufous tinge on the collar and upper breast, which is the most obvious plumage distinction from the smaller European Nightjar.

Nightjar (see below).

European Nightjar *Caprimulgus europaeus*

This is a relatively uncommon pine-forest species in Britain, as a summer migrant from Africa.

European underlines the geographical range.

Nightjar conveys both the bird's nocturnal habits and the noise that it makes. The element *-jar* indicates a somewhat annoying noise, which is a little unjust: the song is often described by the kinder word 'churring'.

Caprimulgus returns to the folklore of Classical times, which maintained that the nightjars were 'goat-suckers'. The word derives from the Greek *capra* (nanny goat) and *mulgere* (to milk). The truth behind that story is that the herds attracted and disturbed insects, which were the real target of these insectivores, but it is not surprising that birds seen around flocks in poor light would be misunderstood. The myth persisted in such a way that the alternative name *Goat Sucker* was long used in everyday English.

The specific name *europaeus* is obvious in its meaning, but is an unusual example of the use of the word in this context: the Linnaean System was originally a Eurocentric device and had little need of such a word for local species. The only other example is the Nuthatch, *Sitta europaea*.

* Alternative names such as *Goat Owl*, *Fern Owl* and *Churn Owl* were also in traditional use. The *Owl* element derives from its nocturnal habits, *Goat* of course relates to the Goatsucker legend, *Fern* to the bracken of its nesting habitats and *Churn* to the churring song.

Egyptian Nightjar *Caprimulgus aegyptius*

This species has been recorded only twice in Britain, in 1883 and 1984. It normally breeds in North Africa and the Middle East.

Egyptian and *aegyptius* tell part of its geographical story.

Nubian Nightjar *Caprimulgus nubicus*

This very small desert-dwelling nightjar ranges into southern Israel, Egypt and Arabia.

Nubian and *nubicus* link the bird to Nubia, a region of the Upper Nile in Egypt.

Golden Nightjar *Caprimulgus eximius*

This Sahel species was added to the Western Palearctic list only in 2019, when it was proved to be breeding in the western Sahara.

The names **Golden** and *eximius* (distinguished) both refer to the unusual plumage colour of this species.

6. APODIDAE: Swifts

The family name is based on *Apus*.

In times past swifts were thought to be a form of hirundine, often going under such names as Black Martin. However, they are far more closely related to the nightjars and, surprisingly, even more closely to the hummingbirds. In short they are not passerines, in spite of their resemblance to the swallow family. This is a good example of convergent evolution, with both families developing similar shape and structure for rapid aerial flight and feeding.

At a level of folklore and superstition, swifts were often treated with some suspicion, as witnessed by such names as Devil Screamer.

White-throated Needletail *Hirundapus caudacutus*

This Central Asian species has been recorded surprisingly often as a vagrant to Britain, but it is known to roam widely.

White-throated is self-evident.

Needletail describes a tail that is composed of short, spiny feathers.

Hirundapus combines *hirundo* (swallow) with *apus* (swift).

The specific name *caudacutus* (literally 'sharp-tail') repeats the 'needletail' image.

Chimney Swift *Chaetura pelagica*

This American species is a rare vagrant to Britain and is the only American swift recorded as crossing the Atlantic.

Chimney refers to the fact that the bird traditionally nests in chimneys. Before the European colonisation of North America the birds used tree hollows, but with forests felled for agriculture and the building of artificial alternatives in the form of houses with chimneys, the birds changed their habits and now rely almost exclusively on such structures. Modern architecture does not suit them so well: in Wolfville, Nova Scotia, a single dairy chimney has been preserved as a refuge for a few of them.

Chaetura literally means 'hair-tail', which is very like 'needletail' above.

The specific name *pelagica* is misleading, because it has nothing to do with sea-going. Instead, it derives from a nomadic Ancient Greek tribe the *Pelasgi*.

Alpine Swift *Tachymarptis melba*

This Mediterranean species has been recorded as a vagrant to Britain.

Alpine is used in the sense of 'montane', though the bird does not necessary require or use only high altitude regions. It generally frequents rocky habitats.

Tachymarptis conveys its speed in a name meaning 'fast catching'.

The specific name *melba* appears meaningless, until it is seen as an accidental compression of *melanoalba* or *melalba* (black and white). The error seems to originate in Linnaeus's notes (see Laughing Gull for a similar compression).

Cape Verde Swift *Apus alexandri*

This Atlantic island species is also recorded in north-west Africa.

Cape Verde is the main base of this species, which is also known as *Alexander's Swift*, with reference to the specific designation, *alexandri*. Lieutenant Boyd **Alexander** (1873–1910) was an English army officer and ornithologist who is also commemorated in Boyd's Shearwater. The use of both his forename and his surname for different birds is unusual, if not unique.

(Common) Swift *Apus apus*

This is a summer migrant in Britain.

Common is not needed until there are more exotic swifts in the air.

Swift is a simple and wholly apt name for this fast-flying bird. This is a good example of a plain English name for a single species that has radiated as a generic form. It is significant that all the other swifts listed here require a descriptive or geographical epithet.

Apus suggests that the birds are 'without feet'. In reality the feet are present, but are extremely reduced because the birds land only to nest. Amazingly, they sleep on the wing and mate in the air during flight.

Plain Swift *Apus unicolor*

This is a small species of Madeira and the Canaries.

Plain and *unicolor* say it all by telling us that this is a featureless bird.

Pallid Swift *Apus pallidus*

This Mediterranean species is a scarce vagrant to Britain.

Pallid and *pallidus* indicate that this is a relatively pale species, the comparison being with *Apus apus*.

Pacific Swift *Apus pacificus*

This species has rarely been recorded in Britain. Like the White-throated Needletail, it tends to wander widely.

Pacific and *pacificus* indicate its normal domain.

Little Swift *Apus affinis*

This North African species is a very rare vagrant to Britain.

Little Swift is distinctly smaller than any of the other swifts listed here, other than Chimney Swift, which is of a similar size and also square-tailed, though the white rump of this bird is very different.

The specific name *affinis* usually means that a bird resembles another: the comparison species is probably the Common Swift. The same device is used with Lesser Scaup in view of Greater Scaup.

White-rumped Swift *Apus caffer*

This is a mainly sub-Saharan species, but small numbers now breed in Spain and North Africa. It has been recorded once in Britain, in 2018.

White-rumped is self-evident, but in this case the bird has a long, forked tail.

The specific name *caffer* is an historic reference to South Africa as *Kaffirland* (from an old Arabic name).

7. OTIDAE: Bustards

The family name is based on *Otis*.

Bustards of any sort are rarely seen in Britain, though Great Bustards were once native in open grassy areas. They were hunted out of existence here by 1832. Other species occur, or have occurred historically, as rare vagrants.

Great Bustard *Otis tarda*

This is a rare vagrant. There are resident populations in some parts of Europe, but those in Russia are migratory. During the twenty-first century, a population has been reintroduced into Wiltshire, and this now appears to be self-sustaining.

The bird is literally **Great** and has the reputation of being one of the world's heaviest flying birds.

Bustard derives from two French forms (*outarde*/*bistarde*), which were imported in the Middle Ages. The key word is the latter, which is a compression of the Latin *avis tarda* to an Italian form *vistarda*, which then morphs via French into *bistarde* (Lockwood).

Otis means 'eared' and is confusingly similar to the owl genus *Otus*. In this case it refers to the prominent facial feathers of the Great Bustard. (Such near duplication of this sort is not unprecedented: the duck genus *Aythya* is echoed by the auklet genus *Aethia*, for instance.)

The specific name *tarda* is particularly odd, since it appears to be the Latin for 'late'. According to the Roman Pliny, however, it was a Latinised form of a local Iberian name for the bird and has nothing at all to do with lateness.

Arabian Bustard *Ardeotis arabs*

This species still has a foothold in Morocco.

Arabian and its Latin equivalent *arabs* relate to the fact that part of the population is found in the Arabian Peninsula, even though most of the population is found in sub-Saharan Africa.

Ardeotis unites *Ardea* (heron) with *Otis* (bustard) to describe a tall, rather rangy form of bustard.

Houbara Bustard *Chlamydotis undulata*

This is a species of North Africa and the Middle East and has colonised Lanzarote.

Houbara was originally the French rendition of the Arabic name for the bird.

Chlamydotis uses the Greek word for 'cloak' to create a name meaning 'cloaked bustard'.

The specific name *undulatus* draws attention to the wavy markings on the mantle plumage.

Macqueen's Bustard *Chlamydotis macqueenii*

This long distance migrant was recorded four times in Britain during the nineteenth century, when it was much more numerous in its native range than it is today. It migrates between breeding areas in Central Asia and wintering grounds in Arabia, Afghanistan and Pakistan.

General T.R. **MacQueen** presented a specimen to the British Museum in about 1832.

Denham's Bustard *Neotis denhami*

This species occurs in south-west Mauritania, as well as widely south of the Sahara.

Major Dixon **Denham** (1786–1828), was an English naturalist and explorer.

Neotis links the Greek *neos* (new) with the traditional name *Otis* (see Great Bustard above) to create a further genus.

Nubian Bustard *Neotis nuba*

This is a species is found in Mauritania, but also in the southern edges of the Sahara and into the northern Sahel.

The vernacular name comes from the **Nubian** region to the south of Egypt.

The specific name *nuba* does the same job, but compare with *C. nubicus* (Nubian Nightjar) above. Two different ornithologists (Phillip Cretzschmar and Martin Lichtenstein respectively) were at work here.

Little Bustard *Tetrax tetrax*

This is a very rare vagrant to Britain.

Little is the smallest of its family in the region: it is the size of a hen Pheasant.

Tetrax is the Greek for a type of gamebird (compare *tetrix* for Black Grouse).

8. CUCULIDAE: Cuckoos

The family name is based on *Cuculus*.

The migratory Cuckoo – the one that says its name – was once much more common in Britain and was widely fêted as a major harbinger of spring. In medieval Europe its two-note song marked the end of the dark winter, of cold, and of food shortages. It heralded the start of a new season of growth and warmth. The most famous manifestation of the joy that it brought is found in the round 'Sumer is icumen in, Llude sing Cuccu!', which was first sung at Reading Abbey in the Middle Ages. Today the bird is far less common and is heard by relatively few people, so its symbolism is largely lost. Like many early folk names, the name of this one bird was eventually used more widely as the generic name for a whole family, but only Common Cuckoo has the call that gave the family the name.

The other side of the bird's reputation is distinctly negative and is based on its actual behaviour. As a nest parasite it was considered to be a cheat, and this reputation gave rise to a number of derogatory terms, such as 'cuckold', 'a cuckoo in the nest' and the sinister modern term 'cuckooing'. It is clearly a high-risk strategy, since it is estimated that the female Common Cuckoo lays more eggs than most other species, simply because a number of her eggs are rejected by would-be hosts. Not all cuckoo species are nest parasites, however: the coucals raise their own young, as do the two American species listed below. Cuckoos are not the only nest parasites, of course – see Brown Cowbird later.

Senegal Coucal *Centropus senegalensis*

This species breeds in Egypt.

Senegal and *senegalensis* remind us that it has a wide range which stretches into West Africa.

Centropus tells that the bird is 'spur-footed' (Greek *kentron*, spur, and *pous*, foot). Like all cuckoos, they are zygodactyl – that is they have two forward-facing toes and two backward – but the name seems to have been earned by the Greater Coucal, *Centropus sinensis*, which has an exaggerated spur on the inner rear toe. In pheasants and jungle-fowl, for example, such a spur is usually used in battles between cock birds for multiple hens, but the monogamous lifestyle of coucals suggests that this must have another purpose.

Great Spotted Cuckoo *Clamator glandarius*

This Mediterranean migrant sometimes overshoots into Britain.

Great and **Spotted** both appear to have been added very early to describe this bird. The size comparison is with the Common Cuckoo, while 'spotted' contrasts its upper markings with the other's barred breast.

Clamator means 'shouter' and seems to be the deliberate antithesis of the Common Cuckoo's 'melodious'.

The specific name is misleading, since *glandarius* echoes the name of the acorn-eating Jay. However, this largely insectivorous cuckoo merely lives in the trees that produce the acorns, in this case cork oaks.

Jacobin Cuckoo *Clamator jacobinus*

This black-and-white sub-Saharan Africa species has been recorded only rarely in the Western Palearctic, most recently in Algeria.

Jacobin and ***jacobinus*** both relate to the bird's plumage and are references to the black-and-white garb of Dominican friars, who were known as Jacobins in France. The bird was first recorded by the French scientist Georges-Louis Leclerc, Comte de Buffon (1707–88).

* The infamous Jacobin Club of the 1789 French Revolution took its name from a meeting place, a former Jacobin monastery. The word Jacobins was originally a nickname for members of the Dominican Order, given when they established a monastery in the Rue St Jacques (*Jacobus* in Latin).

* The imagery of Jacobin reappears in a hummingbird species, while, for similar reasons, versions of Dominican appear in the names of a gull and a plover.

Yellow-billed Cuckoo *Coccyzus americanus*

This is a scarce vagrant from America, as the specific name ***americanus*** reminds us.

Yellow-billed is contrasted with the next species.

The generic name *Coccyzus* comes from a Greek word meaning 'to cry like a cuckoo' (though in this case the voice is more crow-like).

Black-billed Cuckoo *Coccyzus erythropthalmus*

This is a rather rarer American vagrant.

Black-billed is self-evident.

The specific name ***erythropthalmus*** describes the red orbital ring of the eye.

Asian Koel *Eudynamys scolopaceus*

This South **Asian** species is recorded in the Arabian Peninsula.

Koel is an echoic word: like cuckoo it is based on a two-note call, which is haunting and almost owl-like.

Eudynamys refers to the 'fine strength' of the birds, which are bold enough to parasitize much larger species, such as the House Crow.

The specific name *scolopaceus* derives from *Scolopax* (snipe/woodcock) and refers to cryptic plumage patterns, which are exhibited by the female of this species.

Diederik Cuckoo *Chrysococcyx caprius*

This African species has been recorded in Israel and Cyprus.

Diederik is Afrikaans and is said to derive from the bird's call. The spelling was revised by the IOC in 2012/13 from a corrupted form, *Didric*.

Chrysococcyx is based on the Greek *khrusos* (gold) and *kokkux* (cuckoo).

The specific name ***caprius*** means cuprous (copper-coloured).

(Common) Cuckoo *Cuculus canorus*

This summer migrant is far less **common** today than in the past, but the qualifier avoids confusion with other cuckoo species.

The word **Cuckoo** originated in medieval England. It was based on the echoic Old French *cucu* and rooted in the Latin form. That name eventually replaced the rather dour Nordic name *Gowk*, perhaps because it was so strongly associated with the joy and optimism of the above-mentioned popular round (see also *optatus* below).

Cuculus is the Latin for 'cuckoo'.

The specific name *canorus* means 'melodious'.

* *Gowk* was used mainly in Scotland and the North. It was imported by the Vikings and, according to Lockwood, almost certainly derived from a much older two syllable form that was suitably echoic.

Oriental Cuckoo *Cuculus optatus*

This close relative migrates into areas mainly east of the Urals.

Oriental confirms its easterly bias.

The specific name *optatus* describes the bird as 'welcome' (see introductory note above).

African Cuckoo *Cuculus gularis*

This largely sub-Saharan species is migratory within the **African** continent and is recorded in Mauretania.

The specific name *gularis* draws attention to the throat, which is much paler grey than the upper-parts.

9. PTEROCLIDAE: SANDGROUSE

The family name is based on *Pterocles*.

This family is little-known in Britain, particularly over the past 100 years or more. It is represented by only one species – and that one has an extraordinary history.

Pallas's Sandgrouse *Syrrhaptes paradoxus*

This Asian species has been recorded in Britain (see note below).

Peter **Pallas** (1741–1811) was a German naturalist and explorer.

Sandgrouse applies to a family that has evolved some grouse-like characteristics and lives mainly in arid, open areas.

Syrrhaptes means 'sewn together'. The feathered toes are in fact fused, 'rather suggesting the foot of a rabbit', as Coward put it. There is only one other species in this genus.

The specific name *paradoxus* means 'strange' or 'extraordinary'.

* The BTO records that there were a series of extraordinary irruptions of this species into Britain in 1863, 1888 and 1908. There were even breeding attempts. In 1908, Coward phlegmatically reported that he 'just missed seeing a covey in Cheshire'. He, like others, was baffled by such movements. Such irruptions have not recurred since then, probably because the birds' natural range has contracted owing to the expansion of agriculture in its natural breeding areas. There is one record from Shetland in 1990.

Black-bellied Sandgrouse *Pterocles orientalis*

This widespread species breeds in the Iberian Peninsula, North Africa and the Middle East.

Black-bellied is self-evident.

Pterocles means 'splendid-winged'.

The specific name *orientalis* (eastern) apparently did not take into account the fact that a large part of the range is in Western Europe and North Africa.

Pin-tailed Sandgrouse *Pterocles alchata*

This widespread species breeds in Spain, and in France is limited to La Plaine de Crau in Provence.

Pin-tailed (like the Pintail duck) has long, slender tail plumes.

The specific name *alchata* derives from an Arabic name *al kattar*, which appears in Linnaeus's notes as *katta*.

Spotted Sandgrouse *Pterocles senegallus*

This species breeds in North Africa, but has been recorded as a vagrant in Italy.

Spotted could easily apply to most of this genus. In this case the female is finely spotted on a sandy ground.

The specific name *senegallus* relates to Senegal.

Crowned Sandgrouse *Pterocles coronatus*

This species is found in North Africa and occurs in Israel.

Crowned and *coronatus* mean the same, though it seems an oddly chosen name. The male does have a small black mask, but it is difficult to see the link. The bird was named by Lichtenstein (see next species).

Lichtenstein's Sandgrouse *Pterocles lichtensteinii*

This African and Asian species breeds in Morocco, Sinai and Israel.

M.H.C. **Lichtenstein** (1780–1857) was an eminent German zoologist. The dedication was by Coenraad Temminck.

Chestnut-bellied Sandgrouse *Pterocles exustus*

This African species has been recorded in Egypt and as a vagrant to Hungary.

Chestnut-bellied separates this from the Black-bellied form.

The specific name *exustus* describes the bird's dark plumage as 'burnt' or 'scorched'.

10. COLUMBIDAE: Pigeons, Doves

The family name is based on *Columba*.

The distinction between *Pigeon* and *Dove* is linguistic and cultural, but the ornithological application of the two words is erratic and the differences have no biological foundation. The reason we have two words is down to the mixed origins of the English language.

Dove is the older name, with Anglo-Saxon roots and based on an even older Germanic form. It appears to be related to their diving display flight. That name prevailed until the fourteenth century, though Lockwood points out that the word was not greatly favoured by the early Church owing to its Germanic superstitious link with death. For that reason it does not appear much in documentary evidence. That prejudice was eventually turned on its head when the later Church used the dove as a symbol of divine love and peace, an association that still colours our use of the word.

Pigeon is first evidenced in French as late as the thirteenth century, when *pijon* simply meant a young bird – that is, one making piping noses. By the fourteenth century, it had evolved to fit its present usage, albeit with erratic spellings, when introduced into fifteenth-century English. The current spelling was first seen in 1483.

In modern usage doves tend to be the smaller ones and pigeons the larger species, but there are absolutely no rules. The matter does not generally relate to the genus of the bird nor to its colour. In the first species below we see both forms used for different manifestations of the same species and in the second that the Woodpigeon was long known as the Ring Dove.

The Pigeon family is a large one, with well over 300 species worldwide, yet mainland Europe has only five representative species, all of which are found in Britain. The rest of the recorded species are peripheral to the region or occur as vagrants.

Rock Dove/Feral Pigeon *Columba livia*

This species is rare under the first name and common under the second.

Rock Dove names the truly wild form, a cliff-nesting species from which the domesticated forms were raised, probably millennia ago. It still exists in a relatively pure form on remoter cliffs of northern Britain.

Feral Pigeon, or sometimes *Town Pigeon*, refers to the birds that live freely near human habitation and are descended from domesticated Rock Doves. In the past, free-flying birds were encouraged to breed in the loft-space of houses or in dovecotes, which enabled them and their eggs to be harvested for food. These birds show evidence of selective breeding in their colour variation. Today, fully domesticated forms are known by names such as Racing Pigeon, Loft Pigeon and Fancy Pigeon. Such birds as those were studied closely by Darwin as he formulated his theory of evolution.

Columba is the Latin for 'dove'.

The specific form *livia* is more interesting since it is a medieval concoction for the species, translated into Latin from the Greek *pellos* (dark-coloured) and related to *livid* (dark bluish grey, as in bruising).

Common Wood Pigeon (Woodpigeon) *Columba palumbus*

This is an extremely common species that in recent years has become very familiar in suburban habitats and even urban areas. Resident birds in Britain are boosted in winter by flocks from mainland Europe.

In international usage it is **Common Wood Pigeon** in three words. Its name is so well used in Britain that it compresses into the single form **Woodpigeon** in normal usage.

Historically *Wood Pigeon* was used for both woodland species, but Pennant ensured the usage for this one species. It was long rivalled by the alternative name *Ring Dove*, but that faded after 1912, when Ernst Hartert and his allies favoured the current name.

The specific name *palumbus* is the Latin for 'Woodpigeon'.

* In its breeding habits, the Woodpigeon is not dependent on old trees, as is the Stock Dove: it does well in shrubs, scrub and marginal land. Because such habitat is now widely available, it is much more numerous than the smaller Stock Dove.

* The alternative traditional name, *Ring Dove*, which Coward clearly still preferred in 1926 and which lingered for many more years, was for its white collar. It underlines again the interchangeability of *pigeon* and *dove*. That name is still sometimes heard and is probably the older one: Lockwood suggests that it is probably an ancient Germanic form.

Stock Dove *Columba oenas*

This is the least well-known resident species, because it is generally similar in coloration to the 'standard' Rock Dove and the much more numerous Woodpigeon. Consequently it can be easily overlooked.

Stock relates to the bird's preference for nesting in hollow trunks of trees. (One of its alternative local names was *Hole Dove*.) 'Stock', in this sense, is an Old English word for a trunk or a stump. It is at the root of the word 'stockade' (a defensive fence built of logs) and of 'the stocks' (the wooden frame in which criminals were once punished).

The specific name *oenas* is based on a Greek word for 'pigeon'.

Laurel Pigeon *Columba junoniae*

This species is endemic to some of the Canary Islands.

Laurel relates to the laurel forests of these tropical islands with high rainfall.

The specific name *junoniae* relates to the former names of the islands of La Palma and Gomera, as *Junonia Mayor* and *Junonia Menor*.

Bolle's Pigeon *Columba bollii*

This species is also endemic to some of the Canary Islands.

The bird was named by the English ornithologist F.D. Godman (1844–94), a founder-member of the BOU. He visited the Canaries in 1871 and subsequently dedicated this species to the German naturalist, Carl August **Bolle** (1821–1909), who had written a monograph on the birds of the islands after an extended 1856 visit. Bolle succeeded Christian Brehm (1787–1864) as the chairman of the German Ornithological Society.

Trocaz Pigeon *Columba trocaz*

This species is endemic to Madeira.

Trocaz is a local Portuguese dialect name meaning much the same as 'ring dove'. It traces back to the Latin *torques* (collar), with reference to the bird's diffuse silver-grey collar-patch.

Yellow-eyed Pigeon *Columba eversmanni*

This Asian species has been recorded in Russia.

Yellow-eyed actually refers to a yellow eye-ring rather than to the eye itself.

The specific name *eversmanni* is after A.E.F. **Eversmann** (1794–1860), a Prussian explorer and biologist who worked extensively in Russia. His name also appears later in **Eversmann's** Redstart.

Speckled Pigeon *Columba guinea*

This is a sub-Saharan African species that has been recorded on the north-western coast.

Speckled describes an area of white spots on the wings.

The specific name *guinea* was originally widely used to describe a large area of West Africa.

(European) Turtle Dove *Streptopelia turtur*

This once common migrant is now scarce-to-endangered in Britain.

Turtle evolved in English from the echoic Latin form *turtur* (now the bird's specific name). The song is a gentle purring. In the work of Shakespeare and in the King-James Bible, the word tends to stand alone. Its coupling with **Dove** is in evidence as early as 1300, but that usage became more common during and after the seventeenth century.

Streptopelia is a way of adding a collar (*streptos*) to a dove (*pelia*), but in Greek.

* The song of the Turtle Dove was long an iconic part of the British rural summer, but numbers have plummeted in recent years and the sound is now rare. This bird is a victim of intensive farming practices, of hunting along its migration routes and of subsistence hunting in the Sahel. Isabella Tree and Mark Avery, among others, have written in some detail of the reasons for its tragic decline and about what the species needs in order to survive in Britain.

Oriental Turtle Dove *Streptopelia orientalis*

This species is usually found in Asia, breeding mainly to the east of the Caspian Sea, and wintering on the Indian subcontinent. Odd birds appear in Britain every few years and have been noted more frequently since 2000. Their out of season appearances make them more conspicuous. They are generally darker than the European Turtle Dove.

Oriental and *orientalis* are both self-evident.

* The bird was listed as a separate species in Coward's 1926 British list, though under the name *Eastern* Turtle-Dove. There appears to be only one record prior to that time. Note the use of the hyphen, a device that was in much wider use then.

(Eurasian) Collared Dove *Streptopelia decaocto*

This dove colonised Britain spontaneously during the early 1950s. It is now familiar in many gardens and the great wonder of its arrival is largely forgotten.

Eurasian is only required in an international context.

Collared describes a narrow black neck stripe.

The specific name *decaocto* (eighteen) is based on the Greek story of a servant girl who repeatedly complained about her wages being only 'eighteen pieces'. It is said to imitate the three-note song, so the key word would have probably been pronounced *dec'octo*.

* The story of this bird is the converse of the decline of the Turtle Dove. It was only at the start of the nineteenth century that the birds started to expand into Europe from southern Asia. When the Hungarian, Frivaldsky, named the bird in 1838, it was still relatively new to his part of Europe. James Fisher tracked the expansion in *British Birds*, in a 1953 article that recorded a relatively slow movement through the Balkans and Alps until about 1941, after which it took only a decade for the birds to expand through Western Europe. It was first noted in Britain in 1952. Fisher also reproduced a version of the 'eighteen pieces' legend.

Mourning Collared Dove *Streptopelia decipiens*

This is a widespread African species that has been noted in Egypt.

The name **Mourning** is confusingly similar to that of the American species below and underlines the importance of the scientific alternatives. In this case, the broad, black collar would appear to be the reason for the name. (The bird was first described in 1870. In the Victorian tradition, mourning dress was *de rigueur* and is still symbolised by a black armband.)

The specific name *decipiens* implies the fact that the bird is 'deceptive', probably in the sense that it is deceptively similar to the Red-eyed Collared Dove.

African Collared Dove *Streptopelia roseogrisea*

This species of the Sahel, Horn of **Africa** and parts of Arabia has been recorded as far north as southern Egypt.

The specific name *roseogrisea* describes the pink-grey of its plumage.

Laughing Dove *Spilopelia senegalensis*

This small dove is found in North Africa and in the Eastern Mediterranean.

The **Laughing Dove**'s name relates to its song.

Spilopelia refers to the spots (Greek *spilos*) of its one congener and again uses *pelia* (dove).

The specific name *senegalensis* relates to Senegal.

Namaqua Dove *Oena capensis*

This tiny African species breeds in South Israel.

Namaqua is a region in southern Africa.

Oena is based on a Greek word for 'Dove', which also appears in the name of the Stock Dove.

The specific name *capensis* takes us back to South Africa and the Cape provinces.

Mourning Dove *Zenaida macroura*

This American dove has been recorded as a vagrant to Britain.

In this case, the name **Mourning** Dove relates to its rather sad song ('Oh dear! Poor me!') rather than to any external markings.

Zenaida was an invention of Charles-Lucien Bonaparte (1803–57), who named the genus after his wife, **Zenaïde**.

The specific name *macroura* means 'long-tailed'.

Bruce's Green Pigeon *Treron waalia*

This is another widespread African species that has been recorded in Egypt.

James **Bruce** (1730–94) was a Scottish explorer, whose most famous exploit was the discovery of the source of the Blue Nile.

Green is somewhat misleading, since the bird is grey with yellow underparts. The colour depends on carotenoids in the diet, which is exclusively the fruit of a single fig species.

Treron is a Greek word for 'pigeon'.

The specific name *waalia* is an Abyssinian (Ethiopian) name for the species.

Rails to Flamingos

11. RALLIDAE: RAILS

The family name is based on *Rallus*.

Water Rail *Rallus aquaticus*

This is a common but fairly secretive species.

Water Rail was originally a literal translation of the Latin *Rallus aquaticus*, chosen to contrast with the now defunct name *Land Rail* (Corncrake).

Rail comes via the French *râle* (a death rattle) from the Latin *rascula* (referencing the rasping call). It has no linguistic connection with fence rails.

African Crake *Crecopsis egregia*

This African species has been recorded in the Canary Islands.

Crecopsis means that it is 'crake-like' (see *Crex* below) in appearance.

The specific name *egregia* means that it is 'extraordinary', though this does not seem to relate to its appearance, but more probably to its enigmatic taxonomy.

Corn Crake *Crex crex*

This is a scarce migrant species, now found mainly in outlying areas of Britain.

Corn because the bird frequents summer grain and grass crops.

Crake is the anglicised form of the Latin *Crex*, which echoes the call '*crex! crex!*'

Corncrake is often preferred as a single word.

* An alternative traditional name was *Land Rail*, which paralleled the *Water Rail*.

Sora *Porzana carolina*

This American species is a very rare vagrant to Britain.

Sora is generally thought to be a word of Native-American origin – and is therefore very rare among North American bird names.

Porzana (see next species).

The specific name *carolina* relates to early British settlement of the Carolina States (dedicated to *Carolus Rex*, Charles I).

Spotted Crake *Porzana porzana*

This is a secretive and scarce migrant species.

Spotted is self-evident.

Porzana was originally a Venetian name for this species.

* This was such a rarely seen species that Pennant had to name the species from scratch. His first version, the Small Spotted Water Hen, was clearly not satisfactory. In typical fashion, he then named it the *Spotted Gallinule*. Montagu essayed *Water Crake*, but George Shaw's 1824 offering was the one we now use.

Lesser Moorhen *Paragallinula angulata*

This African species is recorded in Egypt.

The English name is self-evident.

In 2017, this species was placed in a unique new genus *Paragallinula*, which means 'near to *Gallinula*'.

The specific name *angulata* means angular, perhaps for the pointed shape of the bill shield.

(Common) Moorhen *Gallinula chloropus*

This is a widespread species, but **Common** is needed only in an international context.

Moorhen is 'a hen of the moors'. That name was first in evidence in 1300. *Moor* was used in a traditional sense, which included lowland marsh (e.g. Otmoor, Sedgemoor). That sense is not so widely understood in modern usage, so the name was rejected by the AOU when the American species were split from the Eurasian form. In America, Pennant's 1768 name, Common Gallinule, was preferred, so today the IOC list shows two closely related species bearing quite different English names.

Gallinula conveys the sense of 'a little hen'.

The specific name *chloropus* means 'green-legged'.

Eurasian Coot *Fulica atra*

This is a widespread and common species.

Eurasian is somewhat inaccurate because it currently includes an Australian subspecies.

Coot is echoic, based on the sharp call of the bird.

Fulica is Latin for 'Coot'.

The specific name *atra* means 'black'.

* This is a familiar species that many people take for granted. It is not generally appreciated that large numbers of Continental birds swell the British population in winter. The BTO gives a figure of 26,000 pairs in summer and of 205,000 birds in winter – a fourfold increase.

Red-knobbed Coot *Fulica cristata*

This Iberian and North African species is not recorded in Britain.

Red-knobbed is for two red tips surmounting the frontal shield.

The specific name *cristata* treats the red knobs as 'crests'.

American Coot *Fulica americana*

This species has been recorded as a vagrant in Britain and in several other parts of Europe.

American and *americana* are self-evident.

The split of the former **Purple Swamphen** *Porphyrio porphyrio* is only a decade or so old. The complex that bore that specific name ranged from the Iberian Peninsula to the Pacific. The former subspecies have now been split into six species. Three of those occur within the relevant area. The most westerly form retains the original specific name, but under the vernacular name Western Swamphen.

These can be confiding birds and were kept as pets in Classical times: one was illustrated in one of the murals exposed at Pompeii.

Sightings of more than one form have been recorded in Britain, but have generally been treated as probable escapes, though the Western Swamphen at Minsmere, Suffolk, appeared to be a genuine vagrant, with the species known just across the Channel, in France.

Western Swamphen *Porphyrio porphyrio*

This species has been recorded once in Britain, at Minsmere in 2016.

Western is the most westerly species – in this case the Western Mediterranean – in the natural range of a former species complex known as the *Purple* Swamphen.

Swamphen defines a larger species that has habits like moorhens and coots.

Porphyrio is derived from the Greek for 'purple'.

* The move away from the name 'Purple' avoids the previous confusion with the smaller Purple Gallinule (below), which is an occasional vagrant to Europe from America.

Grey-headed Swamphen *Porphyrio poliocephalus*

The Middle East and the Indian subcontinent are home to this species.

Grey-headed and *poliocephalus* mean the same, and describe a paler and brighter bird than the dark Western species.

African Swamphen *Porphyrio madagascariensis*

This is a species found widely in **Africa**, but also in parts of the Eastern Mediterranean.

The specific name *madagascariensis* reminds us that it is also found in Madagascar.

Allen's Gallinule *Porphyrio alleni*

This African species has been recorded in Britain.

Allen's and *alleni* commemorate British naval officer Rear-Admiral William Allen (1792–1864), who explored Niger in Africa.

Gallinule is a name based on the Latin *gallinula* (little hen).

Purple Gallinule *Porhyrio martinica*

This American species has been recorded as a vagrant in Britain and in some other parts of Europe.

The name **Purple** befits a species with purple-blue plumage.

The specific name *martinica* relates to the island of Martinique in the West Indies.

Baillon's Crake *Zapornia pusilla*

This is a very rare vagrant to Britain, which is otherwise found from Eastern Europe to Australasia.

Louis-Antoine **Baillon** was a nineteenth-century French collector.

Zapornia is a modified anagram of *Porzana*. It was resurrected in 2020. It was originally created by William Leach in 1816 as a way of naming a closely related new genus.

The specific name *pusilla* means 'very small'.

* Anagrams are also in evidence in the Swallow (*Delichon*) and Kingfisher (*Dacelo*) families.

Little Crake *Zapornia parva*

This is another very rare vagrant to Britain. Most birds breed in Eastern Europe, but there are a few remnants of its much-contracted breeding range in parts of France.

Little is actually slightly larger than Baillon's, though smaller than Spotted.

The specific name *parva* is merely a repetition of 'little' in Latin.

Striped Crake *Aenigmatolimnas marginalis*

This African species has been recorded in Italy, Malta and Algeria.

There are no mysteries in the English name.

Aenigmatolimnas is literally 'enigmatic', in its first part at least. That refers to the fact that this bird puzzles the taxonomists: historically it has been moved around and now is seen as unique in this genus. The second part of the word is based on the Greek *limnas*, which is to do with marshes.

The specific name *marginalis* implies that it stays near the edges of marshes.

White-breasted Waterhen *Amaurornis phoenicurus*

This South Asian species has been recorded as a vagrant to Kuwait.

White-breasted describes a pied bird that is also white-faced.

Waterhen has been used historically as a generalisation for birds of this family, but this species appears to be the only one to bear the name formally (others of this genus are known as bush-hens).

Amaurornis means 'dusky bird'.

The specific name *phoenicurus* is shared with the Common Redstart, since this bird too has a russet red tail.

* Curiously there is a related species known as the Watercock, *Gallicrex cinerea*. As Lockwood pointed out, the root-words of both *cock* and *hen* meant much the same as *fowl* or *bird* today.

12. GRUIDAE: Cranes

The family name is based on *Grus*.

Cranes are generally large and iconic birds, whose performance of elaborate mating dances, and sometimes impressive flocks, get them noticed. Inevitably, they have become part of folklore and legend, including the story of Gerana. Her very name is linked to the Greek root of the word crane, *geranós*. She first offended the goddess Hera and then, in her manifestation as a crane, waged incessant war on the Pygmies. Aristotle peddled the story that a sentinel crane would hold a stone in one claw so that the noise would wake the bird should it relax and drop the stone. Pliny perpetuated that false claim, which became part of the background story of the species. Similar fascination and mythology attended non-European species around the world.

The Common Crane was once well known in Britain, but wetland drainage and hunting caused it to disappear. Small numbers had begun to breed spontaneously in Norfolk in the late twentieth century, but it was decided to boost numbers with reintroduction in Somerset. Since then, there is evidence of re-establishment. Vital attention is needed elsewhere to save such species as the Siberian Crane from extinction.

Siberian Crane *Leucogeranus leucogeranus*

A very small number of this species may still exist within the region in the western part of its Asian range.

Siberian is for its breeding range.

Crane: see below.

Leucogeranus means 'white crane'. Several adjustments have been made in the taxonomy of cranes, and this genus (originally created by Charles-Lucien Bonaparte) was adopted formally by the IOC in 2017.

Sandhill Crane *Antigone canadensis*

This is an American species that was recorded as a vagrant to Britain in 1981, 2009 and 2011. It has also been recorded elsewhere in Europe.

Sandhill seems to relate to the choice of a raised mound on which the birds perform their courtship dance.

Crane: see below.

Antigone was the specific name given by Linnaeus to the Sarus Crane. That formed the basis of this relatively new genus. The name originates in the myth of Antigone of Troy who was turned into a stork by the goddess Hera. Linnaeus confused this with the similar legend of Gerana, who was changed into a crane in a very similar legend. (N.B. This is not the more famous Theban Antigone, daughter of Oedipus, who was the subject of tragic dramas by both Sophocles and Anouilh.)

The specific name *canadensis* relates to Canada, where most of these cranes breed.

Common Crane *Grus grus*

This species was historically resident in Britain until extirpated. Irregular passage migrants occur, but a native population is now being re-established.

Common refers to the European species to distinguish it in a world context.

Crane and *Grus* both derive from a common Indo-European ancestral word based on the call. The former travels via the Germanic form *kranô*, while the related Greek word *geranós* compresses to form the Latin name.

(See 'Linguistic Background' in the introductory notes for more detail on this name.)

Hooded Crane *Grus monacha*

This species was recorded for the first time in the Western Palearctic in October 2023. It was noted in Russia just west of the Urals. It normally winters in China, Korea and Japan.

Hooded describes the white cowl that distinguishes these birds.

The specific name *monacha* (monk) echoes the *monachus* of Cinereous Vulture and Monk Parakeet in referring to the hooded appearance.

Demoiselle Crane *Grus virgo*

The occasional sightings of this species in Britain are considered to be escapes. Asian-breeding wild birds migrate through Israel and may also be seen in Cyprus.

Demoiselle and *virgo* are 'maiden' in French and Latin respectively. This is a more graceful-looking species than the Common Crane.

13. PODICEPIDAE: Grebes

This family name is based on *Podiceps*.

Pennant introduced the name **Grebe** into English in 1760. Prior to that, the vernacular form Arsefoot had been widely used, but Pennant here succumbs to a tendency to favour more 'genteel' French or Latinate names: Gyr Falcon, Gallinule, Avocet and Guillemot are other examples. Latham seems to have turned that on its head a few years later when, in 1787, he created the genus *Podiceps*, which effectively translates the original Anglo-Saxon. It is tempting to see that as a somewhat witty gesture on behalf of our earlier heritage.

Apart from that, grebes and their names have a bumpy history. For a long while they were classed with the divers (loons) and, like those, have a complicated naming history in which British and American traditions have differed and sometimes clashed. As a hangover from the combined classification of divers and grebes, the Americans clung on for many years to the genus *Colymbus*, which was eventually abandoned, but a remnant of that is still found in the generic name of the Pied-billed Grebe.

A much larger tangle in the naming of two other species starts with a slightly ambiguous note in Linnaeus's notebook and grows into a complicated tale involving two continents and many ornithologists. The resultant compromise still retains evidence of the original mistake, while local usages seem destined to resist the international solution. More of that below…

Little Grebe *Tachybaptus ruficollis*

This is a common and widespread breeding species in Britain.

Little is self-evident for the smallest European grebe species.

Grebe is from the French *grèbe*, which is possibly rooted in the Savoyard for a gull species.

Tachybaptus means 'to sink fast'. This describes an escape strategy.

The specific name *ruficollis* is confusing, since it means 'red-necked', which is the vernacular name of another species (below).

* *Little Grebe* was one of Pennant's 1768 creations, but Yarrell later championed the local name *Dabchick* as being more authentic when he used it in 1843. It had a number of regional variants in the spelling. 'Dab' means '*little*', while the '*chick*' part means 'small bird', so it makes the point very clearly. Although it did not appear on Coward's 1926 list as an accepted alternative, the name has always maintained a certain popularity, and it remains one of the best-known alternative names.

Pied-billed Grebe *Podilymbus podiceps*

This American species is a rare vagrant to Britain.

Pied-billed describes a black stripe on the ivory-white bill (though this mark disappears in winter).

Podilymbus combines *Podiceps* with *Colymbus* (a now defunct genus that once included divers and grebes).

The specific name further acknowledges a relationship to the genus *podiceps* (see below).

Red-necked Grebe *Podiceps grisegena*

This more easterly species occurs in relatively small numbers in Britain in winter.

Red-necked applies only in summer plumage, though this may sometimes be seen before the bird leaves in spring.

Podiceps (*podex*, vent, *pes*, foot) is a Latinised form of the old vernacular name 'Arsefoot', which describes the fact that the feet are placed at the rear of the body.

The specific name *grisigena* means 'grey-faced', a feature that stands out more in breeding plumage. (Confusingly, *ruficollis*, red-necked, is used for the Little Grebe.)

Great Crested Grebe *Podiceps cristatus*

This is a widespread species found on both inland and coastal waters.

Great is apt for the largest of the family found in Britain, and originally helped to distinguish it from the two smaller 'crested' grebes that occur. However, no other grebe species actually shares the word 'crested' in its name, so this word is largely redundant.

Crested applies all year round to this species, though the summer head plumage is more elaborate.

The crest features again in *cristatus*.

* Great Crested Grebe was one of the key species involved in the late nineteenth-century exploitation of birds for the fashion industry: whole tippet cloaks were made from their skins and they became seriously endangered. The British protests against such exploitation led to the formation of the RSPB and, vitally, to the protection and recovery that now make this a relatively common bird.

The naming of the next two species is historically complex. The British naturalist George Edwards had already coined the names Eared and Horned when Linnaeus recorded a slightly ambiguous note to that effect. In spite of the fact that all other data in Linnaeus's notes pointed to a link between *eared* and *auritus*, it eventually emerged that the name '*auritus*' was given to the 'wrong' species, because it is the Horned Grebe that now bears that name.

Subsequent efforts to remove the anomaly contributed to the abandonment in Britain of both Horned and Eared during the nineteenth century, and to their respective replacements by the curious form, Slavonian, and the more literal Black-necked. Meanwhile, the Americans retained the original names.

In some ways the simplest solution would be to get rid of *auritus* and Eared altogether: the American subspecies is *P. a. cornutus* and that means 'horned', but the rule of precedence protects the 'nuisance' name.

When the IOC sought an international, twenty-first-century compromise, Horned and Black-necked became the agreed forms, and that seems very reasonable. However, it seems highly probable that Eared and Slavonian will survive in local usage on their respective sides of the Atlantic.

Horned Grebe *Podiceps auritus* (INTERNATIONAL AND AMERICAN NAME)
Slavonian Grebe *Podiceps auritus* (BRITISH NAME)

This small species breeds in the extreme north of Britain, but appears in winter in southern coastal waters. The background to the two names is discussed above.

Horned records the upswept, golden head plumes worn in breeding plumage. This is the traditional American usage that, after its adoption by Pennant, was also current in Britain until the late nineteenth century. It is now the agreed international usage.

Slavonian was introduced by Latham, who noted that the bird was 'known in Sclavonia' [*sic*] (a region of Russia). In 1802 Montagu adopted the name in preference to Horned, and it eventually prevailed.

The specific name is *auritus* (eared) – and that is at the root of the change, as well as the source of much confusion (see notes above).

Black-necked Grebe *Podiceps nigricollis* (INTERNATIONAL AND BRITISH NAME)
Eared Grebe *Podiceps nigricollis* (AMERICAN NAME)

This species breeds in Scotland and Ireland, and is seen further south in Britain in winter.

Black-necked and *nigricollis* mean the same, and separate this species in summer plumage.

Eared reflects the down-swept, golden ear-tufts worn in breeding plumage.

14. PHENICOPTERIDAE: Flamingos

The family name is based on *Phoenicopterus*.

The original vernacular English names of the two Old World flamingos were coined in the context of those two species alone. In the wider context of a world populated by six species in three genera, the names *Greater* and *Lesser* now appear to be ungrammatical.

The resurrection of two retired genera in this family is relatively recent: for many years only the genus *Phoenicopterus* was in use.

Greater Flamingo *Phoenicopterus roseus*

This species occurs fairly widely in Mediterranean Europe.

Greater is used because this is the larger of two Old World species.

Flamingo has a curious origin that goes back to the Middle Ages. It related to pink-skinned, rosy-cheeked Flemish traders (Flemings) who called into Iberian ports and stood out among sallow-complexioned locals. The word was coined in sixteenth-century Portuguese, though the Spanish versions, *flamengo/flamenco*, were in use for the pink bird even earlier. Flamenco dancers, of course, imitate the stately strutting display of the birds.

Phoenicopterus is the Latin for 'Flamingo', from the Greek meaning 'flame-winged'.

The specific form *roseus* is the Latin for 'rosy'.

Lesser Flamingo *Phoeniconaias minor*

This mainly African species occurs in small numbers in Southern Europe.

Lesser and *minor* both indicate that it is the smaller of the two species.

Phoeniconaias adds the crimson element to a water nymph (a naiad). This resurrected genus was originally created by G.R. Gray in 1869.

Two exotic species are established as free-flying, feral birds in north-west Germany close to the Dutch border. What began as a small flock of apparently escaped Chilean Flamingos now includes a few American and Greater flamingos. Some breeding has occurred and some hybrids have been noted, which has implications for the wider integrity of the native population of Greater Flamingo.

American Flamingo *Phoenicopterus ruber*

This is a Caribbean and South American species that has been recorded in small numbers among the German feral flock.

American suggests that it was the first-named of the four New World species. It is also widely known as the *Caribbean* Flamingo.

The specific name *ruber* (red) is for its bright salmon-red plumage.

Chilean Flamingo *Phoenicopterus chilensis*

This species has a foothold in a feral flock living on the German–Dutch border.

Chilean and *chilensis* tell of the origins of this paler species.

Buttonquails to Painted-snipes

This section contains what are commonly referred to as 'waders' in Britain or 'shorebirds' in North America. Strictly speaking, the first and last families included here have not normally been included under those terms, though they are now considered to be fairly closely related. In any case, such generalised terms are conveniences rather than classifications, and both are imperfect: some species do not do a lot of wading and some are rarely seen on shores of any sort, though the majority are certainly associated with lakes and ponds, rivers and streams, marshes and swamps, and beaches of every sort.

15. TURNICIDAE: Buttonquails

The family name is based on *Turnix*.

There are fewer than 20 species of buttonquail worldwide. They share many characteristics and habits with quails, but are not related to them in any way: instead, they are closer to the waders. The similar characteristics of buttonquails and quails are an example of convergent evolution.

Small Buttonquail *Turnix sylvaticus*

This species was declared extinct in Spain in 2018, but small numbers of the same subspecies appear still to exist in Morocco, so reintroductions may occur.

Buttonquail suggests that the bird resembles a small quail, with *button* used to indicate smallness.

Turnix recalls the genus of the quails, *Coturnix*.

The specific name *sylvaticus* implies that this is a woodland species, but it tends to favour a wider range of scrub, woodland edges, grassland and overgrown fields.

* The former name, *Andalusian Hemipode*, which used to appear so ridiculously extravagant in every handbook of European birds, is clearly a scientist's concoction. Andalucia is of course a province of Spain, while the Latinism *Hemipode* means 'half-foot', and that is because it only has three toes. Latham called it the *Andalusian Quail*.

* It has been difficult to prove the extinction of a species that is rarely seen at all, but it has been seen in recent years in Morocco and once in Algeria. Its loss in North Africa would represent the loss of a subspecies, though other forms exist south of the Sahara.

16. BURHINIDAE: Stone-curlews

The family name is based on *Burhinus*.

The exact place of this family in the sequencing of birds was long seen as a puzzle. Although they resemble waders in a general way, most of them tend to favour semi-arid and arid habitats. In spite of the name, the birds are not at all closely related to the curlews. Pennant originally called the British form the *Norfolk Plover*, but revised that a few years later to the *Thick-kneed Bustard*. He had adapted that last name from Pierre Belon's 1555 concoction of *Oedicnemus* (see below), but it was Leach who gave the name further life when he used *Thick-knee* without any trimmings. The one merit of that name is its neutrality, but it is unattractive, inaccurate and of no use in the field. Of the ten species in this family, half carry the name 'stone-curlew' and half are 'thick-knees', though there is no clear rationale for the difference.

Eurasian Stone-curlew *Burhinus oedicnemus*

This summer visitor to Britain is scarce in its downland and Breckland habitats.

Eurasian, because there are other stone-curlews around the world (though some are called *Thick-knees* (see below).

Stone-curlew is so-called partly because it prefers stony habitats and partly because it loosely resembles a Curlew in voice and size (though not in bill-shape). This name was recorded by Merrett in 1667 and also used by Ray in 1678.

Burhinus is from the Greek *bous* (ox) and *rhinos* (nose) to describe the short, stout bill.

The specific name *oedicnemus* comes from the Greek *oidos, oideos* (swelling), *kneme* (leg).

Senegal Thick-knee *Burhinus senegalensis*

This species is found in Egypt.

Senegal and *senegalensis* suggest that the bird has a wider range in Africa.

Thick-knee refers to the thickening of the joint of the leg, which equates more closely to the human ankle.

17. HAEMATOPODIDAE:
OYSTERCATCHERS

The family name is based on *Haematopus*.

(Eurasian) Oystercatcher *Haematopus ostralegus*

This is a common species in Britain.

Eurasian reminds us that there are several other species of oystercatcher around the world.

Though it is not obvious, the very name **Oystercatcher** underlines that last fact. When Catesby returned in 1719 from a personal voyage of discovery to the Americas, he introduced this name, which was based on the feeding behaviour of the American form. Oddly, the name caught on and eventually superseded the most widely used English name, *Sea Pie*, in spite of the fact that the local British birds take mainly cockles and mussels (hence a regional name *Mussel-cracker*).

Haematopus tells us that it is 'blood-footed' for its red legs.

The specific name means *ostralegus* means 'oyster-gathering'.

* The name *ostralegus* appears in the 1758 tenth edition of Linnaeus's *Systema Naturae*, and that is probably the reason why it enters the French language in 1781 as Buffon's *l'huîtrier*. That name then appears in the 1788 edition, which also indicates that '*ostralegus*' had by then been recorded from various parts of the Antipodes, Captain Cook having completed his three voyages and his naturalists having revealed many of their discoveries. It was sometime later that the various pied oystercatchers were formally split into several separate species. Unfortunately, the original, American, bearer of the accepted vernacular name is today rather dully described as *palliatus* (cloaked), while the European form wears a borrowed alias.

* The word oyster*catcher* seems an odd term for a feeding style that involves fairly passive prey. The actual feeding habits have been studied widely and reveal some odd facts: first, that some birds prise open bivalve shells and that others smash them at the base; and secondly, that the bill structure adapts over a period of time to suit periods of feeding on shellfish or on worms.

18. RECURVIROSTRIDAE: STILTS AND AVOCETS

The family name is based on *Recurvirostra*.

Black-Winged Stilt *Himantopus himantopus*

This is a scarce but increasing visitor to Britain, and a few have bred.

Black-winged is self-evident, but the name is confusingly similar to the American *Black-necked* Stilt, with which it was formerly 'lumped'.

Stilt is a comparison of the bird's long legs to the poles used to enable people to walk tall for practical tasks or for amusement.

Himantopus is from the Greek *himas, himantos* (strap) and *pous* (foot), which is an image of the bird folding those gangly legs as it attempts to sit on the nest.

* A long-staying, unringed bird resided at Titchwell, Norfolk, from 1993 to 2005 and was known to birders all over the country as Sammy. Stilt sightings were still extremely uncommon, so it is estimated that he had about 1.5 million visitors during his stay, which was good publicity for the site owners, the RSPB. His eventual disappearance spawned a number of obituaries.

* While the greater recent frequency of the birds in Britain is pleasing to some, to others the change is a major sign of a global warming in which Southern European wetlands are drying out and forcing the birds northwards.

Pied Avocet *Recurvirostra avosetta*

This is now a relatively common species in Britain, having re-established itself during the second half of the twentieth century to become the symbol of the modern RSPB.

Pied is for this wholly black-and-white species. There are other species elsewhere.

Avocet and *avosetta* are seemingly rooted in a Latin form *avis sitta* for 'a sort of magpie'. The words arrive via French and Italian respectively. It seems that in 1760 Mathurin Brisson mistakenly exchanged the *s* for a *c* in the French form, and that eventually found its way into English via Prideaux Selby in 1833 to modify Pennant's choice of *Scooping Avoset*.

Recurvirostra refers to the bird's upturned bill.

* Lockwood lists a number of older local English names: *Crooked Bill, Scooper, Barker, Clinker, Cobbler's Awl* and *Yelper*, in his own words 'none exactly decorous'. On this occasion Pennant's leaning towards a French form seems to have been a wise move.

19. CHARADRIIDAE: LAPWINGS AND PLOVERS

The family name is based on *Charadrius*.

The tradition in Britain has been to use the word 'plover' for most of this family, with the exception of Lapwing, which was a simple vernacular name for *Vanellus vanellus*. The IOC now prefers 'lapwing' to designate all species of the genus *Vanellus*, so the original Lapwing requires the descriptor Northern for the familiar European form.

Some major changes were made to speciation in this family during 2023. It is possible that resequencing may also follow.

(Northern) Lapwing *Vanellus vanellus*

This is a common species in Britain.

Today it is **Northern** to distinguish it from other lapwings.

Lapwing is an ancient name based on an eighth-century Anglo-Saxon form *læpiwince*, which has even older Germanic roots. In the strict, original sense the *lap-* element was a name for a crested bird, while the *wince* element seemed to refer to the crest's movements. The later transition to *Lapwing* is therefore a little misleading, as the *-wing* element is a corruption. There were once more regional names for this bird than for any other, but Lapwing outlasted them all, perhaps because it seems to conjure up an image of the flight. Lockwood also records the later name *Flop Wing*, which he assumes to have grown from the fairly obvious idea of *flapwing*.

Vanellus is from the Latin word for a 'winnowing fan' (a tool used in primitive agriculture to separate chaff from grain). This image conveys the broad 'fingered' wing in a very graphic way.

* The many regional and local names given to this species in Britain reflect the fact that it was once very common and very widespread. Some forms emphasised the crest and others the voice. The best known today of those traditional names is *Peewit*, which was based on the call. The name *Green Plover* started life as a name for the Golden Plover, but was later much used for the Lapwing. It did not fade from general use until well into the twentieth century.

* In the Middle Ages, the bird was considered to be a symbol of treachery, which seems a little odd. This was largely because its crest made it a Northern substitute for the Hoopoe, which had since Classical times been associated with the legend of the treacherous King Tereus, who had been transmuted into a Hoopoe as punishment for his crimes (see also *Swallow* and *Nightingale*).

Spur-winged Lapwing *Vanellus spinosus*

This Middle-Eastern species is unknown in Britain, but ranges throughout the Middle East and into Greece.

Spur-winged and *spinosus* both relate to a small bony spur at the 'wrist' (see also Spur-winged Goose).

Black-headed Lapwing *Vanellus tectus*

This African species has been noted in Israel and Jordan.

Black-headed is self-evident.

The specific name *tectus* means 'with covered head' (referencing Black-headed).

Grey-headed Lapwing *Vanellus cinereus*

This East Asian species was recorded in 2018 on Turkey's Black Sea coast, but the appearance of one in Northumberland in May 2023 caused a major sensation.

The **Grey** is reflected in the specific name *cinereus*. The bird is generally grey-brown above, but the head is a blue-grey.

Red-wattled Lapwing *Vanellus indicus*

This Central Asian species has wandered to Eastern Europe.

Red-wattled is for the patch of red skin near the eye.

The specific name *indicus* relates to India.

Sociable Lapwing *Vanellus gregarius*

This Eastern species is a scarce vagrant to Britain.

Sociable and *gregarius* describe the fact that the birds spend a lot of time in flocks.

White-tailed Lapwing *Vanellus leucurus*

This Asian species is a rare vagrant to Britain.

White-tailed and *leucurus* mean the same thing.

(European) Golden Plover *Pluvialis apricaria*

This is a common British wintering species.

European distinguishes this from two close North American relatives.

Golden distinguishes these birds from the Grey Plover. The name was chosen by Pennant in 1776, based on Brisson's names: Latin *Pluvialis aurea* and French *Pleuvier doré*.

Plover and *Pluvialis* both derive from the Latin *pluvius* (rain). The birds were said to call before rain.

The specific name *apricaria* fancifully suggests that the bird sunbathes to obtain its golden glow.

* This species was once widely known as the Green Plover, but that name was also applied to the Lapwing, hence Pennant's choice of *Golden*.

Pacific Golden Plover *Pluvialis fulva*

This is a scarce American vagrant to Britain.

Pacific indicates a westerly bias in its American range.

The specific name *fulva* speaks of a dark tawny brown. It is generally darker than the Golden Plover.

American Golden Plover *Pluvialis dominica*

This is a relatively frequent stray to British shores.

American has a more widespread range right across the continent.

The specific name *dominica* compares the summer plumage to the robes of a Dominican friar. (That image also occurs with the Kelp Gull, *L. dominicanus*, and in names relating to the Jacobins, a Parisian nickname for the order.)

Grey Plover *Pluvialis squatarola*

This is a common winter visitor to Britain from its Arctic breeding grounds.

Grey is the traditional name in Britain, because the bird generally is overall grey in its winter plumage. That has been preferred as the international name to the American use of '*Black-bellied*', which describes the very striking breeding plumage. The problem there is that all three 'golden' plovers also have similar features.

The specific name *squatarola* is rooted in a Venetian name *sgatarola*.

(Common) Ringed Plover *Charadrius hiaticula*

This is a common species in Britain.

Common separates it formally from *Little* (below).

Ringed is for the band of dark plumage on the breast.

Charadrius is Late Latin for a yellowish bird. It derives from the Greek *kharadrios* (a drab, nocturnal bird living in ravines, *kharadra*).

The specific name *hiaticula* is rooted in the Latin *hiatus* (cleft) and *cola* (dweller), effectively a repetition of *kharadrios* above.

* The generic and specific names here do not suit this sort of bird or the habitat it frequents, so both those names seem to have been selected randomly from a classical lexicon.

Semipalmated Plover *Charadrius semipalmatus*

This is a rare American vagrant to Britain.

Semipalmated and *semipalmatus* mean 'semi-webbed' and describe the small skin extensions between the toes, which may help the birds to swim short distances. The name occurs in the name of a sandpiper and the *Willet* (both below).

Little Ringed Plover *Charadrius dubius*

This was once very rare in Britain, but is now a regular summer migrant.

Little because it is smaller than *hiaticula*.

The specific name *dubius* implies that it can be confused with, or not easily separated from, its larger cousin.

* To Coward in 1926 this was still a 'rare wanderer'. He did not live to see the first one breed in Britain in 1938. The story of its first confirmed breeding is recorded in fictional form in Kenneth Allsop's *Adventure Lit Their Star* (1949).

Kildeer *Charadrius vociferus*

This is a scarce vagrant to Britain from America.

Kildeer is merely an onomatopoeic version of the bird's call and has no sinister implications. The specific name ***vociferus*** suggests that it calls often.

The next species was moved in 2023 from the genus *Charadrius* to the resurrected genus *Eudromias*.

(Eurasian) Dotterel *Eudromias morinellus*

This is a scarce passage species, a few of which breed in mountains in Scotland.

Eurasian is something of a technicality, as a handful of Southern Hemisphere species are also known as dotterels.

Dotterel is usually thought to be rooted in the idea of 'dotty' or 'eccentric' because it was considered too trusting and therefore too easily caught. However, Lockwood felt that an old local name *Dot Plover* might equally have been the root, the word *dot* relating to a single note call.

Eudromias means 'swift of foot'.

The 'dotty' concept appears to be supported by ***morinellus***, which is related to *Morus* (silly) in booby/gannet.

In 2023 all the plovers below were considered 'significantly divergent' from the genus *Charadrius* to be moved into the genus *Anarhynchus*, which is that of the Wrybill and is described as 'the oldest available name for this clade'. Since that word means 'backwards bill' and relates to the uniquely shaped appendage of this odd New Zealand plover, it seems to be an uncomfortable decision and may be provisional.

Kittlitz's Plover *Anarhynchus pecuarius*

This species occurs in Egypt and strays to Israel.

Heinrich von **Kittlitz** (1799–1874) was a Prussian explorer/naturalist who was a friend and colleague of Eduard Rüppell.

The specific name ***pecuarius*** relates to 'a grazier of cattle or flocks'. That may be explained by the fact that it is sometimes found on short, grazed grassland.

Three-banded Plover *Anarhynchus tricollaris*

This African species is sometimes recorded in Egypt.

Three-banded and ***tricollaris*** refer to a distinctive triple 'necklace' of black-white-black.

Kentish Plover *Anarhynchus alexandrinus*

In spite of a name based on an English county, this European species is relatively rare in Britain.

Kentish was chosen by Latham when he received his first specimen from the county of Kent, apparently from the same source that provided him with specimens of what became known as the Sandwich Tern and Dartford Warbler, which are named after two locations in that same county.

The specific name *alexandrinus* is a reference to Alexandria in Egypt.

* The juxtaposition of two widely differing geographical names makes this a curiosity. Linnaeus had previously coined the scientific name, based on Fredrik Hasselqvist's notes. The Swedish academic had responded to Linnaeus's concerns about lack of data on birds of that region by making a voyage of exploration, which proved too much for his weak constitution. After his death at Smyrna, his specimens and notes found their way back to Linnaeus.

Greater Sand Plover *Anarhynchus leschenaultii*

This is a scarce vagrant to Britain.

Before the recent split of Lesser (below), it was the **Greater** of two similar species.

Sand refers to the arid and desert-like habitats in which it breeds.

The specific name commemorates Jean-Baptiste **Leschenault** de la Tour (1773–1828), a French botanist and ornithologist.

In early 2023, the IOC flagged the split of the next species.

(Lesser Sand Plover *Charadrius mongolus*)

This Asian species is a rare vagrant to Britain, with just seven records.

It was originally the **Lesser** of two Sand Plovers.

The specific name *mongolus* means Mongolian, for one of the areas where it is found.

The two new names are:

Siberian Sand Plover *Anarhynchus mongolus*
Tibetan Sand Plover *Anarhynchus atrifrons* (black-fronted)

The seven British records to date are divided almost evenly between the two types, so it seems likely that both new species will eventually be included on the BOU list. The two new vernacular names generalise the regions where these birds occur.

Caspian Plover *Anarhynchus asiaticus*

This Asian species is a rare vagrant to Britain.

Caspian relates to the region of the Caspian Sea.

The specific name *asiaticus* is self-evident.

Oriental Plover *Anarhynchus veredus*

This Asian species (hence **Oriental**) has been recorded in Finland.

The specific name *veredus* is the Latin for a 'swift horse': it is a long-winged, fast-flying species that breeds as far north as Mongolia and winters as far south as Australia.

20. PLUVIANIDAE: Egyptian Plover

The family name is based on *Pluvianus*.

This species is unique in its family. It is now considered to be closer to the plovers than to the coursers and pratincoles with which it was formerly linked.

Egyptian Plover *Pluvianus aegyptius*

This African species has been recorded in Israel.

Egypt and *aegypticus* both suggest that the bird was first identified there, but this is the most northerly part of its normal range.

Pluvianus was coined from a French name *pluvian*, which in turn is based on *pluvius*, the Latin for 'rain' (see Golden Plover above).

* This bird is sometimes called a *Crocodile Bird*, an idea that started with the Greek writer Herodotus, who passed on the story that the bird picked out waste food from the crocodile's mouth. This behaviour has never been observed or verified in modern times.

21. ROSTRATULIDAE: Painted-snipes

The family name is based on *Rostratula*.

This is a small family of three species that exhibit reversed sexual dimorphism, in that the females are more colourful, while the males incubate eggs and raise young.

Greater Painted-snipe *Rostratula benghalensis*

This is a species found in northern Egypt.

Greater separates this from others in the genus.

Painted-snipe tells that the bird is colourful, but the hyphen warns that it is not a true snipe: convergent evolution has caused some structural similarities.

Rostratula implies that this is 'a little bird with a large bill'.

The specific name *benghalensis* features an area of India, but the bird also occurs in marshes in Africa and South-East Asia.

Sandpipers to Pratincoles

22. SCOLOPACIDAE:
SANDPIPERS AND SNIPES

The family name is based on *Scolopax* (see Woodcock).

It is worth noting that a number of waders in this family have been 'tidied up' in recent years and now appear in a different genus: among them are the Ruff, Buff-breasted Sandpiper, Broad-billed Sandpiper and the tattlers. Sadly, there are also some probable cases of extinction.

This is a large family and many species were traditionally caught for food. They were therefore fairly well known in markets, and it seems probable that a number of more generalised names, such as godwit, had a fairly wide usage because of such commerce. Such names are used below as subheadings.

Sequencing of this family is under review and will probably be changed.

CURLEWS: *BARTRAMIA* AND *NUMENIUS*

Upland Sandpiper *Bartramia longicauda*

This American species is a scarce vagrant to Britain.

Upland reminds us that this species breeds in upland plains at over 1,000 m/3,000 ft.

The unique genus *Bartramia* commemorates 'the grandfather of American ornithology', William Bartram (1739–1823), who earned that accolade for his support of Alexander Wilson (1766–1813), the 'father'. The species was once known as *Bartram's Sandpiper*.

The specific name **longicauda** comments on the bird's long tail.

* This species was recorded by Coward (1926) as *Bartram's Sandpiper*.

Eurasian Whimbrel *Numenius phaeopus*

This is a mainly passage species in Britain, but breeds in the Northern Isles.

Eurasian is now needed because the American form has been recognised as a full species (next).

Whimbrel is echoic, having originated as '*whimmerel*' in a Northern dialect form, where *whimmer* means 'whimper'.

Numenius is from the Greek for a new moon, and relates to the crescent-shaped bill of this and other curlews.

The specific name *phaeopus* is the Medieval Latin for 'whimbrel', from the Greek *phaios* (dusky brown) and *pous* (foot).

Hudsonian Whimbrel *Numenius hudsonicus*

This has been recorded as a rare vagrant to Britain.

It was formerly treated as a subspecies, but finally split in 2016. It has a number of plumage differences, especially in lacking the white rump of the Eurasian form.

Hudsonian and *hudsonicus* are to do with Hudson Bay, Canada, which is part of the bird's normal range.

Little Curlew *Numenius minutus*

This north-east Siberian species has been recorded only very rarely in Britain and Western Europe.

Little: it is Greenshank-sized and its specific name *minutus* tells us that it is the smallest curlew.

An alternative name, *Little Whimbrel*, seems to make better sense, since whimbrels are small curlews.

For **Curlew** see Eurasian (below).

Slender-billed Curlew *Numenius tenuirostris*

Tragically, this Asian species, which used to be recorded in Western Europe, has not been confirmed anywhere during the twenty-first century and may well be extinct.

Slender-billed and *tenuirostris* mean the same.

For **Curlew** see Eurasian (below).

Eskimo Curlew *Numenius borealis*

This North American species was recorded historically a number of times in Britain but is now probably extinct.

As the names **Eskimo** and *borealis* imply, it was an Arctic species and once very numerous, but was heavily persecuted.

For **Curlew** see Eurasian (below).

Eurasian Curlew *Numenius arquata*

This is a common species in Britain.

Eurasian describes the range of this one of several species called 'curlew'.

Curlew is from the Old French *corlieu*, based on the sound of the two-note call of this species. It has been in general use since at least the early fourteenth century.

The specific name *arquata* uses the Latin for 'bow-shaped' to underline the shape of the bill.

GODWITS: *LIMOSA*

Separating the two common godwits in the field challenges birdwatchers because the difference is not always that obvious. One useful tip lies in the names. *Limosa* means 'muddy' and the name of the Black-tailed Godwit has twice as much mud, because that one is twice as likely to be found in muddy pools. Bar-tailed prefers the tide-line. Conversely the vernacular names do not help a lot, because the tails are hard to see. What would be more helpful would be the name 'White-winged' Godwit instead of Black-tailed. That is the difference on the wing and, if not an official name, it is an idea to carry into the field.

Black-tailed Godwit *Limosa limosa*

This is a fairly common species on passage through Britain and in winter. Two distinct subspecies can be identified in breeding plumage – the nominate form and *islandica*.

Black-tailed it is, though that is generally not that obvious. The white wing-bars make a more striking diagnostic field mark when the bird is in flight.

The true roots of **Godwit** remain a mystery: it may have an echoic origin, but links have also been suggested to an Anglo-Saxon root related to its medieval status as a culinary delicacy.

Limosa means 'muddy' – and this bird loves soft, wet mud. For that reason it is often found in brackish pools and estuaries, though it will also feed in damp pasture.

* The darker subspecies *islandica* does migrate to Iceland, while the paler form travels into mainland Eurasia.

Bar-tailed Godwit *Limosa lapponica*

This is a coastal passage and wintering species, which breeds widely across the arctic regions of Eurasia and Alaska.

The **Bar-tail** is not at all the bird's most striking feature, but it serves to contrast it with the Black-tailed. The lack of white wing-bars and a long white rump-patch make it look more Curlew-like.

The specific name *lapponica* (from Lapland) is a very narrow representation of a much broader Arctic breeding range.

Hudsonian Godwit *Limosa haemastica*

This American species is a very rare visitor to Europe: it has been recorded in Britain, Ireland and Sweden.

Hudsonian is for Hudson's Bay in Canada, which befits a northerly breeder.

The specific name *haemastica* (bloody) reflects the dark red tinge of the breeding plumage.

SANDPIPERS, *ARENARIA, CALIDRIS*

(Ruddy) Turnstone *Arenaria interpres*

This is a common circumpolar breeder that can be found on almost any beach in the world outside the breeding season.

Ruddy is a partial description of its breeding plumage, and is used because there is one other species of turnstone in North America.

Turnstone because it turns stones (and weeds) to reveal food items.

The Latin *Arenaria* means 'of the sand'. It is a beach-dweller.

Jobling points out that Linnaeus first misinterpreted use of a Gotland dialect name for the Redshank and then confused it with the Swedish for 'interpreter'. His final selection of *interpres* is the Latin for 'messenger', but that seems good for a bird that scuttles about a great deal.

Great Knot *Calidris tenuirostris*

This Siberian breeder has been recorded rarely in Britain and in other parts of Western Europe.

Great indicates that this species is on average larger than Red Knot (but not that much so).

Calidris comes from the Greek *kalidris* (a grey waterside bird mentioned by Aristotle).

The specific name *tenuirostris* means 'slender-billed'.

Red Knot *Calidris canutus*

This Arctic breeder is seen in Britain as a passage and wintering species.

Red distinguishes this from Siberia's Great Knot (see above). They have brick-red underparts in summer (though the female is more subdued), but that is almost irrelevant in Britain, where the red is rarely seen.

Knot and *canutus* are from King Cnut (Canute), who famously proved that he could not control the tides.

* The names may have little to do with tides, because the Knot was said to be one of the king's favourite foods when cooked in milk.

Ruff *Calidris pugnax*

This is a common passage species in Britain.

Ruff describes the neck-feathers of the breeding male, which are likened to the fashionable large Elizabethan/Jacobean collar. The name seems to have come into being during the late Elizabethan period. The males evolved their larger size and splendid breeding plumage (a 'ruff' of flamboyant neck feathers) in the lek, in which they compete for females.

The traditional, Greek-based, generic name, *Philomachus*, meant 'pugnacious', though that was set aside in 2017 when the bird was moved into the genus *Calidris*.

However, the specific name *pugnax* says the same in Latin, referring to the competitive courting style.

* This species is unusual in having a traditional separate name, *Reeve*, for the female, which has now fallen out of use. That may be based on an older medieval name for the species. Both Ruffs and Reeves were caught commercially for the food market: there is evidence that the much larger ruff commanded a higher price than the reeve, so the two names would have been useful in that context.

Broad-billed Sandpiper *Calidris falcinellus*

This Scandinavian breeding species is a surprisingly scarce vagrant to Britain for a bird that breeds relatively close by, in northern Scandinavia. A distinctly south-easterly bias during autumn migration means that vagrancy to Britain is infrequent, especially as strong westerly winds prevail at that season, though the similarity to a Dunlin may make it hard to spot.

Broad-billed is something of an exaggeration.

The former unique generic name, *Limicola* (mud-dweller), for this species was set aside in 2017, when the bird was reclassified in *Calidris*.

Sandpiper originated as a folk name for small birds that ran around on sandy shores and gave piping calls. (The Americans use the colloquial term 'peeps'.) Unlike many other generic names deriving from traditional English, *sandpiper* was probably always a 'catch-all' name.

The specific name *falcinellus* describes the down-curved bill as 'like a little sickle' (Latin *falx, falcis* – see *Falcon, Falcated Duck, Glossy Ibis*, etc.).

Sharp-tailed Sandpiper *Calidris acuminata*

This eastern Siberian breeder is a scarce vagrant to Britain.

Sharp-tailed is self-evident and is echoed in *acuminata* (sharp-pointed). A number of the smaller, long distance migratory sandpipers have an attenuated rear end in profile because they have slim bodies and relatively long wings.

* In 1926, Coward listed this species as the Siberian Pectoral Sandpiper, *Erolia acuminata*. The genus *Erolia* was adapted from an invention of Vieillot for the Curlew Sandpiper. This old name underlines the bird's close similarity to what we now call simply the Pectoral Sandpiper (see below).

Stilt Sandpiper *Calidris himantopus*

This American species is a scarce vagrant to Britain.

Stilt is for a long-legged species and echoes the name of the *Himantopus* (Stilt) genus. That word (meaning 'strap-foot') is used as the specific name here.

Curlew Sandpiper *Calidris ferruginea*

This is a regular passage migrant in Britain.

Curlew recalls the shape of the larger wader because of its curved bill, but this one is only a little larger than a Dunlin.

The specific name *ferruginea* means 'rust-coloured' for the red-brown breeding plumage, some of which may still be showing during autumn passage.

Temminck's Stint *Calidris temminckii*

This is a scarce species in Britain for reasons similar to Broad-billed Sandpiper.

C.J. **Temminck** (1778–1858), was a Dutch ornithologist.

Stint: see Little Stint below.

Long-toed Stint *Calidris subminuta*

This is a rare vagrant to Britain from Siberia.

Long-toed is literal, but the bird also has proportionately long tibia, making it look more gangly than other stints.

The specific name *subminuta* implies that this species is 'comparable to' the Little Stint, *C. minuta*.

Stint: see Little Stint below.

Red-necked Stint *Calidris ruficollis*

This is another bird from north-east Siberia that is a very rare vagrant to Britain.

Red-necked and *ruficollis* mean the same.

Stint: see Little Stint below.

Sanderling *Calidris alba*

This is a common passage and wintering species.

Sanderling echoes *Sandpiper*, but in this case is just 'a small bird living on sand'. (The diminutive suffix *-ling* is rooted in Anglo-Saxon and is used elsewhere, as in Star*ling*. It is truncated in Dun*lin*.)

The specific name *alba* applies particularly to the winter bird, which is then the whitest of all the small waders. Though darker above and red-throated in breeding plumage it still retains clean white undersides.

* This species outdoes even the Ruddy Turnstone in being one of the most worldwide of all the waders outside the breeding season.

Dunlin *Calidris alpina*

This is the commonest of the small waders in Britain, seen as a passage and wintering species.

Dunlin is a small 'dun'. In this case the diminutive *-ling* has been compressed. 'Dun' was probably a dialect name for a Knot, and indeed it does look like a small Knot. 'Dun' is also a brownish-grey, nondescript colour; it occurs with another diminutive suffix in Dun*nock* (*Prunella modularis*).

The specific name *alpina* means 'of high mountains'. The nominate subspecies is seen in Britain.

Two other visually identifiable subspecies are seen regularly in Britain:

– *schinzii*: H.R. Schintz (1777–1862) was a Swiss naturalist.

– *arctica*: breeds in the Siberian Arctic.

* Dunlin is the commonest of the small waders on British shores and therefore serves as a yardstick with which to compare other small waders. In Britain, at least, White-rumped, Baird's, Western, Least sandpipers and the like are usually well hidden among flocks of Dunlin, which themselves can vary considerably in autumn.

Purple Sandpiper *Calidris maritima*

This is a fairly scarce species found wintering on British shores.

Purple describes the tone of the bird's dark grey winter plumage, rather than a distinct colour.

The specific name *maritima* reminds us that this species is never far from the sea.

Baird's Sandpiper *Calidris bairdii*

This American species is an annual vagrant to British shores.

Baird's and *bairdii*: S.F. **Baird** (1823–1877) was a US ornithologist. However, the AOS announced in 2023 that all eponyms in vernacular names of American species will eventually be replaced.

Little Stint *Calidris minuta*

This Arctic species is a passage and wintering visitor to Britain.

Stint originated as an East Anglian dialect word for Dunlin. It is simply another word for a sandpiper.

Logically, the Dunlin's smaller relative became known as the **Little** Stint. The name *Stint* was later adopted for some other small species seen in Britain.

The specific name *minuta* recalls 'Little'.

Least Sandpiper *Calidris minutilla*

This American species is a rare vagrant to Britain.

Least is certainly one of the smallest of this family.

The specific name *minutilla* underlines that idea, because it is 'very small'.

* In the 1923 list and in Coward (1926) this was known as the *American Stint*. In later decisions on such matters, the Americans discarded such obviously British vernacular forms as *stint* and *shank*.

White-rumped Sandpiper *Calidris fuscicollis*

This North American species is an annual vagrant to British shores.

White rumped describes a feature that is useful in the field.

The specific name *fuscicollis* is Latin for 'dark-necked' – a winter identifying feature to use with the vernacular name.

* Coward's 1926 list records this as *Bonaparte's Sandpiper*, a name that was still in use in some mid-century records, but was superseded by *White-rumped* in the 1950s. Older records still have it as *Schintz's Sandpiper* (see Dunlin subspecies).

Buff-breasted Sandpiper *Calidris subruficollis*

This American species is a fairly regular vagrant to Britain.

Buff-breasted rather understates the warm coloration, which, on the other hand, is a little exaggerated in the specific name *subruficollis* (reddish-necked): it is a warm pinkish-buff.

The former unique genus *Tryngites*, created by Cabanis in 1857 to link the bird to the *Tringa* genus, was set aside in 2017 and the species moved into *Calidris*.

Pectoral Sandpiper *Calidris melanotos*

This is a regular migrant from either American or Siberian populations (see note below).

Pectoral relates to the Latin *pectus* (breast) – a reference to the distinct, darker breast pattern, particularly of the breeding male, which resembles a breastplate of armour. The shape of that pattern is a useful identifying feature even in a juvenile bird.

The specific name *melanotos* refers to a 'black-back'. That describes a darker greyish tone to the mantle of the adult, but does not apply to the brighter mantle of the juvenile form, which is the one seen most often in Britain.

* In 1926, Coward listed this as the American Pectoral Sandpiper, *Erolia maculata*. See also Sharp-tailed Sandpiper (above). In Hartert's 2012 list the two were treated as subspecies of *Erolia maculata*. Together those names remind us that the 'taxonomic industry' has long been a dynamic force in bird names.

Semi-palmated Sandpiper *Calidris pusilla*

This American species is a scarce vagrant to Britain.

Semipalmated means 'semi-webbed' and describes the small skin extensions between the toes.

The specific name *pusilla* means 'tiny, very small'.

* It is remarkable how many ways variants on the 'very small' theme were concocted for this family.

Western Sandpiper *Calidris mauri*

This American species is a very rare vagrant to Britain.

Western is for a species that breeds in Western Alaska.

The specific name *mauri* commemorates the Italian botanist Ernesto **Mauri** (1791–1836).

* Among the records of vagrancy, the very appearance of this bird in Britain at any time is worthy of further examination, simply because almost no other American vagrant could have travelled further. To arrive anywhere on our shores these birds will have travelled some 4,400 miles, or 7,000 km, by the shortest possible route. Not surprisingly, very few birds are recorded per decade. To explain how even one might get here, it is necessary to realise that it is probably the most numerous sandpiper species in the word; that small numbers winter on the American East Coast; and that individuals might be caught and blown by the autumn Atlantic storms.

DOWITCHERS, WOODCOCKS AND SNIPES: *LIMNODROMUS, GALLINAGO, SCOLOPAX* AND *LYMNOCRYPTES*

Long-billed Dowitcher *Limnodromus scolopaceus*

This American species is a relatively frequent vagrant to Britain.

Long-billed describes a snipe-like feature, but it is hard to separate this from Short-billed in the field on length alone: the mean difference is about half a centimetre. In Britain, at least, a vagrant is far more likely to be of this species.

Dowitcher is an enigmatic word: it may be a remnant native name, from Iroquois or Mohawk *tawistawis* (snipe), but some claim it to be a '*Deutscher* snipe' (from German/Dutch-American settlers.) It seems possible that the latter form grew as an approximation of the former.

Limnodromus means 'marsh-runner'.

The specific name *scolopaceus* links it to the woodcock/snipe complex (see Scolopax below).

* The more frequent occurrence of this, the more westerly breeder of the two dowitchers, is counter-intuitive and in some ways parallels the case of the two yellowlegs. The explanation lies in migratory patterns and timing.

Short-billed Dowitcher *Limnodromus griseus*

This American species is much rarer in Britain than the Long-billed.

Short-billed would be more accurately Short*er*-billed, since it is not short.

The specific name *griseus* (grey) is not helpful, since both are grey in winter. The summer plumage might be described as 'grey-red from a distance'.

(Eurasian) Woodcock *Scolopax rusticola*

This British breeding species is boosted in winter by large influxes from Europe.

The use of **Eurasian** is appropriate in a world context.

Woodcock underlines the fact that the species lives in woodland.

Scolopax is the Latinised version of a Greek name for a Woodcock and historically included the snipes.

The specific name *rusticola* is appropriate to a 'country-dweller'.

* Wood*cock* implies the existence of the feminine form, wood*hen*, which appears to be the case in Old and Middle English forms. Lockwood, however, points out that both *cock* and *hen*, each originating in different forms of the Germanic languages, simply had the sense of *bird*, and that their attachment to gender in English was a relatively late development.

* The Woodcock is associated with the strange specialist word, to *rode*, which refers specifically to an evening flight and, according to the OED, is of obscure eighteenth-century origins. Male Woodcocks *rode* (patrol) their breeding territory at dusk, by flying just above tree-top height with deep, slow wingbeats. This is one of the best times to see an otherwise secretive and nocturnal bird.

American Woodcock *Scolopax minor*

This North American species has been recorded in France.

As the specific name *minor* implies, it is a smaller species.

Jack Snipe *Lymnocryptes minimus*

This is a common but secretive winter visitor to Britain.

Jack tends to imply 'a bit of a rogue' in many English idioms, which suits the Jackdaw, but that is not the meaning here. Rather, it has a closer relationship with the smaller target ball (the jack) in a game of bowls or with the jack (knave) in a pack of cards, as a lesser figure. In that respect its meaning is matched by the specific name *minimus*.

Snipe is a word of Germanic origins to do with a cutting tool, in this case the bird's extraordinarily long bill. The word 'snip' is therefore related.

Lymnocryptes reminds us that this bird hides itself very effectively in marshland (Greek *limnē*, marsh, and *kruptos*, to hide).

Pin-tailed Snipe *Gallinago stenura*

This Siberian species is very rare in Western Europe.

Pin-tailed describes the 'pin-like' feathers of the outer tail.

The specific name *stenura* means 'thin-tailed'.

Swinhoe's Snipe *Gallinago megala*

This is an Asian species that has been noted in Russia and Finland.

Robert **Swinhoe** (1836–1877) was an English naturalist (see also **Swinhoe's** Petrel).

The specific name *megala* means great: it is the same size as the Great Snipe.

Great Snipe *Gallinago media*

This Eastern European species is quite rare in Britain.

Great is somewhat larger than Common Snipe.

The specific name *media* places this bird midway between Common Snipe and Woodcock in size.

(Common) Snipe *Gallinago gallinago*

This is a common resident species.

Common is sometimes needed to separate this from other snipes.

Gallinago is linked to the domestic hen, *gallina*. Historically the name also included Woodcock.

Lymnocryptes reminds us that this bird hides itself very effectively in marshland (Greek *limnē*, marsh, and *kruptos*, to hide).

* As a difficult target for shooters (it weaves when flushed), the bird's name eventually became linked with marksmen as 'snipers'.
* The name *Full Snipe* was used in the past by shooters to contrast this species with the smaller Jack Snipe.

Wilson's Snipe *Gallinago delicata*

This American species is a very rare vagrant to Britain.

Alexander **Wilson** was a Scottish-American ornithologist, and is considered to be the father of American ornithology. He occurs frequently in the naming of American species. However the AOS announced in 2023 that all eponyms in vernacular names of American species will eventually be replaced.

The specific name *delicata* means 'delicate', having been long applied to what was considered to be a smaller subspecies of Common Snipe.

Terek Sandpiper is unique in its genus.

Terek Sandpiper *Xenus cinereus*

This is primarily an Eastern European and Asian species. It is very scarce as a vagrant to Britain and Western Europe, largely because of the strict orientation of its north–south, overland migratory movements.

Terek is the name of a Caucasian river, where the bird was first named in 1775.

Xenus is Latinised Greek for 'stranger'.

The specific name *cinereus* means 'ash grey'.

PHALAROPES: *PHALAROPUS*

The names **Phalarope** and *Phalaropus* are reserved for three atypical waders that swim freely and whose generic names (from the Greek *phalaris* coot and *pous* foot) mean that they have coot-like projections of skin on their feet for that function.

They are also distinct from other members of the Scolopacidae in showing a reverse sexual dimorphism, in which the female is the more colourful and leaves the male to incubate eggs and raise young. (The Dotterel and Painted-snipe show some similar characteristics.)

Red (Grey) Phalarope *Phalaropus fulicarius*

This Arctic breeder migrates to winter in southern oceans. It is a regular autumn vagrant to Britain, usually appearing after Atlantic storms, with most of such birds being juveniles.

The question whether this bird is **Red** or **Grey** is a genuine dilemma related to culture and season. In its breeding areas it wears red livery, but when it appears on passage in Britain it is in grey plumage. The formal international name is Red, but it seems to make little sense in Britain to call a grey bird 'red', especially when it is easily muddled with the local 'Red-necked' species.

The specific name *fulicarius* reminds us again of the link with the coots (*Fulica*), but this time it does so in Latin.

Red-necked Phalarope *Phalaropus lobatus*

This species breeds in northern Britain, Iceland and Scandinavia, but is generally scarce on passage through the south of Britain.

Red-necked indicates that the colour is limited in the breeding plumage of this species.

The specific name *lobatus* is an alternative reference to skin formations on the foot.

Wilson's Phalarope *Phalaropus tricolor*

This American species breeds mainly in the west of Canada and the United States, so is a rare vagrant to Britain.

Alexander **Wilson** was a Scottish-American ornithologist. However the AOS announced in 2023 that all eponyms in vernacular names of American species will eventually be replaced.

The specific name *tricolor* is an understatement, as the bird in breeding plumage shows white, black, grey and orange-rufous.

SHANKS: *ACTITIS, TRINGA*

The informal term 'shanks' is used by British birders to describe a group of birds of the genus *Tringa* in which the colour of the legs or 'shanks' (in a now archaic usage) is important.

But first, two small relatives of the genus *Actitis*.

Common Sandpiper *Actitis hypoleucos*

This is not a particularly obvious summer visitor to Britain, but here the name **Common** is used regularly to distinguish it from the many other sandpipers.

Sandpiper is a widely used word for a number of small species that are often seen in sandy shorelines and utter piping calls.

Actitis is of Greek derivation and means 'coast-dweller', though these birds are most often seen around inland waters.

The specific name *hypoleucos* means 'white beneath'.

* As a summer visitor, this little bird was popularly known in some parts of Britain as the *Summer Snipe*, a name noted by Alfred Newton in 1896 as the most popular appellation of this species. Its bill is proportionately much shorter than a true snipe, however.

Spotted Sandpiper *Actitis macularius*

This is the American equivalent of the above and is a scarce vagrant to Britain.

Spotted and *macularius* both tell the same story, though the breast-spots appear only in breeding plumage. In winter it is quite similar to the previous species, and is then best separated by its brighter leg colour.

Green Sandpiper *Tringa ochropus*

This is a common passage species, with some wintering in Britain.

Green appears to be a logical name for a bird whose wings were described by Coward as 'greenish brown with a bronze gloss', though Linnaeus's reference to the legs (below) may have also influenced the name.

Tringa: this name was given to this species by Aldrovandi in 1599, though its origins are in Aristotle's use of the Greek *trungas* for a 'thrush-sized water-bird that bobs its tail'. Later the name was transferred to the whole genus.

The specific name *ochropus* means 'yellow-footed' and appears to be an error committed by Gessner in 1555. The legs are in fact grey-green (though those of the related Wood Sandpiper fit Gessner's bird). In spite of his own notes including the words *pedibus virescentibus* green-footed (see above), Linnaeus did not correct the error.

* Lockwood also recorded a Norfolk name, *Martin Snipe*, which is a useful aide-memoire for the birdwatcher: it indirectly draws attention to the white rump – like that of a House Martin.

Solitary Sandpiper *Tringa solitaria*

This American species is a rare vagrant to Britain and can be readily confused with the Green Sandpiper, though it is smaller.

Solitary and *solitaria* mean the same thing, and imply that the bird is not generally seen in flocks.

Grey-tailed Tattler *Tringa brevipes*

There are just two accepted British records of this Siberian species, which normally winters south, as far as Australia.

Grey-tailed is self-evident.

The name **Tattler** originates in North America with the Wandering Tattler, whose calls gave away the presence of a hunter by 'tattling' (telling tales).

The two tattlers were previously classed in the now-retired genus *Heteroscelus* (which related to the shape of the leg bones), but were moved in 2017 into *Tringa*.

The specific name *brevipes* means 'short-footed'.

* For Coward in 1926, this was the *Grey rumped Sandpiper*.

Lesser Yellowlegs *Tringa flavipes*

This is a fairly regular vagrant to Britain.

It is the **Lesser** of two American yellowlegs.

Yellowlegs is now preferred in America and internationally to the Anglicism *Yellowshank* (see note below).

The specific name *flavipes* is another way of conveying 'yellow legs'. (Compare *ochropus* in Green Sandpiper.)

* The name *Yellowshank* was used by John James Audubon for this species, and that unadorned form was considered the standard name in Coward's 1926 list. The advantage of the archaic form was that it was in keeping with the names of the closely related Redshank and Greenshank. For Audubon's generation, the eventual Greater Yellowlegs went primarily under another name (see below), so the qualification Lesser was not deemed necessary until much later – after the two names were aligned.

* The Americans ultimately preferred the plainer name Yellow*legs*.

* Like the Long-billed Dowitcher, the relative frequency in Britain of this, the more westerly of the two yellowlegs, is down to the timing and pattern of migration.

Willet *Tringa semipalmata*

This American species has occurred as a vagrant in a number of European locations, but not in Britain.

Willet is thought to be an echoic name. It may well have had its roots in *Curwillet*, a Celtic-Cornish name for a Sanderling, which is recorded in both Ray and Pennant.

Willet was moved in 2017 from the now-retired genus *Catoptrophorus* to *Tringa*. (The extensive white wing-patches were responsible for the old generic name, which means 'mirror-bearing'.)

The specific name *semipalmata* means that the bird has partially webbed feet.

* There are two quite distinct subspecies, *Eastern* and *Western*, which are different in plumage, structure and habits. There seem to be good reasons for them to be split.

Common Redshank *Tringa totanus*

This is a very **common** species in Britain.

Redshank is a traditional name that uses the archaic word '*shank*' for legs. In this case the red legs traditionally contrast with the green ones of the closely related Greenshank (see also Yellowlegs).

The specific name *totanus* derives from *tótano*, an Italian name for the species.

* The term *shank* is found in 'shank's pony' an ironic term for walking, and in the soubriquet of King Edward I, Edward Longshanks.

Marsh Sandpiper *Tringa stagnatilis*

This is a scarce vagrant to Britain.

Marsh and *stagnatilis* tell much the same story about the bird's habitat preferences, the latter meaning 'belonging to pools and fens'.

Wood Sandpiper *Tringa glareola*

This species is mostly seen on passage in Britain.

The name **Wood** Sandpiper is somewhat misleading: the bird is not particularly associated with woodland, so Pennant seems to have made a rather vague assumption when he coined this name.

The specific name *glareola* associates the bird with gravel.

Spotted Redshank *Tringa erythropus*

This is fairly scarce as a wintering species in Britain, where it is found mainly on the East and South coasts.

Spotted applies in the breeding season when the black plumage is speckled with white, making the bird look rather like a large Starling.

The specific name *erythropus* simply means 'red-footed'.

Common Greenshank *Tringa nebularia*

This is a moderately scarce wintering and common passage species in Britain.

Common is virtually unused, since the only other 'greenshank' is the Asian species, Nordmann's, which has not been recorded in the Western Palearctic.

Greenshank is an obvious name for a bird that has green-grey legs.

The specific name *nebularia* states that the plumage is 'cloud-grey'.

Greater Yellowlegs *Tringa melanoleuca*

This is much rarer in Britain than its smaller cousin, in spite of the fact that its American range is closer. The differences of migratory routes and timing make the larger species less likely to be storm-carried.

It is the **Greater** of the two yellowlegs species.

Yellowlegs is an obvious descriptive name: see also *Lesser Yellowlegs* (above).

The specific name *melanoleuca* means 'black and white', which relates to the fine chequering of the wings.

* This species was known to Audubon as the *Tell-tale Godwit* or *Tell-tale Snipe*: like the Tattler (above), it warned of the hunter's approach. He also acknowledged that it was known in 'the Western Country' as the *Great Yellow-Shank*.

* Coward's 1926 version of the British list has it as the *Greater Yellowshank*.

23. DROMADIDAE: CRAB PLOVER

This bird is unique in its family. It is currently considered to be closer to the coursers and pratincoles than to the plovers.

Crab Plover *Dromas ardeola*

This bird of the Persian Gulf region has been noted in countries of the Eastern Mediterranean.

The bird takes its name from the **crabs** that it eats.

Plover was originally a name of convenience: the bird is plover-like in a general way.

Dromas means 'running', which is a useful clue perhaps to its relationship to the coursers.

The specific name *ardeola* reminds us that the bird can have an upright, heron-like posture (*Ardea* = heron).

24. GLAREOLIDAE:
COURSERS AND PRATINCOLES

The family name is based on *Glareola*.

Birds of this family generally frequent warmer climes and are not often seen in Britain.

Cream-coloured Courser *Cursorius cursor*

This is an occasional vagrant to Britain.

Cream-coloured is self-evident.

Courser, *Cursorius* and *cursor* all tell us that the bird is a 'runner' (Latin *currere*, to run).

Collared Pratincole *Glareola pratincola*

This is an occasional vagrant to Britain.

Collared is an odd name, since the black line, worn only in the breeding season, runs down from the eye and is also a feature shared with Black-winged.

Pratincole and *pratincola* describe this as a 'meadow-dweller'.

Glareola derives from the Latin for 'gravel' (see Wood Sandpiper above).

Oriental Pratincole *Glareola maldivarium*

There are a few confirmed records in Britain.

The **Oriental** Pratincole is found in South and Southeast Asia.

The specific name *maldivarium* derives from the fact that the first specimen was caught and named near the Maldives by Johann R. Forster, the biologist on Cook's second voyage (see Forster's Tern).

Black-winged Pratincole *Glareola nordmanni*

This is an occasional vagrant to Britain.

Black-winged is obvious for the darker species.

The specific name *nordmanni* commemorates the Finnish naturalist Alexander von Nordmann (1803–1886), whose name is also attached to a species of greenshank (see above).

Gulls to
Tropicbirds

25. LARIDAE:
NODDIES, GULLS AND TERNS

The family name is based on *Larus*.

Laridae has broadened its scope, because it formerly included only the gulls, but now embraces noddies, skimmers and terns. Sequencing in particular will probably be changed in the near future as the result of ongoing reviews.

Birdwatchers often use the vernacular collective, **Larids**, to refer particularly to gulls.

From that come the terms **Laridophile** and **Laridophobe**. Gulls are so difficult that they have become something of a specialism among some birdwatchers (the Laridophiles), while others (the Laridophobes) steer away from tackling their complexities.

> **Noddies** were formerly treated as members of the retired tern family, Sternidae, but are now included in the Laridae and sequenced separately before gulls.

Brown Noddy *Anous stolidus*

There is one historical record from Germany of this otherwise tropical species.

Brown is self-evident.

Noddy is more intriguing: it was originally a slang word for a senile person (i.e. with a nodding head). Because they did not flee, the birds were considered stupid by the mariners who raided their colonies for eggs or birds as food.

The specific name *Anous* continues that theme by meaning 'foolish' or 'silly', and that is further underlined by *stolidus*, for 'stupid'.

> **Skimmers** too have now been subsumed into the Laridae.

African Skimmer *Rhynchops flavirostris*

This species has been recorded in Israel and Morocco, though never in Western Europe.

There are three species of skimmer, and this one comes from **Africa's** tropical regions.

Skimmer describes the feeding action in which the bird skims low over the water to trawl the surface with its elongated lower mandible.

Rhynchops inaccurately describes a 'cut-off bill': the upper mandible is shorter than the lower and gives that impression, but of course it is the lower mandible that has evolved to be longer to enable the fishing style.

The specific name *flavirostris* tells us that the bill is yellow, though that too is somewhat inaccurate: only the tip of the bright orange-red bill is yellow. It seems likely that the name is intended to contrast this feature with the black-tipped bill of the American species, *Rhynchops niger*.

GULLS

The gull family has always been difficult, not only because all gulls change appearance according to age and season, but also because it has been difficult to unravel the complex question of speciation, particularly in the Larus complex. The late twentieth and early twenty-first centuries have been characterised by a tumult of redefinitions and splits. Not all is stable in 2024: the ink is hardly dry on the split of Mew Gull and some other species may yet follow the same route. In any case, there are a few traps inherent in the names themselves, so one must tread warily.

The word **Gull** has Celtic roots, and came into general English usage via the Cornish language, in about 1430, though it did not come into more formal use until the seventeenth century. Previously the word *Mew* served more widely (see Preface and below). Lockwood also records the use of *Sea Cob* (1530) or *Cob*, particularly on the East and South coasts, where its links with the Dutch *kobbe* would not be surprising.

The popular use of the catch-all **seagull** has become less accurate than it used to be. Thanks to greater urbanisation and infrastructure developments, many gulls now live and breed inland: landfill sites provide food, while artificial waterbodies, such as reservoirs and disused gravel pits, offer bathing and roosting facilities.

(Black-legged) Kittiwake *Rissa tridactyla*

This species breeds widely but patchily around the coasts of Britain.

Internationally this is **Black-legged Kittiwake** because there is a Red-legged species in the Pacific.

Kittiwake is an echoic name based on the distinctive three-note call of the bird.

The Icelandic name for the species, *Rita*, is the origin of the generic name *Rissa*.

Unlike other gulls, the nominate form has no rear-facing toe, hence it is 'three-toed' or *tridactyla*. (Curiously, the Pacific subspecies *pollicaris* (thumbed) has a vestigial rear toe.)

* This species normally requires rocky cliffs for nesting, but may also use man-made structures, as at Sizewell, Suffolk, where an offshore platform provides nesting ledges.

Ivory Gull *Pagophila eburnea*

This is a scarce winter vagrant to Britain from its Arctic breeding grounds.

Ivory seems an appropriate name for an all-white gull.

Pagophilia (a genus unique to this species) describes it as a 'sea-ice lover'.

The word *eburnea* is from the Latin *ebur* (ivory).

* The bird was originally described by naval officer, Constantine Phipps, in 1774 as *Larus eburneus*, but Johann Kaup recognised its unique character and created the new genus in 1829.

Sabine's Gull *Xema sabini*

This bird migrates down the Atlantic from Arctic Canada and is a regular autumn vagrant to Britain's coasts. Such birds are often juveniles.

General Sir Edward **Sabine** (1788–1883) is commemorated in the vernacular and specific names. He was a British Arctic explorer and, later, President of the Royal Society. However, the AOS announced in 2023 that all eponyms in vernacular names of American species will eventually be replaced.

Xema is another unique genus name. It was apparently a fanciful invention of Leach in 1819.

During the restructuring mentioned above, the genus *Chroicocephalus* was resurrected for a subset of gulls long classified in *Larus*. The complicated name came as a shock to the early twenty-first century. Few people realised that the genus had first been introduced in 1836 by the English ornithologist Thomas Campbell Eyton to distinguish the 'masked' gulls. For many years the official blanket use of *Larus* had obscured the subtle differences found within the family. *Chroicocephalus* loosely means 'coloured or stained head', and that applies to the species that develop black, brown or grey hoods during the breeding season. There are exceptions, though: Europe's Slender-billed Gull and the Antipodean forms are white-headed all year-round. However, these gulls do share a general similarity of wing pattern.

Black-headed Gull *Chroicocephalus ridibundus*

Probably the best known of all British species, the Black-headed proliferated as an inland gull during the twentieth century and also colonised part of the east coast of America.

Black is a misnomer: though appearing black from a distance, it is actually sooty brown when seen close to (but Brown-headed Gulls exist in Asia).

More potential confusion exists here, since *ridibundus* means 'laughing' – and the Americans have a Laughing Gull (see below).

Brown-headed Gull *Chroicocephalus brunnicephalus*

This Central Asian species has been recorded in Israel, but never in Western Europe.

Brown-headed and *brunnicephalus* mean the same, but the brown is distinctly paler than that of the Black-headed Gull.

Slender-billed Gull *Chroicocephalus genei*

This Mediterranean species is very rare in Britain.

Slender-billed is self-explanatory, though the illusion is partly to do with the shape of the head and length of the bill.

The specific name *genei* commemorates Giuseppe **Gené** (1800–1847), a nineteenth-century Italian naturalist.

Bonaparte's Gull *Chroicocephalus philadelphia*

This small gull is an occasional vagrant from North America.

Charles-Lucien **Bonaparte** (1803–1857) was a French ornithologist who spent six years in America. The AOS announced in 2023 that all eponyms in vernacular names of American species will eventually be replaced.

The North American town of **Philadelphia** lends its name for the specific form.

* C-L. Bonaparte was exiled to America for a time with other members of his family after the 1815 fall of his uncle, the Emperor Napoleon I of France. He was both helped and hampered by his famous heritage, but was a prodigious talent in the field of nineteenth-century ornithology. He worked in America from 1822 to 1828. In that short time he had a huge impact on the ornithological scene there. Among his creations was the dove genus *Zenaida*, which he named after his wife. He eventually continued his work in Italy and France, between assorted political ventures, and was finally director of the Jardin des Plantes in Paris.

* The naming of this species took a bumpy road. It was first described in 1815 by George Ord as *Sterna philadelphia* and then moved to the genus *Larus*. In Audubon's account of the species we find the name *Larus bonapartii*, deriving from Swainson and Richardson's *Fauna Boreali-Americana* (1831): this appears to be the point at which Bonaparte is linked with the bird. We can assume that the rule of precedence then overruled the Latin form, leaving Bonaparte only in the vernacular name.

Grey-headed Gull *Chroicocephalus cirrocephalus*

This is a species with separate South American and African races. It has been recorded in Spain, North Africa and the Eastern Mediterranean.

Grey-headed is self-explanatory and that is reinforced by *cirrocephalus*, which means the same.

Little Gull *Hydrocoloeus minutus*

This is a regular passage species through Britain.

Little and *minutus* point to the fact that this is the smallest of all gulls.

Hydrocoloeus was originally used in 1829 by Johann Kaup, but was discarded for many years while *Larus* embraced most gulls. The name was resurrected for use a few years ago. It has two elements: *hydro* (water) and *koloios* (a web-footed bird).

Ross's Gull *Rhodostethia rosea*

Ross's is a scarce winter visitor to Britain from the High Arctic.

Ross's Gull records Rear Admiral Sir James **Ross** (1800–1862), a polar explorer who is also remembered in the name of an Antarctic sea. (Note that a different Ross appears in the name Ross's Goose.) However, the AOS announced in 2023 that all eponyms in vernacular names of American species will eventually be replaced.

Rhodostethia is another unique genus and refers to the 'rosy breast', which is underlined for its pinkness by *rosea*.

* The original specimen was collected by the young Ross on one of several Arctic voyages. William MacGillivray originally named it in 1824 as *Larus roseus*, but revised that in 1848 when he created the new genus, *Rhodostethia*.

Laughing Gull *Leucophaeus atricilla*

This is a scarce vagrant to Britain from the Americas.

Laughing Gull refers to one of the bird's calls but, confusingly, the Latin form *ridibundus* (laughing) belongs to the Black-headed Gull.

Leucophaeus means 'ash-coloured'. The genus contains just five dark-backed American species.

The specific name *atricilla* is a curiosity, because it tells us, erroneously, that this white tailed bird has a black tail. This was a mistake by Linnaeus. He apparently meant to write *atricapilla* (black-headed), but the omission of that particular syllable upends his logic.

* Linnaeus made a few such errors, but it has to be remembered that his collection of data on living things included very much more than just birds. In the context of a body of work that underpins all modern classification of plants and animals, he must be forgiven for the odd error. It simply proves that this scientific superman was only human after all. He was, in any case, far from being alone in making that sort of slip.

Franklin's Gull *Leucophaeus pipixcan*

This is yet another scarce vagrant to Britain from the Americas.

The British explorer Sir John **Franklin** (1786–1847) is remembered in this name. The first specimen was taken during his 1825 voyage to Canada. However, the AOS announced in 2023 that all eponyms in vernacular names of American species will eventually be replaced.

It was later realised that the sixteenth-century Spanish explorer-naturalist Francisco Hernandez de Toledo had left a very accurate description of this species during exploration of Mexico, where the bird is found in winter. Johann Wagler acknowledged this when he named the bird formally in 1831, using the specific name *pipixcan*, which is derived from a local Amerindian form recorded by the Spaniard.

Relict Gull *Ichthyaetus relictus*

This species was once treated as a subspecies of Mediterranean Gull. It is sometimes also called the *Central Asian Gull*, but has been recorded in European Russia.

Relict and *relictus* relate to the fact that it breeds in an isolated area. It has a small and threatened population.

For *Ichthyaetus*, see Pallas's Gull (below).

Audouin's Gull *Ichthyaetus audouinii*

This Mediterranean species has been recorded only very rarely in Britain.

J.V. **Audouin** (1797–1841) was a French entomologist and ornithologist.

For *Ichthyaetus*, see Pallas's Gull (below).

Mediterranean Gull *Ichthyaetus melanocephalus*

The Mediterranean Gull now breeds in Britain and has long been more widespread than just the **Mediterranean**, with which it has only loose associations.

For *Ichthyaetus*, see Pallas's Gull (below).

The specific name *melanocephalus* is another one to provoke confusion, since it means 'black-headed'. That is more appropriate to this species, which in Coward's 1926 list was in fact called the *Mediterranean Black-headed Gull*. In the breeding season it has a more extended and darker area of true black on the head and upper neck than that of the Black-headed Gull, *C. ridibundus*.

Pallas's Gull *Ichthyaetus ichthyaetus*

This Asian species is a very rare vagrant to Britain.

The preferred international name, **Pallas's Gull**, commemorates Peter Pallas (1741–1811), a German explorer and naturalist.

Ichthyaetus means 'fish eagle', which seems oddly misleading. It made better sense when this species was *Larus ichthyaetus*, where the specific name clearly describes the 'eagle-like' characteristics of a large gull. It certainly seems quite inappropriate as a generic name for the smaller relatives of this genus.

* The alternative name *Great Black-headed Gull* was widely used for many years, but has now been set aside.

White-eyed Gull *Ichthyaetus leucophthalmus*

This species of the Red Sea strays to Israel.

White-eyed and *leucophthalmus* mean the same.

Sooty Gull *Ichthyaetus hemprichii*

This species of the Red Sea and Persian Gulf also strays to Israel.

Sooty is self-evident for a dark grey gull.

F.W. **Hemprich** (1796–1825) was a German naturalist and explorer.

MEW GULL AND COMMON GULL

The American name Mew Gull prevailed until 2021 on the IOC list over the name Common Gull in spite of the fact that the European form was the nominate one and the American form was designated as a subspecies. Common Gull was never abandoned in general use in Britain, where Mew was largely ignored. In 2021 the IOC announced a definitive split of the two forms, with *Common Gull* as *Larus canus* and the American form as *Larus brachyrhynchus*, but under the vernacular name of *Short-billed Gull*.

The apparent 'new' name is a version of one created by the Scottish naturalist and explorer John Richardson in 1831, when he described the 'short-billed mew gull', *Larus brachyrhynchus*.

It appears that the name *Mew Gull* was permanently set aside because there was a need to separate the newly split species from the complications of older, mixed records.

From the British perspective, it may seem to be a good idea to be rid of an unused Americanism, but in some ways the loss of Mew is regrettable. The name is in fact one of the oldest and simplest of bird names, which once served for all gulls. It derives originally from a Norse form and has many other related Germanic forms. It had faded from general use in Britain by the end of the seventeenth century and had been replaced by the word *gull*. However, the word had been transported earlier by those migrating to North America and preserved into the twenty-first century. Sadly, a piece of linguistic history, older than the name of England itself, has been quietly discarded.

Common Gull *Larus canus*

When, in 1768, Pennant chose **Common Gull** to replace *Sea Mall* (a variant of *Sea Mew* via *Sea Maw*) he considered it to be the most numerous of its genus. He may have been wrong in that assumption, as that is certainly not the case today. Discomfort with that anomaly may have provoked the suggestion that the species may be so-named because it frequents 'commons' (i.e. grassland), but this seems somewhat fanciful.

The specific name *canus* simply means 'grey'.

Short-billed Gull *Larus brachyrhynchus*

This North American species has a bias to the north-west of the continent. The only recorded appearance in the Western Palearctic is in the Azores.

Short-billed and ***brachyrhynchus*** both convey the fact that the bill is shorter than its Eurasian cousin's.

The original name, *Short-billed Mew Gull*, was created in 1831 by its finder, the Scottish explorer and naturalist, John Richardson, who was an associate of John Franklin (as in Franklin's Gull). For some years the name Short-billed Gull was used formally in American records, but was later replaced by Mew Gull.

* Richardson is also remembered in *Richardson's Goose*, as well as in the now archaic form *Richardson's Skua* (*Arctic Skua*/*Parasitic Jaeger*).

Ring-billed Gull *Larus delawarensis*

This American species is a fairly regular vagrant to Britain.

The name **Ring-billed** derives from the broad dark band on the bill, which is present at all ages.

The Delaware River, where the species was first recorded, gives rise to the specific name *delawarensis*.

Great Black-backed Gull *Larus marinus*

This is a common enough species in Britain, but is less-frequently found inland than some other species.

Great: this is simply the largest gull (the largest in the world, in fact), and with a truly **black back**.

The specific name *marinus* implies that this species is closely associated with the sea.

Kelp (Cape) Gull *Larus dominicanus vetula*

Kelp (from the seaweed) is a species of the coasts of the Southern Hemisphere. In recent years the subspecies *vetula* (known popularly as **Cape Gull**) has bred in north-west Africa and has been recorded in Portugal, Spain and France. One eventually appeared in Britain in 2022 and was reported again in 2023. This form may eventually be treated as a separate species.

Cape is for its connection to South Africa's Cape of Good Hope.

The specific name, ***dominicanus***, reflects the fact that this is 'cloaked like a Dominican friar', a reference to its black mantle.

The subspecific name *vetula* means 'old woman', which probably refers to the 'cloaked' appearance, this time with a shawl in mind. (Compare *viduata*, with its link to *widow*, in the name of the White-faced Whistling Duck.)

A few species are known colloquially as 'white-winged' gulls because they lack black feathers in the wing-tips. Some, like Glaucous-winged, are intermediate, with grey wing-tips.

Glaucous-winged Gull *Larus glaucescens*

This is an accidental American species from the Pacific Coast that has been recorded in Britain, the Canaries and Morocco.

Glaucous-winged describes the pale blue-grey of the wings, which are also tipped with grey.

The specific name *glaucescens* means the same as glaucous.

Glaucous Gull *Larus hyperboreus*

Small numbers of this Arctic species winter in Britain.

Glaucous describes the pale blue-grey mantle, which is emphasised by the pure white of the wingtips.

The 'very northerly' breeding range accounts for the specific name *hyperboreus*.

Iceland Gull *Larus glaucoides glaucoides*

This is a regular, though scarce, winter visitor to Britain.

Iceland is something of a misnomer, because the species breeds in Greenland and Canada, though it is a passage and winter visitor to Iceland. That name is generally used for the nominate form, since the current subspecies (Kumlein's and Thayer's) have previously been treated as full species.

The fact that it 'resembles a Glaucous Gull' accounts for the specific name *glaucoides*.

Kumlien's Gull *Larus glaucoides kumlieni*

This form has been recorded as a rare vagrant to Britain.

Its status has changed since its first identification as a full species: since 2017 it has been considered to be a more westerly subspecies of the Iceland Gull. It generally has darker wingtips.

It was originally named as a full species in 1883, commemorating the ornithologist Ludwig **Kumlien** (1853–1902), son of the Swedish-American naturalist Thure Kumlien.

Until the end of the twentieth century, Herring Gulls were treated as a varied complex, but genetic analysis has since revealed a need to split the complex into a number of separate species. This resulted in some considerable shuffling and has led to a number of new names.

European Herring Gull *Larus argentatus argentatus*
Larus argentatus argenteus

European is now a very generous range-description of a species that is mainly confined to Atlantic Europe. The two subspecies can often be separated visually and both are found in Britain in winter.

Herring Gulls used to follow the old herring fishing fleets, hence the name.

The genus *Larus* was previously used for most gull species, but is now limited to a smaller part of the complex. The Latin form is derived from the Greek *laros* (gull).

The specific name *argentatus* means 'silvery'. The nominate form is the more northerly and is sometimes known as the **Scandinavian** Herring Gull.

The name *argenteus* (the southern subspecies that breeds in Britain) means a different sort of 'silvery': it is in fact paler.

American Herring Gull *Larus smithsonianus*

This species, particularly the dark first year form, is recorded as a scarce vagrant to Britain.

American distinguishes this species from *European* Herring Gull.

The specific name *smithsonianus* commemorates James Smithson (1765–1829), the American philanthropist and founder of the Smithsonian Museum of Natural History in Washington.

* An alternative name, *Smithsonian Gull*, is sometimes used in America, since that was how Elliott Coues named this form in 1862 when he described it from specimens held in the museum. It was first treated as a full species and then as a subspecies of Herring Gull. The wider use of the alternative name might be preferred to avoid confusion with the European form, but with the American abandonment of eponyms that will not now happen. In any case there seems to be some need to convince all stake-holders of its separate status. Curiously, the AOS is the one organisation that does not recognise this split, in spite of what appeared to be convincing evidence published in 2002.

Vega Gull *Larus vegae*

This Far-Eastern species was recorded in Ireland in 2016 as a first for the Western Palearctic.

The name **Vega**, and the specific name *vegae*, relate to Cape Vega in northern Siberia.

That said, the trail to the real roots of that name reads like a game of consequences: the cape itself was named after the Norwegian ship that made the first voyage through the North-East Passage in 1878–9. The ship in turn was named after the brightest star in the constellation Lyra. The ultimate root of that name lies in an Arabic word that was translated into Latin as the 'falling vulture'. It is interesting that such an odd chain has a bird at each end.

* This is another unresolved split of the Herring Gull complex.

Caspian Gull *Larus cachinnans*

Caspian Gull is a scarce but regular visitor to Britain.

Many of this species breed around the **Caspian** Sea.

The specific name, *cachinnans*, means 'laughing raucously'.

* For a time during the early twenty-first century, this species was classed with the Yellow-legged Gull, both of them under the specific name *cachinnans*, which, confusingly at the time, stayed with Caspian Gull when the two were split.

Yellow-legged Gull *Larus michahellis*

This Mediterranean species is now seen in Britain regularly.

Yellow-legged is self-evident (though the leg colour is not unique to this species).

The specific name commemorates K.W. **Michahelles** (1807–1834), a German zoologist.

* Even in its much-reduced scope, this species still embraces some difficult issues, because the exact definition of Azorean and Atlantic forms remains complicated (see Stoddart and McInerny, 2017).

Armenian Gull *Larus armenicus*

This Caucasian species has not been recorded in Britain.

Considered by some to be a race of Yellow-legged Gull, this is given separate species status by most authorities because of a number of behavioural and physical differences.

Its native range is in and around **Armenia**, hence also *armenicus*.

Slaty-backed Gull *Larus schistisagus*

This Far-Eastern accidental has been recorded once in Britain, as well as in Latvia and Lithuania.

Slaty-backed is self-explanatory.

The specific name *schistisagus* is made up of two Latin elements meaning 'slate-cloaked'.

Lesser Black-backed Gull *Larus fuscus fuscus*

Larus fuscus intermedius
Larus fuscus graellsii
'Heuglin's Gull' *Larus fuscus heuglini*

The visual separation of the subspecies is possible, owing to mantle colour and other structural differences. In Britain, the common form is *graellsii*, with *intermedius* also occurring in winter. The occurrence of *fuscus* is much rarer and is the subject of much debate among laridophiles. To date there are no proven examples of *heuglini* closer than Russia and Finland.

This species is smaller than Great Black-backed, therefore it is the **Lesser** of the two.

The specific name, *fuscus*, means 'dark grey', which is true of all the subspecies below, though the nominate form (aka the **Baltic Gull** for its distribution) is truly black.

The form *intermedius* is intermediate in shade between *fuscus* and *graellsii* – just pale enough for the black wingtips to remain distinct.

The charcoal-grey form, *graellsii*, commemorates M. de la Paz Graells y de la Agüerra (1809–1898), a Spanish zoologist.

The form *heuglini* is known in English as **Heuglin's Gull**. M.T. Heuglin (1824–1876) was a German explorer and ornithologist.

TERNS

Terns long had their own family, the Sternidae, but they have now been subsumed into the Laridae.

The word **Tern** is based on the medieval English form, **stern** (variants stearn, starn), which in turn have Nordic-Germanic roots.

William Turner gave this the Latin form *Sterna* in 1544. That was later adopted as the formal generic name.

Tern (or tarn), was noted by Ray in 1678, though he preferred Sea Swallow. Pennant formalised the word '**tern**' by adopting it in 1768.

A more poetic generalisation, **Sea Swallow**, is of course a simple descriptive idea, based on the long-winged shape and forked tail, which together resemble the profile of a large swallow. That idea is echoed in the specific name of the Common Tern. Some other species incorporate Greek forms to convey the same idea.

Most terns favour warmer climates, but a small number of species breed in Britain, all of them migrating from further south to do so. A number more species are recorded as vagrants, some well away from their normal ranges.

Gull-billed Tern *Gelochelidon nilotica*

This bird is recorded as a vagrant to Britain. It breeds in warmer zones, including the Mediterranean.

Several aspects of this bird are **gull**-like, including the shape of the bill.

Gelochelidon is literally a 'laughing-swallow' (based on the alarm call plus shape).

The specific name *nilotica* is a link with the River Nile.

* Until 2019 this species was unique in its genus, but with several subspecies. The Australian form (*macrotarsa*) is now treated as a full species in its own right.

Caspian Tern *Hydroprogne caspia*

This, the largest of the terns, is a scarce vagrant to the British Isles.

It was originally named by Pallas from a specimen from the **Caspian Sea**, hence both the vernacular and specific names, though its distribution is very much wider. The nearest breeding area to Britain is around the Baltic Sea.

This is the sole member of its genus. *Hydroprogne* is a 'water-swallow' (after the Greek legend of Procne/Progne, a tragic victim of the tyrant Tereus.)

Royal Tern *Thalasseus maximus*

This is another American species that is recorded as a rare vagrant to the British Isles.

Royal and *maximus* both indicate that this is the largest of the genus.

This tern and its allies used to be classified in the genus *Sterna*, but have, since 2011, been classed as *Thalasseus*, which derives from the Greek for 'fisherman'.

Greater Crested Tern *Thalasseus bergii*

This is another rare vagrant to the British Isles. Its normal range is from the Indian Ocean to the Pacific.

This is the **Greater** of the original two species bearing the name **Crested Tern** (see also *Lesser*, next).

The specific name, *bergii*, records a German naturalist, C.H. **Bergius** (1790–1818).

Lesser Crested Tern *Thalasseus bengalensis*

This is another rare vagrant to the British Isles from its breeding areas along the coasts of North Africa.

This was the **Lesser** of the two crested terns that were recognised prior to the description of the following species.

The specific name *bengalensis* reflects the fact that this species appears in the Bay of Bengal in the Indian Ocean (it is also much more widespread).

West African Crested Tern *Thalasseus albididorsalis*

This is a recent split of a former subspecies of Royal Tern. It breeds as far north as the west coast of Morocco.

The rather cumbersome English name is typical of so many modern splits (see the introductory note on English nomenclature).

The specific name *albididorsalis* tells us that the bird's back is 'whitish'.

Sandwich Tern *Thalasseus sandvicensis*

This species breeds widely at British coastal sites and is sometimes seen along the South Coast in winter.

In 1797, the ornithologist John Latham received specimens of this species from a collector who lived in the coastal town of **Sandwich** in Kent and named it after that town.

The specific name, *sandvicensis*, is a Latin version of Sandwich.

Cabot's Tern *Thalasseus acuflavidus*

This is a rare vagrant to Britain from North America. One confirmed record of the American form was from a ringed recovery. It is probably under-recorded because it was long seen as a subspecies of Sandwich Tern. This split is not universally accepted and some authorities still treat Cabot's as a subspecies of Sandwich Tern.

This shorter-billed form is named after an American naturalist, Samuel **Cabot** (1815–1885). However, the AOS announced in 2023 that all eponyms in vernacular names of American species will eventually be replaced.

The specific name, *acuflavidus* (*acus*, needle, and *flavidus*, yellowish) is more appropriate to the more southerly yellow-billed subspecies generally known as *Cayenne Tern*, since the nominate form has a black bill with only a yellow tip.

Elegant Tern *Thalasseus elegans*

Very surprisingly, a small number of this Pacific American species now breed on the Atlantic Coast of Europe and one or two have wandered to British shores. This colonisation appears to be a new phenomenon.

Elegant and *elegans* both describe a bird that is slender and elongated, the bill in particular.

Little Tern *Sternula albifrons*

This species breeds widely on British coasts.

Little is obvious: it is the size of a thrush. (All terns of this genus are small and several carry a vernacular name that reflects that, examples being *Least* Tern and *Fairy* Tern.)

Sternula is a diminutive form of *Sterna* (see introductory note above).

The bird has a white forehead, hence the specific name **albifrons** (*alba*, white, and *frons*, forehead).

Saunders's Tern *Sternula saundersi*

This close relative of the Little Tern is mainly an Arabian Sea species, which occurs in the north of the Red Sea and is now confirmed as breeding in Egypt.

Howard **Saunders** (1835–1907) was a British ornithologist who specialised in gulls and terns.

* The bird was originally described by Hume as the *Black-shafted Tern*, for the strong black of some of the primary feathers that helps to separate it from the Little Tern.

Least Tern *Sternula antillarum*

This American equivalent of the Little Tern has been recorded in Britain a few times, but is easily overlooked because of its similarity.

The specific name *antillarum* links it to the Antilles, the island chain on the eastern side of the Caribbean.

Aleutian Tern *Onychoprion aleutica*

This is a Pacific species that has been recorded just once in Britain, on the Farne Islands, in 1979, which Lees and Gilroy (2021) highlighted as a truly exceptional example of tern vagrancy.

It is named for the **Aleutian Islands**, which stretch out from Alaska towards Asia.

Onychoprion means 'saw-clawed', relating to the deep indentations of the webbing between the claws.

The specific name *aleutica* refers to the islands, which in are turn named after the native Aleut people.

Sooty Tern *Onychoprion anaethetus*

This is a rare vagrant to Britain. It breeds in the West Indies and in the Red Sea, eastwards.

Sooty is self-evident, referencing the mantle colour.

The specific name *anaethetus* means 'stupid, senseless' (compare such names as noddy and booby). These are all names reflecting the fact that the nesting birds did not flee from humans when their nest sites were raided for food.

Bridled Tern *Onychoprion fuscata*

This species has been recorded occasionally in Britain. It breeds in the Caribbean, in Mauretania and eastwards from the Persian Gulf.

Bridled: the bird has a black facial pattern reminiscent of a horse's bridle.

The specific name, *fuscata*, means 'very dark', though it is paler than Sooty.

Roseate Tern *Sterna dougallii*

This species breeds in small numbers at off-shore sites to the west and east of Britain.

In the breeding season the birds develop a rose-coloured tinge to the breast feathers, hence they are **roseate** (pinkish) at this season.

Sterna is explained in the header note for this section.

Dr Peter McDougall (1777–1814) sent a type specimen of the bird to the ornithologist George Montagu in 1813, and was rewarded by the naming of the species as *Sterna **dougallii***.

Common Tern *Sterna hirundo*

Common Tern, as the name suggests, is widespread in the British Isles and in America in the breeding season, both at inland and coastal sites. The name was first used by Pennant to replace *Greater* Tern, the latter contrasting with *Little Tern* at a time before the Arctic and Roseate Terns had been distinguished.

The specific name, *hirundo*, is Latin for 'Swallow'.

* Common and Arctic Terns, especially when on the wing, can be hard to separate. Baffled birders sometimes report *Commic* Terns. This wonderfully ironic term harps back to the days of the music halls, which presented mixed acts or 'turns' of song, dance, acrobatic skill and comedy. A comic turn is today's stand-up comedian.

White-cheeked Tern *Sterna repressa*

This species occurs in Egypt, the Red Sea and the Persian Gulf. It is not recorded in Britain.

In breeding plumage, the **white cheek** stands out from a grey body.

The specific name *repressa* means 'restrained', though its significance for the tern is not obvious.

Arctic Tern *Sterna paradisaea*

This species breeds in the British Isles and northwards as far as the **Arctic** Circle.

The specific name *paradisaea* relates to the bird's long tail streamers, as in Birds of *Paradise* and the *Paradise* Kingfisher, where *Paradise* equates to exotic beauty.

* After spending the summer in the long days of the Northern Hemisphere, this bird 'winters' in the seas around Antarctica during the Antarctic summer. It is therefore reputed to experience more daylight than any other bird species.

Forster's Tern *Sterna forsteri*

This North American species is a scarce vagrant to Britain.

It commemorates Johann R. **Forster** (1729–1798) and was, according to Choate, so dedicated in 1834 because Thomas Nuttall admired his treatise on the birds of Hudson Bay. However, the AOS announced in 2023 that all eponyms in vernacular names of American species will eventually be replaced.

* Forster's greater claims to fame were his publication of *A Catalogue of the Animals of North America*, which assured his recognition in Britain, and his subsequent participation in Captain Cook's second Pacific voyage.

The three following species are often generalised as 'marsh terns' for their inland habitat preferences.

Whiskered Tern *Chlidonias hybrida*

These are scare visitors to Britain from the Mediterranean and Eastern Europe.

Whiskered refers to the white cheeks of the breeding plumage, which give the appearance of side whiskers against the black cap and dark-grey body.

Chlidonias (from the Greek *khelidonios*) is another way of saying that the bird is 'swallow-like' (see also swallows and *Chelidon*).

The specific name here is enigmatic: *hybrida* seems to imply that it is between 'sea' and 'marsh' terns in characteristics.

White-winged (Black) Tern *Chlidonias leucopterus*

This is a scarce but regular vagrant to Britain when it strays westwards during migration.

BOU retains the word **Black**, which is omitted in the International name. The bird has a black body and head in breeding plumage, with strongly contrasting white wings, hence **White-winged** and the equivalent, *leucopterus*.

Black Tern *Chlidonias niger*

This is a regular passage species, which does not breed in Britain, but breeds widely near fresh water in Europe.

Black and *niger* reflect the colour of the breeding plumage on the head and body. In this case the wings are grey and do not contrast so strongly as the previous species.

* A juvenile of its American subspecies *surinamensis* (from Surinam) was first recorded in Britain in 1999, and since then adult forms have been identified occasionally as *American Black Tern*.

26. STERCORARIIDAE: Skuas

The family name is based on *Stercorarius*.

This is an area where there have long been differences between the AOU and the BOU on the question of naming.

Taxonomists seem now to have agreed that all skuas should be classified in the genus *Stercorarius*, where previously the larger species had their own genus, *Catharacta* (rapacious). The concept behind this division appears to be that the larger species were seen as predators, while the smaller ones were treated as kleptoparasites. The truth is that all skua species show aspects of both feeding styles and are scavengers too.

The original division seems to be the reason why the Americans call large species **skuas** and smaller ones **jaegers**, a tradition that has now been accepted by the IOC as the standard.

This seems an odd decision, since the Americans are clearly familiar with the name skua, while jaeger is completely alien to the British tradition and so seems more of an imposition than a compromise.

Jaeger (hunter) reflects the influence of German-American immigrants on American English.

South Polar Skua *Stercorarius maccormicki*

This close relative of the Great Skua has been recorded just once in Britain, but is known in the North Atlantic more generally.

The name **South Polar** is self-evident: this Southern Hemisphere species was certainly well off course.

For **Skua** see next species.

R. McCormick (1800–1890), a British naval surgeon, is commemorated in the specific name *maccormicki*.

* It is interesting to compare the treatment of this name with those of McDougall in Roseate Tern (above) and MacQueen in the bustard (below). The various species' authors had no set rules in this matter, as witnessed also by the variants in the use of Rüppell's name, among others.

Great Skua *Stercorarius skua*

This species breeds in the north of the British Isles and is a passage species further south.

It is **Great** because it is the largest species in the Northern Hemisphere.

Skua comes from the Faroese *skúvur*, via Norse. It was anglicised by Ray in 1678 and then Latinised by Pennant in 1768. This name has been traditionally used in Britain for all members of the family.

Stercorarius means 'dung eater' and relates to the fact that skuas scavenge and also eat food regurgitated by their victims.

* One of the most familiar and persistent traditional names for the species is *Bonxie*, a Shetland dialect word that was greatly popularised by the television presenter Bill Oddie in the late twentieth century and so gained great currency among modern birders.

Historically, in Cornwall, it was called *Tom Harry*, which may have in part been a play on the verb *to harry* (pursue) but also represents a maritime example of the use of personal names, another of which occurs (obscurely) in Guillemot (see below).

Pomarine Jaeger *Stercorarius pomarinus* (INTERNATIONAL NAME)
Pomarine Skua *Stercorarius pomarinus* (BRITISH NAME)

This species breeds in Arctic regions and is a passage species in Britain and Ireland.

The strange names **Pomarine** and *pomarinus* derive from two Greek words for 'nostril cover'. For obscure reasons they draw attention to the cere, the waxy covering found at the base of the bill in some species, such as pigeons.

* Strictly speaking, Pomarine is an inaccurate rendition of the Greek. It was created by Stevens in 1826, but the BOU made a rather pedantic attempt in 1883 to replace it with the cumbersome form *Pomatorhine*. That did not catch on, but nonetheless appeared in the respective lists of Hartert (2012) and Coward (1926) and even as late as the BOU's own checklist of 1952. It was formally dropped sometime later.

Parasitic Jaeger *Stercorarius parasiticus* (INTERNATIONAL NAME)
Arctic Skua *Stercorarius parasiticus* (BRITISH NAME)

The Arctic Skua breeds in northern Britain and in sub-Arctic regions, and is a regular passage species further south.

Both names came into formal use in the nineteenth century, but **Arctic Skua** is the older by several years. It was first used by Fleming in 1828 to replace Pennant's *Arctic Gull*. There is no rule about allowing the older name to prevail in the vernacular, as it does in scientific names. The name Arctic Skua may have been influenced by its association with the Arctic Tern as a favourite target.

Parasitic and *parasiticus* are a little inaccurate: the species is in fact kleptoparasitic – that is, it harasses other species in order to steal their food.

* Coward (1926) has the name *Richardson's Skua* (see also *Richardson's Goose* and *Mew Gull* above).

Long-tailed Jaeger *Stercorarius longicaudus* (INTERNATIONAL NAME)
Long-tailed Skua *Stercorarius longicaudus* (BRITISH NAME)

This skua breeds in Arctic regions and is a scarce passage species in Britain.

Long-tailed and *longicaudus* mean the same thing.

* Coward (1926) has the name *Buffon's Skua*, after the French ornithologist the Comte de Buffon.

27. ALCIDAE: Auks

The family name is based on *Alca*.

Names within this family to some extent reflect a Northern bias, with a dense cluster showing their Scandinavian origins. That of course reflects the distribution of the species concerned. Several local British names are also preserved, particularly in dialect forms that travelled to America. The real oddity is the imported French name, Guillemot, which now seems so comfortably 'British' alongside the 'Americanism', Murre, which is not what it seems.

While the name of the Great Auk, *Pinguinus impennis*, remains on the IOC list, it exists only as a reminder that this large but flightless North Atlantic auk was finally extirpated by hunters by the middle of the nineteenth century, having been exploited historically for its meat and feathers. The name *Pinguinus* derives from a popular vernacular alternative name that was similar in a number of languages and is of disputed roots. What does appear certain is that it survives as the word *penguin*, the generic name for a number of Southern Hemisphere species to which it was transferred in the late Middle Ages. The specific name *impennis* (wingless) was not literal, but conveyed the Great Auk's flightlessness. Its closest living relative is the Razorbill.

The appearance in relatively recent years of several Pacific alcid species in North Atlantic waters is still an unresolved mystery, which is discussed by Lees and Gilroy (2021). Climate change is one possibility, but not all species involved fit that hypothesis.

Little Auk *Alle alle*

This Arctic pelagic species occurs around British coasts and sometimes seen inland during or after storms.

Auk derives from the Norse *alka* for this family of seabirds.

Little is self-evident for the smallest of the family.

Alle comes from the Lapp name for Long-tailed Duck, and was confused by Linnaeus.

* The preferred vernacular name in America is the *Dovekie*, which name was transposed from a Scottish dialect word for the Black Guillemot (see below).

Thick-billed Murre *Uria lomvia* (INTERNATIONAL NAME)
Brünnich's Guillemot (BRITISH NAME)

This species breeds in Arctic regions and is rare in British waters.

Thick-billed is something of an exaggeration: the larger bird has a shorter and somewhat stouter bill than the Common Guillemot.

For **Guillemot** see next species.

Murre was originally a Cornish word, noted by Ray in 1662 as a name for the Razorbill. It was taken to America by colonists, but not used widely or generally in Britain (though Lockwood records occasional examples of its use between 1602 and 1845). He also links the word to a Welsh form *morr*, which was probably echoic, referring to the *urr* sound made by nesting birds. Lockwood points to a similar Scottish word, *Marrot*, also of echoic origins. Other traditional Welsh names include *moriah* for Guillemot and *morra* for Razorbill, which appear to be related forms.

M. **Brünnich** (1737–1827) was a Danish ornithologist.

Uria has Greek roots that refer to a 'waterbird'.

The specific name *lomvia* is a Swedish word meaning 'auk' or 'diver'.

Common Murre *Uria aalge* (INTERNATIONAL NAME)
Common Guillemot *Uria aalge* (BRITISH NAME)

This bird breeds on rocky cliffs around Britain.

For **Murre** see above.

Guillemot originates in French and was first used by Ray in 1678 before being formalised by Pennant a century later. It is probably a diminutive based on Guillaume, the French version of William. It has been so long in use and is so familiar that it now seems natural to defend it against the international preference for *murre*.

Common is self-evident in the context of birds breeding closer to inhabited areas, and is therefore most familiar.

The specific name *aalge* has similar roots to *alka*, above.

* The use of the familiar name *Guillaume* here is in keeping with the widespread medieval traditions, in both English and French, which attach personal names to animals or birds; examples are *Reynard the Fox* and *Mag the Pie*.
* This species is famed for nesting at the closest density of any nesting species, but also for the intriguing shape of its egg (see Birkhead, *The Most Perfect Thing*).
* 'Bridled' Guillemot is not a subspecies – rather, it is a morph.

Razorbill *Alca torda*

This species breeds on rocky coasts in Britain.

Razorbill compares the broad bill of the bird to an open-bladed or cut-throat razor.

Alca is yet another variant of *alka* (above).

The Swedish Gotland dialect furnished Linnaeus with *torda*, a local name for this species.

* The shape of the bill is reflected in a different image in the Scottish name *coulter neb*, which likens the bill to the setting blade of a plough (from the Latin *culter*, knife).

Black Guillemot *Cepphus grylle*

This species is found mainly on western and northern rocky coasts of Britain and sometimes favours harbours, where it is often quite confiding.

It is mainly **black**, but with a distinctive white wing-patch.

Curiously, the Americans also use the vernacular **Guillemot** for birds of this genus.

Cepphus is a word derived from the Greek name of an unidentified waterbird.

The specific name *grylle* is another name taken by Linnaeus from the Gotland dialect.

* A Scottish dialect name, *Tystie*, is still in minority use. In 1966, the British ornithologist James Fisher recommended its general use to underline the fact that this bird was not that closely related to the Common Guillemot, but this idea did not take root.

* A second familiar name, *Dovekie*, was originally an affectionate term created by Scottish fishermen, who likened the Black Guillemot's courting behaviour to that of doves. The word was preserved by Scottish emigrants to the New World and somehow became the American name for Little Auk (see above). The original usage is reflected in both the vernacular and specific names of a Pacific relative, the Pigeon Guillemot *Cepphus columba*.

Long-billed Murrelet *Brachyramphus perdix*

This East Asian species has been recorded only once in Britain.

Long-billed appears to be self-evident, but it is apparently contradicted by the generic name *Brachyramphus*, which means 'short-billed'. The matter becomes clear when one sees that all three of the genus have short bills, but that this one's is relatively longer.

Murrelet is a contrived diminutive for a smaller species of *murre*.

The specific name *perdix* is from the Greek for 'partridge', since Pallas considered this bird to be of a similar size: his notes include the comment: *Magnitudine Perdicis*. (It seems that the Quail might have been a more accurate comparison.)

Ancient Murrelet *Synthliboramphus antiquus*

This Pacific species has been recorded once in Britain as a long-staying vagrant.

It is **Ancient (*antiquus*)** because its mantle is like an old-person's grey shawl.

Synthliboramphus means that the bird has a 'compressed bill' – another way of naming a different genus with a word that complements the previous genus *Brachyramphus*.

Parakeet Auklet *Aethia psittacula*

The normal habitat of this species is the Bering Sea, but is has been recorded in Sweden.

Aethia comes from the same roots as the duck genus *Aythya*. That word was originally used for an unidentified seabird mentioned by various classical authors. (Such potential for confusion is not unprecedented in bird-family names: there are several such cases.)

Parakeet and *psittacula* both draw attention to the (inverted) parrot-like broad round bill of this species.

Auklet clearly implies that this is a small auk.

Crested Auklet *Aethia cristatella*

The normal habitat of this species is the Bering Sea, but it has been recorded in Iceland.

Crested and *cristatella* record the frontal tuft of feathers, which is present to a greater or lesser extent in all plumages.

Atlantic Puffin *Fratercula arctica*

This species breeds on both sides of the Atlantic.

The Americans need the qualifying word, since they have other puffin species on their Pacific coasts. The simple name Puffin generally suffices in Britain for the one species.

Puffin originated to describe the fat (puffed up) young of the Manx Shearwater (now, confusingly, *Puffinus puffinus*), which were once harvested for food. The word became associated by chance and misuse with the unrelated bird that now bears the vernacular name. The story becomes even more curious because young Puffins now have their own diminutive forms as *pufflets* or *pufflings*.

Fratercula is an anthropomorphic idea, because the small upright figure looked like a tiny cloaked friar (Latin *frater*, brother, plus a diminutive ending).

The specific name *arctica* is misleading, since the birds also breed in temperate zones, including Britain. The names of the discarded subspecies of Atlantic Puffin tended to reflect such detail.

Horned Puffin *Fratercula corniculata*

This Pacific species has been recorded just once in the Western Palearctic, off Norway.

Horned describes a distinctive black mark above the eye.

The specific name *corniculatus* derives from *corniculum*, a little horn.

Tufted Puffin *Fratercula cirrhata*

A single Tufted Puffin, another Pacific species, appeared on the Kent Coast in 2009. Its inclusion on the British list has not changed the habits of most birders, who are still happy to talk of Puffins in the traditional way.

In breeding plumage, the **Tufted** (or Crested) Puffin wears curling white adornments on an otherwise black head. That is also the meaning of *cirrhata*.

28. PHAETHONTIDAE: TROPICBIRDS

The family name is based on *Phaethon*.

There are three extant members of this family, all living in tropical waters. Some wander widely, but sightings off Europe are extremely rare.

Red-billed Tropicbird *Phaethon aethereus*

This species has been recorded in British waters.

Red-billed is self-evident, but distinguishes this from other related species of its genus.

Tropicbird recognises that this species in found in Equatorial regions.

Phaethon is the Greek for 'the sun', Phaeton in legend being the son of Helios, the sun god, and of the Oceanid, Clemene. This was Linnaeus's tribute to this exotic bird.

The specific name *aethereus* (in the air) associates this bird with the freedom of flight.

White-tailed Tropicbird *Phaethon lepturus*

This species has been recorded off the Azores.

The specific name *lepturus* tells that the **white tail** is also slender. (Greek *leptos*, slender, and *ouros*, tail.)

Divers to
Shearwaters

29. GAVIIDAE: Divers/Loons

The family name is based on *Gavia*.

The names: *diver* and *loon* represent a further cultural and historical difference between Britain and America. Historically these birds are 'divers' in Britain, while the Americans have traditionally called them 'loons'. So it remained, until the IOC tried to harmonise the names used by all English-speaking nations. Along with the skuas, this family proved to be one of the most difficult to resolve: current compromises do not always feel comfortable, in spite of the fact that so few species are involved.

Red Throated Loon/Diver *Gavia stellata*

This species breeds in Scotland and Northern Ireland, is relatively common in winter around some British coasts, but is scarce inland.

Red-throated applies only in breeding plumage, but to both males and females. In 1747, Edwards had recorded this species as the *Red-throated Ducker or Loon*, which, Lockwood feels, was the matrix for Pennant's version.

Loon was originally associated with this species as an English dialect name. That was rooted in *lómr* (Old Norse 'to moan'). The bird's eerie wailing call readily allowed that original element of its meaning to blend neatly with the Latin-rooted *loon*, a word based on *luna* (referencing moon-madness as in *lunatic*). That folk name travelled to the Americas with settlers.

Diver was made the formal usage by Pennant, but the word was in evidence as early as 1500. It is self-evident for a bird that dives to feed.

Gavia was originally the Latin for an unidentified seabird.

The specific name *stellata* (starry) refers to the fact that in winter the upper plumage is slightly spangled with white.

Black-throated Loon/Diver *Gavia arctica*

Small numbers breed in Scotland, but only a few hundred of these birds are noted in British waters in winter.

Black-throated is self-evident for the breeding plumage of both sexes, though the throat is white in winter. This was another of Pennant's creations: in this case, Lockwood doubts that there was any colloquial precedent, simply because this is a scarce species.

The specific name *arctica* is self-evident.

* *Arctic Loon* has long been the preferred American vernacular for this species, but in this case the British name is the official international preference.

Pacific Loon *Gavia pacifica*

In recent years a small number of this mainly North American species have been recorded wintering in British waters.

Pacific and *pacifica* perhaps underline the uncomfortable fact that the ice-melt in the Arctic Ocean may now give access from the Pacific to Western Europe. However, Lees and Gilroy (2021) feel that the explanation for such vagrants is more complicated.

Common Loon *Gavia immer* (INTERNATIONAL AND AMERICAN NAME)
Great Northern Diver *Gavia immer* (BRITISH NAME)

This species is relatively frequent in British coastal and some inland waters in winter.

It is **Common** only in an American context (though it is also the most common diver species on inland British waters in winter).

Great Northern clearly relates to the bird's size on the one hand and to its geographical distribution on the other. Lockwood traces the origins of Pennant's concoction to Brisson for the first element and to Linnaeus for the second.

The Norwegian language gave the specific form *immer*.

* A compromise IOC name, Great Northern Loon, lasted only a short while. It was one of their most controversial ideas to date.

Yellow-billed Loon *Gavia adamsii* (INTERNATIONAL AND AMERICAN NAME)
White-billed Diver *Gavia adamsii* (BRITISH NAME)

This High Arctic breeder is a rare species in British waters in winter.

Traditionally the Americans have used the name **Yellow-billed** and that is now the formal international use. **White-billed** remains the preferred form in Britain. The exact perception of the colour depends on light, but the bird is generally so unfamiliar to most birders that the detail hardly seems to matter.

The British explorer Edward **Adams** (1821–1856) is commemorated.

ALBATROSSES, PETRELS AND SHEARWATERS

The above terms are historically used in the names of a group of oceanic species that are often referred to as the Procellariiformes or 'tubenoses'. These birds extract all their water needs from the sea, removing a high proportion of the salt via glands in their heads. Surplus salt solutions then drain away through tubes in their bills, hence the odd vernacular collective name.

Birds of the next four families are among the least known to the average birder and are still subject to much research and discovery. They have always been a difficult group, because many of these birds spend most of their lives in the vastness of the world's oceans and only come to land in the remotest locations. What is more, the smallest species often come ashore to their nests after dark to avoid predation.

Science has acquired vital new tools in recent years. As a result, many new species have been designated, so it seems likely that there will be further revelations.

The use (or non-use) of the word 'storm' with 'petrel' is another example of the difficulties in harmonising names. In British usage, only the species that originally bore the name (the Storm Petrel, *Hydrobates pelagicus*) is so-designated. In international terms, the addition of 'storm' to the names of a number of Petrels in Oceanitidae and Hydrobatidae serves to underline their difference from those in Procellariidae (though such niceties do not apply in other duplications, such as warblers or sparrows). In the British tradition, *petrel* is treated as a generalisation. For that reason, where the bird is on the British list, *Storm* is bracketed in all but the original name of *Hydrobates pelagicus*.

30. OCEANITIDAE:
AUSTRAL STORM PETRELS

The family name is based on *Oceanites*.

Wilson's (Storm) Petrel *Oceanites oceanicus*

This Southern Hemisphere species is recorded in small numbers in British waters.

Wilson's commemorates Alexander **Wilson**. However the AOS announced in 2023 that all eponyms in vernacular names of American species will eventually be replaced.

The word *Oceanites* derives from the Greek sea-nymphs, the *Oceanides*.

The specific name *oceanicus* means 'sea-dwelling'.

White-faced (Storm) Petrel *Pelagodroma marina*

This species breeds on Madeira and Selvagens and has been recorded in British waters.

White-faced considerably understates the amount of white on this bird.

Pelagodroma means 'an ocean racer'.

The specific form *marina* underlines its pelagic habits.

Black-bellied Storm Petrel *Fregetta tropica*

This Southern Hemisphere species has been recorded off Madeira.

Black-bellied is self-evident.

Fregetta derives from the German *Fregatte* (frigate).

The specific name *tropica* simply means tropical.

31. DIOMEDEIDAE: ALBATROSSES

The family name derives from *Diomedea*.

Tristan Albatross *Diomedea dabbenena*

This Southern Hemisphere species was recorded once off Italy. It was initially recorded as a Wandering Albatross, but the former subspecies concerned is now split from that.

Tristan takes its name from Tristan da Cunha, a group of islands in the South Atlantic where the bird breeds. In this case the bird only indirectly bears the name of the Portuguese explorer who gave his name to the main island in 1506.

Linnaeus based the genus *Diomedea* on the Greek legend of King Diomedes, whose companions were so upset by his death that they all turned into seabirds (see also the Calonectris petrels).

The specific name *dabbenena* commemorates the Italian-Argentine ornithologist, Roberto **Dabbene** (1864–1938).

* The species is unusual, if not unique, in carrying the names of three different people of three different nationalities and times.

Black-browed Albatross *Thalassarche melanophris*

Occasional errant individuals of this Southern Hemisphere species have been recorded alone or in company with Gannets in British colonies.

Black-browed is for the dark eye-line.

Albatross may be rooted in an Arabic word for bucket. The Spanish *alcatraz* evolved from that, and seems to have been applied initially to pelicans (for the large bill – and from them to the notorious Californian prison island). In the international culture of seafarers, the word eventually embraced other large seabirds and, via the form *algatross*, became the modern vernacular name of this family.

Thalassarche conveys the Greek for 'commander of the seas', a tribute to their size and wide-ranging habits.

The specific name *melanophris* means 'with black eyebrows'.

* When a long staying bird appeared regularly among Northern Gannets in Scottish waters around the millennium, it became known among birders as *Albert Ross*, both a play on words and an echo of the tradition of attaching familiar names to bird species.

Shy Albatross *Thalassarche cauta*

This species breeds off Tasmania, but, amazingly, one was recorded in Israel and Egypt.

Shy and *cauta* mean much the same thing: it implies that the birds are unapproachable, but two of the three islands on which they breed are in any case inaccessible.

* The species is also known by some authorities as the *White-capped* Albatross and, along with other medium-sized albatrosses, is familiarly called a *Mollymawk*, an old mariners' name originating in Dutch and meaning a 'foolish gull'.

Yellow-nosed Albatross *Thalassarche chlororhynchos*

An individual of the Atlantic subspecies was recorded in Britain in 2007.

Yellow-nosed refers to the colour of the bird's bill, while ***chlororhynchos*** renders the same fact in Greek.

* The circumstances in which this bird occurred and reappeared over Sweden caused Lees and Gilroy (2021) to suggest that the bird had a 'cognitive abnormality', which is generally a good reason to get lost.

32. HYDROBATIDAE:
Northern Storm Petrels

The family name is based on *Hydrobates*.

The IOC prefers to use **Storm Petrel** in the vernacular names of petrels in this family, to distinguish them from the Procellariidae.

There have been wholesale changes in classification in this area, with all species that were formerly in the genus *Oceanodroma* now moved to *Hydrobates*.

(European) Storm Petrel *Hydrobates pelagicus*

This mainly pelagic species breeds around British and other European coasts.

European now distinguishes this from the other storm petrels in an international context.

This was the original **Storm** or *Stormy* Petrel. For mariners these birds became associated with bad weather superstitions and were often known as *Mother Cary's Chickens*, a name thought to be rooted in Latin prayer to the *Madre Cara*, the Holy Mother, in times of danger.

Petrel is associated with the 'pitterel' or 'pitter-patter' of the birds' feet as they 'walk' on the surface during flight. A link with St Peter has also been suggested, but seems less likely.

Hydrobates means 'water-walking'.

The specific name *pelagicus* means 'sea-going'.

Swinhoe's (Storm) Petrel *Hydrobates monorhis*

This Pacific species has been recorded several times in the North Atlantic.

The English naturalist Robert **Swinhoe** (1836–1877) described this species in 1867.

The specific name *monorhis* suggests that the bird has a single nostril.

Leach's (Storm) Petrel *Hydrobates leucorhous*

This species breeds in rocky cliffs in the North Atlantic.

The name commemorates British zoologist William **Leach** (1790–1836).

The white rump is noted in *leucorhous* (Gr. *leukos* white and *orrhos* rump). The form *leucorhoa* has recently been modified to agree grammatically with its new genus. It was formerly in *Oceanodroma*.

> The former **Madeiran Petrel** has now been split three ways, as Band-rumped, Monteiro's and Cape Verde Storm Petrels.

Band-rumped (Storm) Petrel *Hydrobates castro*

Band-rumped is for the broad white rump.

The specific name *castro* appears to mean castle, but that is unlikely: Alec Zino, with close local knowledge, felt that it was a local rendition of its odd call.

Monteiro's (Storm) Petrel *Hydrobates monteiroi*

Luis **Monteiro** (1962–1999) was a Portuguese ornithologist.

Cape Verde (Storm) Petrel *Hydrobates jabejabe*

The name **Cape Verde** relates to the archipelago, which is situated off west Africa.

The strange specific name is simply the Latin rendering of a local name *Jabe jabe*, which may be echoic.

33. PROCELLARIIDAE: Petrels, Shearwaters, Diving Petrels

The family name is based on *Procellaria*.

Southern Giant Petrel *Macronectes giganteus*

This Antarctic species has been recorded off Italy just once.

Southern distinguishes this from the Northern form. Confusingly, both qualifiers relate to ranges that are normally within the Southern Hemisphere.

Giant and *giganteus* both reflect the huge size of this species.

Macronectes derives from the Greek for 'great swimmer'.

(Northern) Fulmar *Fulmarus glacialis*

This species was long restricted to sub-Arctic waters, but expanded spontaneously into more temperate waters during the twentieth century to become a common British species. Its fortunes may relate to discards from the fishing industry, which is now less robust.

The name **Fulmar** evolved in St Kilda, a remote island north-west of Scotland, from a Norse name for a 'foul gull': the bird ejects a foul-smelling oil over humans and other predators at the nest site. St Kildans traditionally harvested nesting birds from the cliffs as a subsistence food source.

Southwards expansion of the bird's range during the nineteenth and twentieth centuries made the name more widely familiar.

A sister species (*F. glacialoides*) was discovered in 1840 in the Southern Hemisphere, which necessitates the qualifier **Northern** in certain contexts.

Fulmarus is a Latinised version.

When named formally, the species still had a distinctly northerly bias into icy regions, hence *glacialis*, but its southwards expansion during the twentieth century brought it into more temperate waters.

* In the name of the Southern Hemisphere species *glacialoides*, the addition of *-oides* to the base-word *glacialis* indicates 'a resemblance to'.

Cape Petrel *Daption capense*

This Southern Hemisphere species was recorded once off Gibraltar.

Cape and *capense* both refer to the Cape of Good Hope, where the type specimen was collected.

Daption is a nineteenth-century anagram of *Pintado*, the Portuguese name for that 'painted' species, with its blotchy plumage.

Birds of the genus *Pterodroma* are often called the 'gadfly' petrels because of their sudden soaring movements, which reminded observers of the startled leap of a horse that had been painfully stung by a gadfly.

Atlantic Petrel *Pterodroma incerta*

This is a **South Atlantic** species, but was recorded off Israel and Jordan.

Pterodroma means ocean-racer, a name that attempts to capture the athletic flight pattern. The word is another of Bonaparte's creations.

The specific name *incerta* (uncertain or doubtful) perhaps reflects the difficulties that still pertain in identifying some of this family.

Soft-plumaged Petrel *Pterodroma mollis*

This species breeds in Tristan da Cunha, but has been recorded as far away as Norway and Israel.

Soft-plumaged is echoed by the specific name *mollis* (soft).

Zino's Petrel *Pterodroma madeira*

Alec **Zino** (1916–2004) was a British-educated Madeiran businessman who devoted much of his life to conserving Madeiran shearwaters and petrels. In 1969, he rediscovered this species, which was thought to be extinct.

The specific name *madeira* links them to their breeding-base.

Fea's Petrel *Pterodroma feae*

This species breeds in Cape Verde but is sometimes noted in British waters.

Leonardo Fea (1852–1903) was an Italian zoologist.

Desertas Petrel *Pterodroma deserta*

This species breeds on the **Desertas Islands** off Madeira and has been identified in British waters.

Bermuda Petrel *Pterodroma cahow*

This species breeds on offshore outcrops of **Bermuda** and has been recorded off Ireland.

The specific name *cahow* is based on a local name that echoes the call.

Black-capped Petrel *Pterodroma hasitata*

This is a species of the West Indies. It has been reported once off British shores and once off Spain.

Black-capped is self-evident.

The specific name *hasitata* means 'faltering', which is almost certainly a reference to the 'gadfly' flight.

Trindade Petrel *Pterodroma arminjoniana*

This South Atlantic species has been recorded in the English Channel.

Trindade, where it breeds, is a rocky outcrop about 1,100 km off the coast of Argentina.

The specific name *arminjoniana* commemorates Vitoria Arminjon (1830–1897), an Italian admiral, explorer and writer, who was captain of the ship that collected the first specimen of the species.

White-chinned Petrel *Procellaria aequinoctialis*

This is a species of the Southern Ocean that was noted in 2020 in Orkney.

This dark species has a largely white bill, with white extending onto the 'chin', hence **White-chinned**.

Procellaria tells us that the bird belongs in storms or gales.

The specific name *aequinoctialis* (of the equinox) is a name created by Linnaeus, based on notes by Edwards, for a specimen collected at the Cape of Good Hope. The word is synonymous here with 'equatorial'.

Streaked Shearwater *Calonectris leucomelas*

This Pacific and Indian Ocean species has been recorded very rarely off Israel and Jordan.

Streaked is self-evident.

Shearwater was created to describe the flight style of this long-winged group of birds as they cut over the crests of waves.

Calonectris means 'noble shearwater', which is a reflection of their relatively large size.

The *-nectris* element of the word is from a disused genus and means 'swimmer'.

The specific name *leucomelas* means 'white and black' It is white underneath, but is better described as smudgy dark brown above, though undoubtedly appearing blacker from a distance.

Scopoli's Shearwater *Calonectris diomedea*

This Mediterranean species has been identified in British waters.

This species was first described in 1769 by Giovanni **Scopoli** (1723–1788), but the link was obscured for many years, when the bird was considered to be part of a Great Shearwater complex and later conspecific with Cory's. In 2015, the Scopoli's/Cory's complex was formally split and the Cape Verde Shearwater separated from the latter.

The specific name *diomedea* echoes a genus of albatross (see above), which is based on the name of Diomedes, a Greek king. A Mediterranean species seems to be a more appropriate home for that name.

Cory's Shearwater *Calonectris borealis*

This large species is seen fairly regularly off British coasts.

The American Charles B. **Cory** (1857–1927) established in 1881 that this species was distinct from the Great Shearwater. However, the AOS announced in 2023 that all eponyms in vernacular names of American species will eventually be replaced.

The term *borealis* describes this as the northerly form of the genus.

Cape Verde Shearwater *Calonectris edwardsii*

This species breeds in the **Cape Verde** Islands. It was originally identified in 1883 by the French naturalist J-F. Emile Oustalet, but was subsequently subsumed into the Cory's complex before being reinstated in 2015 alongside Scopoli's.

The dedication in this case is to a French naturalist, Henri **Milne-Edwards** (1800–1885), who was Oustalet's predecessor at the French Natural History Museum.

Sooty Shearwater *Ardenna grisea*

This is also a Southern Hemisphere species seen regularly in the North Atlantic.

Sooty and *grisea* (grey) both describe the dark plumage.

Short-tailed Shearwater *Ardenna tenuirostris*

This Australian species wanders widely in the Pacific outside the breeding season, but has been recorded twice in France and once in Ireland.

Short-tailed is self-evident.

The specific name *tenuirostris* means 'slender-billed'.

* In Norfolk Island and Tasmania, the fat young of this species were traditionally harvested for food under the name *Muttonbird*. Compare the origins of the name *Puffinus puffinus*, below.

Flesh-footed Shearwater *Ardenna carneipes*

This species of the Indian and Pacific Oceans has been recorded off Israel and Jordan.

Flesh-footed describes the pink colour of the legs, which is echoed exactly in the specific name *carneipes*.

Great Shearwater *Ardenna gravis*

This Southern Hemisphere species is recorded in the North Atlantic.

Great is self-evident (though Cory's, formerly treated as a subspecies, is larger).

Ardenna was a term used by the Italian Ulisse Aldrovandi (1522–1605) for the large shearwaters.

The specific name *gravis* means 'heavy' or 'weighty'.

Manx Shearwater *Puffinus puffinus*

This species breeds around British coasts.

Manx derives from the Isle of Man, where the bird no longer breeds. The original shearwater colony was on the Calf of Man, its main offshore island.

Puffinus reflects the fact that the fat (puffed up) young were known locally as 'puffins' when collected for food. This word later jumped to the more familiar name of the auk (see Atlantic Puffin).

Mediterranean Shearwater *Puffinus yelkouan.*

Yelkouan and Balearic Shearwaters were split some years ago, but current research recommends that they should be reunited as two subspecies under this new specific name.

Yelkouan Shearwater *Puffinus yelkouan*

This Eastern Mediterranean species has been noted in British waters.

Yelkouan is unique in deriving from a Turkish word meaning 'wind-chaser'.

Balearic Shearwater *Puffinus mauretanicus*

This is a fairly regular visitor to British waters.

The breeding area is in and around the **Balearic** Islands of the Mediterranean.

The specific name *mauretanicus* reflects the ancient kingdom of Mauretania, which once incorporated a large part of north-west Africa, rather than to the modern country of that name.

Conversely, the Audubon's Shearwater 'complex' was extensively reviewed during the first two decades of the twentieth century and spilt into several species, including the next three.

Audubon's Shearwater *Puffinus lherminieri*

The Caribbean form retains the original name. There is one authenticated historical Western Palearctic record of this form, which was a bird collected in Denmark in 1912.

John James **Audubon** (1785–1851) was a French-American naturalist and artist, whose name was adopted by the largest bird protection group in North America.

The specific name commemorates the French naturalist Félix Louis **L'Herminier** (1779–1833) (making this one of the few bird names bearing two eponyms).

* The AOS announced in 2023 that all eponyms in vernacular names of American species will eventually be replaced. Audubon was their main target, because of his links to slavery.

Persian Shearwater *Puffinus persicus*

This is a species of the Persian Gulf and Northern Indian Ocean, formerly part of the Audubon's Shearwater complex. It has been recorded off Egypt and Israel.

The names are self-evident.

Tropical Shearwater *Puffinus bailloni*

This is another of the species split from the Audubon's complex. It has been recorded in Israeli waters.

The name **Tropical** is self-evident for this Indian Ocean form.

The specific name *bailloni* commemorates the French naturalist Louis-Antoine **Baillon** (1778–1855) (see also *Baillon's Crake*).

The former Macaronesian Shearwater was split in 2015 into the two species below, and that split is upheld by the 2022 report that recommended the 'lumping' of Yelkouan and Balearic as the Mediterranean Shearwater, *Puffinus yelkouan*.

Barolo Shearwater *Puffinus baroli*

This species is named for the Italian philanthropist, the Marchese di **Barolo** (1782–1838), whose family wines also bear his name.

Boyd's Shearwater *Puffinus boydi*

Lieutenant **Boyd** Alexander was a British collector and explorer who died in 1910.

This is an unusual example of the use of a forename in such a dedication (see Pechora Pipit, *A. gustavi*, for another). A further curiosity is that his surname is used in the name of the Cape Verde Swift, *Apus alexandri*.

Bulwer's Petrel *Bulweria bulwerii*

This mainly tropical species wanders widely and has been recorded several times off Europe.

The artist and naturalist, the Rev. James **Bulwer** (1794–1879) worked on Madeira and is recalled in three aspects of this name. The specific name came first as *Puffinus bulwerii*, while the genus was later creation of Bonaparte.

Jouanin's Petrel *Bulweria fallax*

This Indian Ocean species has been recorded in Kuwait.

Christian **Jouanin** (1925–2014) was a French ornithologist working for the Muséum national d'histoire naturelle in Paris, and a specialist in petrels.

The specific name *fallax* means deceitful, presumably because its identity deceived ornithologists for many years. Jouanin described it in 1955.

Storks to Pelicans

34. CICONIIDAE: Storks

The family name is based on *Ciconia*.

African Openbill *Anastomus lamelligerus*

This sub-Saharan **African** species has been noted once in Egypt.

There are two **Openbills** (the other is Asian), which derive their curious name from the space between the mandibles when the bill is closed. This is an adaptation to manipulate the snails that form their diet. The name is a condensed version of Open-billed Stork.

Anastomus derives from Greek to mean 'with mouth wide open'.

The specific name *lamelligerus* refers to 'thin metallic plate', which is a way of describing the unusually formed ribbon-like iridescent display feathers found on the wings and breast.

Marabou Stork *Leptoptilos crumenifer*

This African species has also been recorded in Israel.

Marabou originates in the French form, *marabout*, which was adopted by Temminck at the start of the nineteenth century. That derives from an Arabic name for the bird, *murābiṭ* (a *holy man*, *saint* or *hermit*).

Stork: see White Stork below.

The bird was formally named by the French ornithologist René P. Lesson in 1831, who also created the genus *Leptoptilos*, which means 'delicate feather'. That comes as a surprise because this is a quite ugly bird that feeds with the vultures. The reason for the flattering image appears to lie in the fact that undertail down, in particular from its congener, the Greater Adjutant Stork was particularly valued in the fashion trade under the name *marabou*.

The specific name *crumenifer* relates to the throat-sac of the male, which resembles a leather money pouch (Latin *crumena*) worn round the neck. The *-fer* element is from *ferre*, to bear. Variants of the spelling of this word are confusing, but the current IOC usage is the one given here.

* An alternative name used in English-speaking Africa is the *Undertaker Bird*, which is in keeping with the anthropomorphic element of the formal vernacular name. This name seems suitable to both its dark-cloaked appearance and to its carrion-eating habits.

Yellow-billed Stork *Mycteria ibis*

This African species has been recorded in Spain and in the Eastern Mediterranean.

Yellow-billed is self-explanatory.

Stork: see White Stork below.

Mycteria derives from the Greek *muktēr* (snout).

The specific name *ibis* links this to birds of a generally similar structure with which it was once classed.

Abdim's Stork *Ciconia abdimii*

This sub-Saharan species has also been recorded in Tunisia.

Stork: see White Stork below.

This species commemorates Bey El-Arnaut **Abdim** (1780–1827), a Turkish governor in Sudan.

Black Stork *Ciconia nigra*

This species migrates into Europe to breed but is a rare vagrant to Britain.

Black Stork is mainly black (but with white under).

Stork: see White Stork below.

The specific name *nigra* also means black.

* When he originally classed this as *Ardea nigra*, a member of the heron family, Linnaeus was probably taking into consideration the fact that the bird is mainly a fish-eater, spending much time foraging in freshwater habitats. What is more of a mystery is why he ignored the decisions of his predecessors and contemporaries, who almost universally classed this as a stork.

White Stork *Ciconia ciconia*

White Storks migrate into Europe to breed and are occasional vagrants to Britain, though some sightings may now be of reintroduced birds.

White is self-evident (though the bird has much black on the wings).

Stork is from the Old English *storc*, a word with Germanic roots that simply means 'stick' (with reference to the nest of sticks). Lockwood felt that there is also a link between a Germanic dialect's usage of *Storch* as *penis* and the legend that Storks bring babies.

Ciconia is simply the Latin for 'stork'.

* Over the past few years, captive-bred White Storks have been released on the Knepp estate in Sussex, which is subject to 'wilding'. The birds have now bred there for several consecutive years and a wild population is beginning to establish itself.

35. FREGATIDAE: Frigatebirds

The family name is based on *Fregata*.

There are several forms of frigatebird, which are widely distributed around the tropics, all of them kleptoparasites and all very dependent on land. Unusually for birds that spend much of their lives patrolling the seas, they cannot settle on water. Instead their very light bodies are carried tirelessly for huge distances over the oceans on long, broad wings; it is claimed that they can stay on the wing for as long as two months at a time. Their plumage has little natural oil and becomes useless if saturated, while their long wings would hamper take off.

Ascension Frigatebird *Fregata aquila*

This South Atlantic vagrant has been recorded very rarely in Britain.

The bird takes its name from the **Ascension** Islands, so-called because of their discovery on Ascension Day, 1503.

Frigatebird and *Fregata* are derived from a type of fast and manoeuvrable warship introduced by the French in the sixteenth century. Albin anglicised the French usage in 1738. An alternative name was *Man O' War bird*, which had similar roots. Such names suit a kleptoparasite, which robs other birds of their food.

The specific name *aquila* (eagle) says something of the bird's size and style: it has a 2 m/6.5 ft wingspan.

Magnificent Frigatebird *Fregata magnificens*

A single 2007 record earned this species a place in British records.

Magnificent and *magnificens* underline the general appearance of this bird: the wingspan of this species has been recorded at 2.4 m/8 ft.

Lesser Frigatebird *Fregata ariel*

This Tropical Oceans species was recorded off Israel and Jordan.

The name *Lesser* reflects the fact that this is the smallest frigatebird.

The specific name *ariel* is that of a sprite found in medieval folklore, who also features in Shakespeare's *The Tempest*.

36. SULIDAE: Gannets and Boobies

The family name is based on *Sula*.

The key names used in this family are **Gannet** and **Booby** in the vernacular and *Sula* and *Morus* in the scientific forms, but their respective histories follow a convoluted course, which seems eventually to invert the logic behind the names.

Gannet evolved from the Old English *ganot* and has common roots with gander, as underlined by the use of the alternative name, Solan Goose. As a species that breeds in the north of Europe, the root of the Solan element lies in the Old Norse name *súla*.

During the sixteenth century, Gessner made the link with the largest colony, that on the Bass Rock, by coining the name *Anser bassanus*.

Linnaeus later felt that the birds were a form of pelican, and so modified the name to *Pelecanus bassanus*.

In 1760 Brisson created a more appropriate new genus, *Sula*, but that was superseded in 1816 by Vieillot's use of *Morus*. He had been influenced by Brisson's use of the French name *Fou de bassan*, because Morus means 'a foolish creature'. The word *fou* (fool) had clearly absorbed a tradition rooted in the experience of Spanish mariners, who, when raiding tropical gannet colonies for fresh food, found that the 'silly' birds sat still. As a result, the Spanish called this sort of gannet *bubo* and that term became anglicised as **booby**, while being translated into French as *fou*.

In short, the scientists have left us with an illogical inversion, since the Northern Gannet now bears the 'stupid' genus that originated in the tropics, while the more southerly boobies wear a Norse label.

(Northern) Gannet *Morus bassanus*

This is a widespread North Atlantic species, with a sizeable slice of the world population breeding on British islands and coasts.

It is **Northern** in the context of two other gannet species.

Gannet (from Old English *ganot*) is related to gander, hence Solan Goose, from *súla*.

Morus (*stupid*) started life with the boobies and ended up with the wrong half of the family.

The specific name *bassanus* has its roots on the Bass Rock (in the Firth of Forth, Scotland), historically the largest colony (see the introduction to this section).

Cape Gannet *Morus capensis*

This South Atlantic species has a rather shaky status in the Western Palearctic, with a number of claims from as far north as Britain, but only one firm record from the Azores.

Cape and *capensis* both refer to the Cape of Good Hope in South Africa.

Masked Booby *Sula dactylatra*

This tropical species has been recorded in France, Spain, Morocco and Israel.

Masked describes the bird's black face-patch.

The specific name *dactylatra* describes the bird's black primary feathers with a name that translates as 'black fingers'.

* The term 'fingers' is used in a similar way in English to describe the varying patterns of the spread primary feathers of soaring birds of prey.

Red-footed Booby *Sula sula*

This bird, which is normally found in tropical waters, was recorded in Sussex in 2016.

Red-footed distinguishes this from other booby species, notably *Blue-footed*.

Booby originates as a slang word for someone stupid.

Sula originally meant 'gannet' via the Norse *súla* (see the introduction to this section).

Brown Booby *Sula leucogaster*

Three birds were recorded off the coast of Britain in 2019.

The bird is mainly **brown**.

The belly is white, hence *leucogaster*.

37. ANHINGIDAE: ANHINGAS AND DARTERS

The family name is based on *Anhinga*.

African Darter *Anhinga rufa*

This **African** species has been recorded in Israel and Morocco, and has bred in Turkey.

The word **Darter** describes the bird's habit of impaling fish with a swift thrust of the head and using its fine spear-shaped bill.

Anhinga is a word hailing from the Tupi people of South America, which was adopted in the Americas for a cousin species, *Anhinga anhinga*. That name means 'little head', a name associated with an evil forest spirit. The birds do have small heads in relation to their long neck and large wings, and all four Anhinga/Darter species have a similar structure.

The specific name *rufa* describes the reddish-brown found on the neck of this species.

38. PHALOCROCORACIDAE:
Cormorants

The family name is based on *Phalacrocorax.*

Pygmy Cormorant *Microcarbo pygmaeus*

This is a species found in the Eastern Mediterranean and beyond, but is predicted to spread westwards.

Pygmy and *pygmaeus*: this is a miniature cormorant, the size of a Coot.

Cormorant is a corruption of *Corvus marinus*, the Latin for 'sea crow', obviously because of its size and blackness. The name was evolved primarily for Great Cormorant (see below).

Microcarbo (small and black) revived in 2021 an older genus for the five smallest extant species.

Reed Cormorant *Microcarbo africanus*

This is a widespread sub-Saharan species (hence *africanus*) that has been recorded in Cape Verde.

Reed refers to a habitat preference, though well-vegetated freshwater banks and lakesides would be a more accurate description.

Socotra Cormorant *Phalacrocorax nigrogularis*

This species is found in the Arabian Peninsula and as far north as Kuwait.

Socotra is an island off the coast of Yemen.

Phalocrocorax maintains the crow element in a Greek form, *korax,* but this time the crow is 'bald-headed' (*phalakros*). That makes little sense in itself, because most cormorants have well-feathered crowns. However the name seems to have been created for the white-headed breeding plumage of the *sinensis* race of the Great Cormorant. The Bald Eagle gets its name for similar reasons.

The specific name *nigrogularis* means black-throated.

White-breasted Cormorant *Phalacrocorax lucidus*

This is an African species, still considered by some to be a subspecies of the Great Cormorant. It breeds in the Cape Verde Islands.

White-breasted refers to the adult plumage.

The specific name *lucidus* refers to the whiteness as 'bright' or 'clear'.

(Great) Cormorant *Phalacrocorax carbo*

These cormorants are common in both coastal and inland waters of Britain, and are the most widespread of all cormorant species in the world.

There is generally only one species in Britain, but **Great** now distinguishes this from other world species.

The specific name *carbo* means 'carbon black'.

* Two subspecies occur in Britain: the Atlantic form, *P. c. carbo*, and the European form *P. c. sinensis* (Eastern). Most inland birds are the latter, but not all. With practice these can be distinguished visually. Plumage differences are most obvious in the breeding season.

(European) Shag *Gulosus aristotelis*

This is found mainly on rocky coasts of Britain, but sometimes inland.

Other crested species bear the name Shag, so **European** may be needed.

Shag describes a prominent crest, which is most obvious in the breeding season. Since there are normally only two cormorant species in Britain, these were traditionally separated by the fact that one wore a *shag* (an unruly hairstyle). The name stuck and later became a vernacular collective name for other related species.

The genus *Gulosus*, first used by Montagu in 1813, was revived in 2021 as the result of new genetic studies. It means 'gluttonous'.

The specific name *aristotelis* commemorates the Greek philosopher and naturalist Aristotle.

Double-crested Cormorant *Nannopterum auritus*

This is a common North American species that rarely crosses the Atlantic, but has been recorded in Britain.

In breeding plumage there are small twin tufts on the head, hence the name **Double-crested**.

Nannopterum means 'dwarf-winged', a name that derives from the related Flightless Cormorant of the Galapagos, which has much-reduced wings. That description does not fit this bird, which is fully winged.

The specific name *auritus* inaccurately refers to the tufts as 'ears'. The word is similarly used in the contentious name of grebe, *Podiceps auritus*.

39. THRESKIORNIDAE:
Ibises and Spoonbills

The family name is based on *Threskiornis*.

African Sacred Ibis *Threskiornis aethiopicus*

This African species has a free-flying introduced population in France, where it is now very unwelcome because of the harm done to native species. It has been noted in Britain, where it is notifiable and potentially subject to eradication.

Sacred: it is possible that ibises, like cats, earned a special status in the eyes of the Ancient Egyptians for the simple mundane reason that they were welcomed as pest controllers and scavengers around human habitation and in the fields.

Ibis is the Greek name given to the three species known in the Ancient World (Sacred, Glossy and Bald).

Threskiornis means 'sacred bird' – from the Greek *thrēskeia* (worship) plus *ornis* (bird).

The specific name *aethiopicus* means Ethiopian or, by extension, African.

Northern Bald Ibis *Geronticus eremita*

With only remnant and endangered populations in Syria and Morocco, this bird has been bred and reintroduced into Turkey, Spain, Italy and Austria in an attempt to save a species that was at once widespread and was also worshipped by the Ancient Egyptians.

Northern is used because the related species Southern Bald Ibis is found in South Africa.

Bald underlines the fact that, apart from some long crest feathers, this species has a naked head and face.

Geronticus relates to the baldness, because it means 'old man'.

The specific name *eremita* means 'of the desert', because this is a species of arid rocky deserts. The word hermit has similar roots, and as a result the bird is sometimes called *Hermit Ibis*.

* The birds once bred in areas of Central Europe where they acquired a Germanic name that is still widely used: *Waldrapp* means a 'wood raven'.

* The reasons for the bald head are not explained by a messy feeding habit, as with carrion-feeding vultures or the Marabou Stork. The most likely reason is to do with breeding display, when the bill and face develop a more intense red.

Glossy Ibis *Plegadis falcinellus*

This species is now seen increasingly and annually in Britain. It has now been confirmed that a pair bred successfully in Cambridgeshire in 2022.

Glossy relates to its highly iridescent plumage.

Plegadis is from the Greek *plēgas, plēgados* (sickle) for the curved bill.

The specific name *falcinellus* offers a similar meaning, this time from the Latin (*falx*, *falcis*). This image links the ibis to the falcons and to the Falcated Duck.

* Owing to its similar shape, the Glossy Ibis was sometimes known in Norfolk as the *Black Curlew*.

Eurasian Spoonbill *Platalea leucorodia*

This species has become much more familiar in Britain over the past two decades, particularly on the East and South coasts. It now breeds in small numbers.

Eurasian: this qualifier is only needed when other spoonbill species are concerned. It is unique of its kind in Europe

Spoonbill is an obvious enough descriptive name, which was adapted by the seventeenth-century naturalist Ray from the Latin and Greek names.

Platalea is the Latin for Spoonbill (from the Greek *platea*, broad).

The Greek *leukerōdios* (from *leukos*, white, and *erōdios*, heron) gives the specific name *leucerodia*.

African Spoonbill *Platalea alba*

Occasional records from France, Austria and Denmark were probably escapes, though there may be more doubt about records from Spain, since its largely sub-Saharan **African** range extends into Mauretania.

The specific name *alba* (white) is fairly unhelpful: the key factors are its red face and pink legs.

40. ARDEIDAE:
HERONS, EGRETS AND BITTERNS

The family name is based on *Ardea*.

For the older generation of British birdwatcher, this family illustrates climate change more than any other. Just 40 years ago the Grey Heron was the default heron in Britain and needed no qualifier, while Bitterns were very rare indeed. Since the 1990s, three species of egret have become familiar in Britain. These are already proliferating, and other species seem set to swell the ranks.

Three key names are important here: Bittern, Heron and Egret. They are typical of many of the bird names now used in international English, which evolved in England as names for the most familiar species seen there. All three of those names show no vestige of an Anglo-Saxon heritage, since they all derive from Old French forms imported after the Norman Conquest.

In practice, the words *heron* and *egret* appear to be used very erratically, but that is down to a linguistic curiosity: the fact that both words have a common ancestor in a Old French word *haigron*. That word readily evolves into *heron*, of course, but the leap to *egret* is a little more complex: a small heron becomes *haigrette*, from which the *h* disappears to leave *aigrette*, and hence to the Anglicised form *egret*. (That explanation, of course, compresses several hundred years of change.)

From that it could be argued that herons are large and egrets are small, but in reality usage is quite inconsistent and shows no more logic than in the use of the words pigeon and dove.

Eurasian Bittern *Botaurus stellaris*

This secretive species breeds in relatively small numbers in Britain, and is boosted in winter by birds from the Continent.

Eurasian is needed because there are other bittern species. For a time the qualifying word was *Great* Bittern.

Bittern derives from the Latin, *butire* (to boom like a bittern) and evolves through Old French *butor* (from the Latin *butaurus*) and several Medieval English variants to *bitterne*.

Botaurus is a Medieval Latin form, based on the Latin *butaurus* (bittern).

The specific name **stellaris** (starred or spotted) describes the strong black markings that spangle the mantle, throat and breast.

American Bittern *Botaurus lentiginosus*

This is a very rare vagrant to Britain, with fewer than 20 records.

The significance of the specific name **lentiginosus** is that it merely means 'freckled' and is therefore plainer than the heavily marked Eurasian form, *stellaris* – a telling feature, if and when it occurs as a vagrant.

Least Bittern *Ixobrychus exilis*

This is the American equivalent to Little Bittern. It has been recorded in Iceland, in Ireland in 2019 and Shetland in 2022.

Least is a tiny bittern, generally similar in size to the next species.

The specific name *Ixobrychus* derives from the Greek for 'a bird that bellows in the reeds'.

The specific name *exilis* (slender) underlines its smallness.

Little Bittern *Ixobrychus minutus*

This widespread European species is a scarce vagrant to Britain, but has also bred.

Little is no exaggeration: it is considerably smaller than a Little Egret. That idea is reinforced by the specific name *minutus*.

Yellow Bittern *Ixobrychus sinensis*

This south-Asian and Eastern species has been recorded in Egypt.

Yellow is somewhat exaggerated: the adult is a pale brownish-yellow.

The specific name *sinensis* means eastern.

Von Schrenck's Bittern *Ixobrychus eurhythmus*

This Far-Eastern species was recorded in Italy.

Leopold Von **Schrenck** (1826–1894) was a Russian zoologist and geographer.

The specific name *eurhythmus* is based on the Greek *eurhuthmos* (graceful, well proportioned).

Dwarf Bittern *Ixobrychus sturmii*

This is a species of Tropical Africa that has been recorded in the Canary Islands and in mainland Spain.

Dwarf describes an even smaller close relative of the Little Bittern.

The specific name *sturmii* commemorates German bird artist and collector J.H.C.F. **Sturm** (1805–1862).

(Black-crowned) Night Heron *Nycticorax nycticorax*

This is a widespread species that is found on both sides of the Atlantic. Vagrants are rare in Britain and some sightings may be of escaped birds. Breeding has been recorded.

The qualifier **Black-crowned** separates this species from a number of other close relatives. In America there is also a Yellow-crowned species, so the use of Black-crowned is essential on that side of the Atlantic. Since only one species is recorded in Britain, the qualifier is not generally used.

For **Heron** see the introduction to this section.

Night Heron is no longer hyphenated, but the two words together designate a specific type of heron. Night Herons are largely nocturnal, roosting by day and feeding at night.

Nycticorax is the Greek for a 'night raven' and, because of its nocturnal habits was considered a bird of ill omen in ancient times.

Yellow-crowned Night Heron *Nyctanassa violacea*

This American species has been recorded as a vagrant in several Atlantic islands, including Madeira, and on mainland Portugal.

Yellow-crowned is self-evident.

Nyctanassa means 'queen of the night'.

The specific name *violacea* means 'violet-coloured', a reference to the distinctive shade of grey of the neck and undersides of the bird.

Green Heron *Butorides virescens*

This North America species has only rarely been recorded in Britain. It was once considered to be part of the Striated Heron complex, so historic records of 'Striated' in Britain were of this species. **Green** refers to the bird's upper plumage.

The use of **Heron** in the name of this very small species emphasises the looseness of the term in relation to the origins of 'heron' and 'egret'.

Butorides means 'resembling a bittern' – and especially the juveniles.

Virescens means 'greenish'.

* Choate records one charming folk name for the species: *Fly-up-the-creek* reflected the bird's habit of following the stream as it flew away startled.

Striated Heron *Butorides striata*

This very widespread species occurs in the Red Sea and in southern Israel, but not in Europe.

Striated and *striata* are related to the streaked plumage.

* This species used to include the Green Heron (above) and the Lava Heron. With 21 subspecies spread right across the world, it seems very likely that further splits could be made.

* This species has been recorded in the past under the names *Green-backed Heron* and *Mangrove Heron*, both self-explanatory.

Squacco Heron *Ardeola ralloides*

This European species is a rare but regular visitor to Britain.

Squacco derives from an Italian dialect name *sguacco*, which was first introduced into English by Willughby in 1672 and later modified by John Hill (1753) to the present spelling. The Italian word was based on the bird's voice.

Ardeola is a diminutive of *Ardea* (heron) and applies to a group of small herons often called the 'pond-herons'.

The specific name *ralloides* implies that the birds are 'rail-like' – a reference to the structure and posture of birds such a Water Rail.

Indian Pond Heron *Ardeola grayii*

This Asian species has been recorded in Egypt.

Pond Heron: a vernacular collective term for the four species of the genus *Ardeola*.

The specific name *grayii* commemorates J.E. **Gray** (1800–1875), an English ornithologist working at the British Museum.

* This bird is popularly known in the subcontinent as the *Paddybird* for its habit of frequenting rice paddies.

Chinese Pond Heron *Ardeola bacchus*

Quite amazingly, this Asian species has been recorded more than once as a vagrant to Britain and to other parts of Europe.

Chinese: the breeding range is mainly in China, though in some neighbouring countries too.

The vinous (wine-coloured) plumage gives rise to the specific name *bacchus*, after the Roman god of wine.

(Western) Cattle Egret *Bubulcus ibis*

After many years classed as a rare vagrant, this species has recently colonised the south of England: it first bred in Somerset in 2008.

The **Western** and Eastern forms of cattle egret have now been given full species status. There is no need for the qualifier in local contexts: in this case 'west' and 'east' divide around the Persian Gulf. **Cattle** relates to the fact that the birds are often seen in the company of livestock, which disturb food items sought out by the egrets.

For **Egret** see the introduction to this section.

Bubulcus was a Medieval Latin word for a 'cowherd'.

The use of the name *ibis* resulted from an error. An Egyptian guide convinced the eighteenth-century Swedish traveller-naturalist Hasselqvist that a Cattle Egret was a Sacred Ibis. The deception was recorded for posterity when Linnaeus later used his deceased pupil's notes.

* Coward's 1926 name for this species was the *Buff-backed Heron*. Birds in breeding condition develop areas of pastel-buff wash, while the Eastern form is distinguished by a much more orange-tan coloration at that season.

Eastern Cattle Egret *Bubulcus coromandus*

This Asian species has been recorded in Kuwait, far west of its normal range.

The specific name *coromandus* refers to the Coromandel Coast of India. The term was coined by Portuguese traders and mariners in the early sixteenth century.

Grey Heron *Ardea cineria*

This is a common and widespread species.

Grey is self-evident.

Heron was first used in about 1300 for the Grey Heron, but there were many dialectical variants before the name was stabilised in its current form by Pennant in 1768.

The Latin *Ardea* for 'heron' is from the Greek legend of the burned town of Ardea, from the ashes of which rose a grey bird.

Unsurprisingly, *cineria* means 'ash grey'.

* The many traditional names include *Hernshaw* and *Handsaw*: the latter explains the apparent nonsense of Shakespeare's Hamlet needing to 'tell a hawk from a handsaw'.

Great Blue Heron *Ardea herodias*

This North American equivalent of the Grey Heron has been recorded as a vagrant in Britain.

Great is used because there is also *Little* Blue Heron in North America.

Blue is somewhat misleading: its grey mantle is arguably blue-tinged and is generally darker than Grey Heron.

The specific name **herodias** comes from a Greek legend in which the entire family of Anthus, a disastrously inefficient horse-minder, is transmuted into birds. One of his brothers, *Erodias*, becomes a heron.

Black-headed Heron *Ardea melanocephala*

This sub-Saharan species has been recorded in Cape Verde.

Black-headed is echoed by the specific name **melanocephalus**, which means the same.

Goliath Heron *Ardea goliath*

This is a species of tropical Africa with a small population in eastern Iraq. It has been noted in Israel, Jordan and Syria.

Goliath relates to the giant Philistine warrior of the biblical *Book of Samuel*. It is an appropriate name for the largest extant heron species.

Purple Heron *Ardea purpurea*

This is a scarce vagrant to Britain, but a pair bred successfully at Dungeness in 2010.

Purple and **purpurea** both relate to plumage with a mix of blue-grey and rufous colours. Like the name of the *Great Blue*, it seems a little exaggerated.

The following two species were long classified as *Egretta*, but were moved into the genus *Ardea* without a change of vernacular name. There is, in any case, no consistency in the use of *heron* and *egret* within the genus *Egretta*.

Great (White) Egret *Ardea alba*

Until 2012, this species was a very rare vagrant to Britain, but has now established itself as a breeding species and is fast becoming familiar.

Great and **White** are self-evident for a bird that is as large as a Grey Heron and is pure white.

The specific name **alba** (white) seems more discriminatory in the context of the genus *Ardea* than it did in *Egretta*, from which it was moved in the first decade of the twenty-first century.

* The BOU's retention of the word 'white' in the vernacular name seems to be more of a habit than a necessity. Until well into the twentieth century (e.g. in Coward's 1926 list) it was known as the *Great White Heron*, in which case the 'white' was helpful. The shorter IOC form seems perfectly adequate and is in keeping with *Little* Egret.

* The name *Great White Heron* is often used in America for a form of *A. herodias*, which is a possible split.

* Great Egrets have suffered considerable taxonomic turbulence in recent years: they were split into three species for a time, and then re-lumped as subspecies. They have been variously labelled as *Casmerodius* and *Egretta* but are now *Ardea*. The matter still seems to be unresolved, with the American form flagged as potential re-split. The form seen in Britain is the nominate one.

(Intermediate Egret *Ardea intermedia*)

Intermediate anglicised Wagler's unimaginative 1827 use of **intermedia**, for a bird which is partway between Great and Little Egret in size. However, this name was set aside in 2023 as the result of a three-way split. *Ardea intermedia* is now reserved for the Asian form, Medium Egret. The Australian form now becomes Plumed Egret, *Ardea plumifera*. The relevant form here is now the Yellow-billed Egret, *Ardea brachyrhyncha* (below).

Yellow-billed Egret *Ardea brachyrhyncha*

This species is found in sub-Saharan Africa. It has been noted in Italy and in the Eastern Mediterranean. Until 2023 it was treated as a sub-species of the Intermediate Egret.

Yellow-billed is self-evident and adopts a widely used African name that might readily describe other species.

As Intermediate Egret, it was moved from the genus *Egretta* to **Ardea** in 2020.

The specific name **brachyrhyncha** means short-billed (the Greek equivalent of the Latin *brevirostris* of the former Mew Gull).

Black Heron *Egretta ardesiaca*

This African species has been recorded in Israel.

Black: the bird is in fact slate-coloured, a fact that is recorded in the Modern Latin construction **ardesiaca**, which is derived from the French *ardoise* (slate). This is another of Wagler's names, again from 1827.

Tricolored Heron *Egretta tricolor*

This American species has been recorded in the Canary Islands.

Tricolored and **tricolor** describe a bird with zones of purple-blue, green and white. The IOC uses the American spelling of the vernacular form.

* In the United States it is sometimes still called the *Louisiana Heron*. It breeds mainly in Louisiana and Texas, but has a much wider range outside the breeding season.

Little Blue Heron *Egretta caerulea*

This American species has been recorded in Ireland.

Little is contrasted with the *Great* Blue Heron (above).

Blue and *caerulea* both refer to the purple-blue plumage of the adult.

* Juveniles are white, transitioning in their second year through a blotchy phase.

Snowy Egret *Egretta thula*

This American species was first recorded as a vagrant to Britain in 2001. It is still extremely rare, but other examples could be overlooked owing to its close similarity to the Little Egret.

Snowy is an obvious enough name for a white bird.

The specific name *thula* seems to have been borrowed (probably in error) by Juan Molina from a South American word for the Black-necked Swan.

Little Egret *Egretta garzetta*

This is now firmly established as a British breeding species, having crossed the Channel only 30 years ago.

The species is truly **Little** alongside the Grey Heron and Great Egret and is most probably the species that earned the diminutive French name *aigrette*, which is the source of **Egret** (see the introduction to this section).

The specific name *garzetta* derives from an Italian name for the species. This seems to be a variant of the Spanish name *garceta*, which also happens to be a long lock of hair. In that way, the name seems to relate to the bird's extravagant breeding plumes.

Western Reef Heron *Egretta gularis*

The nominate form is native to sub-Saharan Africa, but has been recorded in parts of southern Europe and North Africa. The Middle Eastern subspecies occurs in Egypt.

Western separates this species from its Pacific equivalent.

Reef describes its coastal habits.

It is interesting here to note the use of **Heron** in the vernacular name and *Egretta* in the scientific form – an illustration of the looseness of use of both words. It is also often called the Western Reef Egret.

The specific name *gularis* relates to the throat. The bird occurs in blue-grey and white morphs, the former having the white throat patch that gives the name.

41. PELICANIDAE: PELICANS

The family name is based on *Pelecanus*.

Pelicans have long been used in Christian iconography as symbols of self-sacrifice: this derives from the fact that young birds reach into the parent's gullet to feed and were once thought to eat the parent's heart. Conversely, modern advertising exploits the appeal of their gawky shape, which lends them a cartoon-like attractiveness.

Some species are popular in zoos and collections from which the occasional escape occurs. The most well known in Britain are those of St James's Park, which were first represented by a gift from Russia in 1664. Their modern-day counterparts became notorious when one swallowed a live pigeon on camera.

Otherwise the birds themselves are almost absent from the British scene. Two species are native to south-east Europe, but a third species has been recorded as a vagrant to Europe.

(Great) White Pelican *Pelecanus onocratulus*

Sightings of this species in Britain are normally treated as escapes. Its normal range is inland Eastern Europe, eastwards to Kazakhstan.

Great distinguishes this species internationally from the *American* White Pelican.

Pelican and *Pelecanus* are both rooted in the Greek word *pelekys* (axe, referring to the massive bill).

The specific name *onocratulus* is the Latinised version of the Greek for 'Pelican'.

Pink-backed Pelican *Pelecanus rufescens*

This sub-Saharan African species has been recorded in Spain, Italy and some other Mediterranean areas.

Pink-backed describes a dusky grey-pink plumage.

The specific name *rufescens* also describes the 'pink' as 'a reddish tinge'.

Dalmatian Pelican *Pelecanus crispus*

A 2016 vagrant to Britain was accepted as a genuinely wild bird. Its European range is similar to that of the White Pelican.

Dalmatia today is a region of modern Croatia, though that is probably the most westerly area of the bird's range.

The head has an untidy appearance, which seems to be the reason for the specific name *crispus* (curly headed).

Osprey to Mousebirds

'RAPTORS' AND 'BIRDS OF PREY'

Collective words for birds with hooked bills and claws seem straightforward enough, though owls seem to sit more comfortably as 'birds of prey' than as 'raptors'. The parameters have been changed fairly radically in recent years, since taxonomists have shown that falcons are more distantly related to the eagles and hawks than are the owls. Seeing as this guide is deliberately organised according to current taxonomical sequencing, the falcons will be found a bit further on.

Nonetheless, hundreds of years of culture and tradition underpin this area, and it is very unlikely that the taxonomic shifts will affect the way these birds are seen in the wider world.

Vernacular names

In this area we see the effects of tradition and culture probably more than in any other set of bird names. This is simply because birds of prey have been caught and trained for hawking and falconry throughout history. The Romans venerated eagles as symbols of power, and Genghis Khan was reputed to carry a fearsome Goshawk to suit his image. In the feudal society of Medieval England, which spawned many of the names which we now use, birds of prey were a status symbol: they had their own hierarchy, and their ownership and use was governed by laws that excluded the poorer classes from hunting with birds. Because of that, the French-led culture of the medieval ruling classes introduced a considerable amount of French-based usage: words such as eagle, buzzard, falcon, kestrel and lanner joined the Anglo-Saxon elements of English, such as kite and hawk. The falconers themselves brought technical terms that still pertain among bird enthusiasts: while the *falcon* is the female, the male is called a *tiercel*, from the French *tiers* (one third), because the male is one-third smaller than his mate. That difference mattered in a world where these birds were seen as tools of a trade, with the female more powerful and the male perhaps more agile.

For ordinary folk it made no difference whether a bird was a Peregrine Falcon to the falconer: if it pursued ducks it was a Duck Hawk, hawk being a bona fide Anglo-Saxon word, widely used for all birds of prey. It was much later that ornithologists steered vernacular names into the more disciplined pattern that we use today. Some of the looseness of terminology travelled with the settlers to North America and accounts for a different evolution of language, when most 'hawks' are of the genus *Buteo* and 'buzzard' is a colloquial term for a vulture. Fortunately the IOC has now harmonised the name of the Northern Harrier, so that when one is identified in Britain it is no longer a Marsh Hawk. In a quotation noted in this section, Benjamin Franklin refers to the Osprey as the fishing-hawk. Such looseness of usage accounts for the fact that the Hobby (another of those French names and a falcon to boot) acquires an improbable link to a buzzard in the specific name *subbuteo*.

42. PANDIONIDAE: Osprey

The family name is based on *Pandion*.

Osprey is once again unique in its family. For a decade or so Western and Eastern Ospreys were treated as two species, but in 2022 the IOC announced that they were to be reunified as a single species, since there was insufficient genetic difference to justify the split.

Osprey *Pandion haliaetus*

This is a regular breeding migrant in Britain.

Osprey evolved from the Latin *ossifraga* (bone-breaker), via the Old French *ossifrage*. That concept was probably the result of an historical confusion with Bearded or Egyptian Vultures, because the Osprey eats very little other than fish.

A medieval transcript error (*Osfrai* to *Osprai*) is the probable bridge to the modern form. Prior to the invention of the printing press, texts were hand-copied and any such mistake would be perpetuated.

Pandion is a further confusion, since it was Nisus, son of Pandion II of Athens, who was said to have metamorphosed into a hawk (see Sparrowhawk below). Marie Savigny made the error in 1809 when he seems to have misremembered the legend.

The specific form *haliaetus* is Greek for 'Osprey' or 'sea-eagle'. That is a difference of spelling from the form used for White-tailed Eagle – *Haliaeetus* – and its congeners. The spelling here was first used by Aldrovandi in 1599)

* The Osprey has long proven to be problematic for the taxonomists: while some early figures classified it as a *Falco*, Willughby, in 1676, used the name *Balbusardus*, which he based on the English popular name *Bald Buzzard*. That name applied loosely to Osprey, Kite and Marsh Harrier for their white or pale heads. The use of 'bald' is akin to that in *Bald Eagle*.

* In a curious reversal of traditions, the Modern French for Osprey is *balbuzard*, which was derived from the English. Buffon adopted that form in 1770 because he rightly felt that the usual name *Aigle de mer* (sea-eagle) was inaccurate.

43. ACCIPITRIDAE:
Kites, Hawks, Eagles, etc.

The umbrella name for this family is based on that of the genus *Accipiter*, which is now just one of many genera in this family.

New World vultures, Osprey and Secretarybird are classified outside this family, though it is possible that further adjustments will be made in future.

ELANINE KITES

These relatively small kites have a wide distribution. They hunt mainly rodents. Only one species is represented in the region.

Black-winged Kite *Elanus caeruleus*

This Mediterranean species had not been recorded in Britain until April 2023, when one was reported in Powys.

Black-winged exaggerates somewhat: there are black shoulder patches above and black primaries below.

The term **Kite** is now used for a number of genera other than the *Milvus* kites for which the name was created (see Red Kite below). The common factor is their lightness and slow-speed agility. The *Elanus* kites tend to hover, while those of the *Milvus* complex prefer to soar.

Elanus is a Greek word for 'kite'.

The specific name *caeruleus* refers to the blue tone of the grey on the mantle and upper wings.

GYPAETINE VULTURES

Two Old World vulture species are classed as a separate branch of the Accipitridae and are not very closely related to the other vulture species below, but are fairly closely related to each-other.

Bearded Vulture *Gypaetus barbatus*

A Bearded Vulture was recorded in Britain in 2020. This species is rarely found north of the Pyrenees or the Alps.

Bearded and *barbatus* both describe the tufts of feathering on the bill.

Vulture evolved via Old French from the Latin *vultur*.

Gypaetus combines the Greek *gyps* (vulture) with *aetos* (hawk).

* This species is also widely known by the German name *Lammergeier*, which means 'lamb-vulture'. That name prevailed in Britain for many years and is still preferred by many.

* In spite of that name, the birds are mainly scavengers and specialise in the consumption of bones. Their habit of dropping bones onto rocks in order to get at the marrow gave rise to another name of French origins: *Ossifrage* derived from the Latin *ossifragus* (bone-breaker), which evolved (misleadingly) into the name Osprey (see above). The long-held assumption that one of these bird dropped the tortoise that killed the Greek playwright Aeschylus has been updated by twenty-first century observations that Golden Eagles in that region sometimes feed by dropping tortoises onto rocks to break them open.

Egyptian Vulture *Neophron percnopterus*

Surprisingly, this migratory species has been recorded in Britain. There are two historical records from the nineteenth century, but subsequent records have often been considered to be escapes. A bird seen in the Scillies in 2021 was thought to be an authentic wild bird.

Egyptian is misleading, as it does occur much more widely, with some breeding in Mediterranean Europe.

Neophron was a character in Greek legend who cruelly duped his friend Aegypius (see Cinereous Vulture).

The specific name *percnopterus* is the Greek for 'black-winged'.

* An alternative popular name for this species, *Pharaoh's Chicken*, appears to date from the British occupations of Egypt between 1882 and 1956.

* This species exhibits one of the rare examples of tool-use. Birds of the African population have been recorded dropping large stones to break open the shells of Ostrich eggs to get at the contents.

HONEY BUZZARDS AND ALLIES

European Honey Buzzard *Pernis apivorus*

This is a scarce breeding migrant in Britain.

European is sometimes used because there are other Honey Buzzard species.

Honey was attached to this buzzard-like species by Willughby, who found honeycomb in a nest. The bird targets bee and wasp grubs rather than the honey itself.

This species resembles the *Buteo* **buzzards**, but is not at all closely related (see *Common Buzzard* below).

Pernis is taken from the Greek name for a type of hawk that was noted by both Aristotle and Hesychius.

The specific name *apivorus* means 'bee-eating': the bird digs out the nests of various wasps, bees or hornets to consume primarily their grubs, though adult wasps or bees are eaten by adults, as well as a variety of other insect prey and small birds and mammals.

* At the conclusion of *The Amazing Mr Willughby* (2018), Tim Birkhead underlines the fact that all elements of this bird's vernacular name are questionable: it is Eurasian, rather than European, in its breeding range; it does not eat honey; and it is not a true buzzard. While the name Willughby is missing from the Pantheon of formal bird names, that little piece of honeycomb does in fact symbolise the story of a much under-estimated great field naturalist.

Crested Honey Buzzard *Pernis ptilorhynchus*

This Eastern species is recorded in the Middle East on passage.

Crested is for a small and insignificant crest.

A 'feathered bill' is implied by *ptilorhynchus*, but that seems no more significant than the crest.

Swallow-tailed Kite *Elanoides forficatus*

This American species has been recorded in Madeira.

Swallow-tailed is for the long, forked tail.

Elanoides suggests that the bird resembles kites of the *Elanus* genus.

The specific name *forficatus* likens the tail to a pair of scissors.

OLD WORLD VULTURES

These Old World species are not closely related to the American species: the phenomenon of convergent evolution accounts for their similarities.

Hooded Vulture *Necrosyrtes monachus*

This sub-Saharan species has been recorded in Morocco.

Hooded refers to feathers on the back of the head and nape.

Vulture evolved via Old French from the Latin *vultur*.

Necrosyrtes is grimly suited to the job of a carrion-eater: it means 'corpse dragger'.

Confusingly, the specific name *monachus* (monk-like, for the hood) is also used for the Cinereous Vulture (see below).

White-backed Vulture *Gyps africanus*

This African species has been recorded in recent years in Morocco, Portugal and Spain.

The names are all self-evident.

Rüppell's Vulture *Gyps rueppelli*

Singles have sometimes joined Griffon Vultures in Spain.

Wilhelm **Rüppell** (1794–1884) was a German zoologist.

The modification here of Rüppell's name to *rueppelli* deals neatly with the umlaut.

* In the name of the warbler it is spelt *ruppeli* (with no *e* and a single *l*) which is a reversion, under the rule of precedence, to a looser spelling used when that bird was first named.

Griffon Vulture *Gyps fulvus*

This species has been recorded in Britain, but very rarely. It is normally restricted to Southern Europe and North Africa.

Griffon seems to stem from some confusion of the Greek words *Gyps* and *Gryps*, the latter for 'Griffon', a legendary beast.

Gyps derives from the Greek *Gups* (vulture).

The specific name *fulvus* is Latin for 'tawny-brown'.

Cinereous Vulture *Aegypius monachus*

This bird breeds in Spain in small numbers and has been recorded as far north as Norway, though not in Britain.

Cinereous literally means 'ash-grey', though this seems to apply to the hood and neck of the adult, since the rest of the bird is in fact far darker (see *Black* below).

Aegypius appears as a character in the same Greek legend as Neophron (see Egyptian Vulture).

The specific name *monachus* means 'monk', an image created by the hooded appearance of the adult.

* This bird was formerly known as the *Black Vulture* (sometimes with *Eurasian* attached), but the current name is now preferred in order to avoid confusion with the American species, which now formally owns that simple form.

* The name *Monk Vulture* appeared for a while, obviously based on *monachus*, but with the Hooded Vulture also carrying that specific name (though in a different genus) it had potential to confuse.

Lappet-faced Vulture *Torgos tracheliotos*

This is an African species, but has been recorded in southern Israel.

Lappet-faced refers to folds of skin that hang on each side of the face. A *lappet* is more generally a decorative fold that hangs from a human headdress.

Torgos is another Greek word for 'vulture'.

The specific name *tracheliotos* means 'cartilage-eared'.

SERPENT EAGLES

This is a group of raptor species that feed mainly on snakes.

Short-toed Snake Eagle *Circaetus gallicus*

This migratory species is fairly widespread in parts of Europe, but has been recorded only very rarely in Britain.

Short-toed is self-evident.

Snake Eagle is now the formal structure for birds of this genus to indicate their main diet.

Circaetus is from the Greek *kirkos* (hawk) and *aetos* (eagle).

The specific name *gallicus* refers to Gaul (an ancient name for France), but the actual breeding range is much wider.

The classification of the next species in somewhat uncertain: it is considered to be related to serpent-eagles, but it has a very broad prey-base and is also a carrion feeder.

Bateleur *Terathopius ecaudatus*

This African species has been recorded in northern Egypt and Israel.

Bateleur is a name given by the French naturalist François Levaillant (1753–1824). It likens the aerobatic expertise of the bird to a tumbler (a circus or street acrobat).

Terathopius is composed from two Greek words, which together mean 'wondrous in appearance'. This bird is the unique member of the genus.

The specific name *ecaudatus*, means 'tailless'. This is not literally true, but the small tail is dwarfed by the impressive, broad wings.

EAGLES

As a collective name, the word Eagle is complex. In a strict sense, 'true' eagles are those of the genus *Aquila*, since the Latin *Aquila* morphed via the French into the English word. However, the term also covers a number of the larger related birds of prey, some of which bear compound names such as Hawk-Eagle and Snake Eagle.

Lesser Spotted Eagle *Clanga pomarina*

This is an Eastern European species not recorded in Britain.

This is the **Lesser** of two 'spotted' eagles.

Spotted in this case describes two very distinct white patches on the upper wing.

Eagle is rooted in the French *Aigle* which in turn evolved from the Latin *Aquila*.

Clanga is from *klangos*, a Greek word for 'eagle'.

The specific name *pomarina* derives from the former Baltic dukedom of Pomerania (and has no connection with the roots of the name *Pomarine Skua*).

Greater Spotted Eagle *Clanga clanga*

This species ranges from Eastern Europe into Asia and has been recorded historically in Britain.

It is the **Greater** of two 'spotted' eagles.

Spotted describes two indistinct white patches on the upper wing.

Wahlberg's Eagle *Hieraaetus wahlbergi*

This sub-Saharan African species has been recorded in Tunis, Egypt and Mauretania.

The dedication is to Johan A. **Wahlberg** (1810–1856), a Swedish naturalist and collector in Africa, who was killed by an elephant in the Okavanga. The bird was named from one of his specimens by his fellow Swede, Carl Jacob Sundevall (1801–1875).

Hieraaetus contains the Greek *hierax* (hawk) and *aetos* (eagle). This is one of several names that combine two elements in that way to indicate that the bird is as least 'eagle-like'.

Booted Eagle *Hieraaetus pennatus*

This small migratory species breeds in Southern and Eastern Europe. The one British record was eventually treated as a probable escape.

Booted is because of heavy feathering on the legs.

The specific name *pennatus* is the Latin for 'feathered' (legs), so that name echoes the sense of 'Booted'.

The Aquila eagles became symbols of the power of great empires: Rome, Austria and Russia among them. The two Imperial Eagles (Eastern and Spanish) still reflect those empires in their very names. Inevitably these eagles became associated with social power in such a way that the size and prowess of various birds of prey flown in falconry reflected the rank of the user. Such rules were particularly strict in feudal England after the Norman Conquest when the whole culture became a ritualised reflection of social control.

Tawny Eagle *Aquila rapax*

Small numbers of this African and Asian species are recorded in extreme North Africa.

Tawny describes its fawn colour.

The specific name *rapax* means rapacious, as befits an efficient predator.

Steppe Eagle *Aquila nipalensis*

This is a largely Asian species, which ranges into south-east Europe during migration.

Steppe describes the central plains of Asia.

The specific name *nipalensis* relates to the Himalayan kingdom of Nepal.

Spanish Imperial Eagle *Aquila adalberti*

This endangered species is limited to the Iberian Peninsula and has been recorded in parts of North Africa.

Spanish: this is now treated as a full species in its own right.

Imperial: the Spanish form was formerly lumped with Eastern Imperial Eagle as the symbol of the Austro-Spanish Empire(s).

The specific name of the Spanish form is *adalberti*, after Prince Adalbert of Bavaria (1828–1875), sponsor of Reinhold Brehm (1830–1891).

* Some sources make an alternative link with a contemporary Prussian Prince Adalbert, but this seems to be incorrect. Adalbert of Bavaria married into the Spanish royal family and seems to have been a friend of Reinhold Brehm, who had settled in Spain. Brehm first described the eagle and sent that material to his father, Christian, to whom the formal description was attributed in 1861.

(See also Thekla's Lark.)

Eastern Imperial Eagle *Aquila heliaca*

Eastern reflects the Eastern European/Middle Eastern range of this species.

Imperial: this species was formerly lumped with Spanish Imperial Eagle as the symbol of the Austro-Spanish Empire(s).

The specific name *heliaca* relates to the Greek *hēlios* (sun), which is possibly rooted in a 'hawk' name, *heleios*, which is found in Hesychius.

Golden Eagle *Aquila chrysaetos*

This species is widespread in Eurasia and North America, but is a scare breeding species in Britain, found mainly in Scotland and, in the recent past, in England's Lake District.

Golden is for the rufous-gold of the head and other areas of plumage. In his addition of this word in 1678, Ray seems to have been influenced by Aldrovandi.

Eagle: Prior to the Norman invasion, the Anglo-Saxon word *Erne* was used for all eagles, but with the evolution of *Eagle* from the French *aigle*, this word sufficed alone in Medieval English to distinguish this bird from *Erne*, the White-tailed Eagle.

The specific name *chrysaetos* derives from the Greek *khrusos* (gold) and *aetos* (eagle).

Verreaux's Eagle *Aquila verreauxii*

This African species has been recorded in Israel.

The dedication in both names is generally attributed to Jules Pierre **Verreaux** (1808–1873), who was a French natural history dealer and collector working with his brother, Jean-Baptiste Edouard. A number of species bear their name, as they were prodigious collectors, with a family taxidermy emporium in Paris. This eagle was named by Lesson, while similar dedications were made for other species by naturalists such as Temminck and Bonaparte. However, those apparent gestures of respect were wholly tainted by one grotesque episode in which the brothers disinterred the corpse of an African warrior to mount it as a stuffed 'specimen', which was later sold to be exhibited in Spain.

* An alternative name for this species is the *African Black Eagle*.

Bonelli's Eagle *Aquila fasciata*

This is a resident Mediterranean species, never recorded in Britain.

Franco Andrea **Bonelli** (1784–1830) was an Italian ornithologist.

The bird was 'promoted' to the genus *Aquila* in 2012, having been previously classed as a hawk-eagle.

The specific name *fasciata* is Latin for 'banded', for the broad, black band on the tip of the adult's tail.

THE ACCIPITERS

This group of raptors is characterised by a word that means 'to grasp': they generally hunt by pursuit and by snatching with their talons.

Gabar Goshawk *Micronisus gabar*

This African species has been recorded in Egypt.

Gabar is a curious name. Jobling explains that it is a compression of the French *garde barré* (a striped guard). The 'guard' is appropriate to the alert, upright posture of this bird of prey, while the 'striped' element is down to the breast-barring.

Micronisus links to the Eurasian Sparrowhawk (*A. nisus*) below, but the *micro* element confirms that it is smaller.

* A similar colonial-French corruption to *Gabar* occurs in *podobe* (*peau daubé*) – see Black Scrub Robin.

Dark Chanting Goshawk *Melierax metabates*

This African species is found in south-west Morocco.

Dark distinguishes this species from *Pale* and *Eastern* forms. This is a dark, blue-grey bird.

Chanting refers to the repeated calls, though this species is reputedly less vocal than its congeners.

Goshawk: it is at least 'goshawk-like'.

Melierax derives from the Greek *melos* (song) and *hierax* (hawk).

The specific name *metabates* means a 'leaper' or a 'vaulter'. Its hunting style includes the act of killing snakes with its feet, in a balletic style similar to that of the Secretarybird.

Shikra *Accipiter badius*

This African and Asian species has been recorded in Israel.

The name **Shikra** derives from an Urdu word meaning 'hunter'.

Accipiter: see below.

The specific name *badius* is Latin for 'chestnut-brown', though that applies only to juvenile birds: adults are pale grey above.

Eurasian Sparrowhawk *Accipiter nisus*

This is a relatively common resident in Britain.

Eurasian now distinguishes this from other sparrowhawks.

Sparrowhawk is a hawk that specialises in hunting small birds such as sparrows. The word is now used generically for several species, as in the case of the next entry. The word started life as two separate words before being hyphenated and finally compressed.

Accipiter derives from the Latin *accipere* (to grasp, though this originally meant something closer to 'catch on to an idea'). This species snatches fleeing prey with outstretched talons.

In Greek legend, *Nisus* was the son of Pandion II of Athens, and was supposedly changed into a hawk when betrayed by his daughter (cf. note on Osprey).

* In the name of the Barred Warbler, *Curruca nisoria*, the specific name links the barring of the warbler to the barring of the hawk (see also Gabar above).

Levant Sparrowhawk *Accipiter brevipes*

This species is found in south-east Europe and eastwards.

Levant means 'eastern', since the sun rises in the east (Latin *levare*, to rise).

The specific name ***brevipes*** means short-footed. This species eats mainly small mammals and reptiles and has not evolved the long central toe used by *A. nisus* to reach out for fleeing birds.

Northern Goshawk previously separated the next two birds from other, mainly Southern Hemisphere species. Following a 2023 split, the qualifiers **Eurasian** and **American** replace **Northern**.

Eurasian Goshawk *Accipiter gentilis*

This species was driven into extinction in Victorian Britain by relentless persecution, but has recovered over the past few decades as escaped or released birds have re-established in the wild. These were probably boosted by vagrants from the Continent.

Goshawk apparently meant 'Goose Hawk', though geese, which are generally birds of open terrain, are not the natural prey of this ambush-hunting sprinter. Lockwood points to the Old English form *gōshafoc* as the name's origin. In pre-Norman times the *hafoc* (hawk) element of the word served for all birds of prey and would have included what became known much later as *falcons*. The large European falcons, Peregrine and Gyrfalcon, do sometimes prey on geese. The Goshawk is also a large bird, so it seems probable that the *Gos* element became attached to the wrong 'hawk' during a period of confused transitional culture after the Norman Conquest.

The specific name ***gentilis*** is the Latin for 'noble'. In medieval falconry this bird was only flown by the nobility, and it was reputedly the favoured bird of Genghis Khan.

* Writing in 1926, in the period between the bird's extirpation in Britain and its modern recovery, Coward listed the Goshawk as 'a rare wanderer from Europe. May have bred.' This supports the probability that not all modern birds originate from escapes or releases.

American Goshawk *Accipiter atricapillus*

The split of this species was announced in 2023: there appear to be good reasons for doing so, since known differences in morphology and behaviour are now supported by DNA evidence. That decision may accelerate action to clarify the bird's status in British records, but there does seem some justification for its inclusion here (see below).

American is a self-evident name. *American Goshawk* was first used by Nuttall in 1832, in preference to Wilson's *Ash-coloured* or *Blackcap Hawk*.

The specific name ***atricapillus***, which echoes *Blackcap* above, was first used for this form in 1812 by Wilson (who actually called it *Falco atricapillus* – the genus was amended much later.) The head of the male is certainly darker than that of the Eurasian form.

* In 1926, Coward listed it as American Goshawk, *Accipiter gentilis atricapillus*, making reference to occasional British records. In *Looking for the Goshawk* (2013), Conor

Jameson discusses historic records from Scotland and Ireland that would support this claim, and there have been other records since Coward's time. It seems likely that the BOURC will be revisiting the matter.

THE HARRIERS

The next six species all exhibit elements of sexual dimorphism: the differences once confused observers into thinking that males and females were separate species, as a result of which they sometimes had separate names. Partly owing to English naturalist George Edwards' 1743 illustration and naming of the Ring-tail'd Hawk (a vagrant Northern Harrier, it appears) it became a habit to refer to female and juvenile forms of several related species as '**ring-tails**'. A common misconception is that the term refers to the very distinctive white rump, which is the striking feature in the field, but it does in fact refer to the barring on the tail.

The Marsh Harrier is the exception in not having the ring-tailed form, but aspects of its recent international name changes have been challenging. For some time at the start of the twenty-first century, it was *Northern* Marsh Harrier because the Australasian form was called the *Southern* Marsh Harrier. When the Australians reclaimed their preferred name, *Swamp* Harrier, Northern was no longer needed, but almost immediately that species was split into *Eastern* and *Western* forms.

The literal reorientation of those names was opportune, because *Northern Harrier* was already in use for the American 'Marsh Hawk', which was finally split from Hen Harrier in 2016.

With a carousel like that, it seems appropriate that the genus is *Circus*.

(Western) Marsh Harrier *Circus aeruginosus*

This species now breeds in moderate numbers in Britain, having come very close to failing totally in the latter part of the twentieth century.

Western is the result of a recent split from an Eastern subspecies.

Marsh describes the bird's preferred wetlands habitats.

Harrier: see next species.

Circus is from the Greek *kirkos*, a part-mythical hawk.

The specific name *aeruginosus* describes the rust-coloured patches in the plumage.

Hen Harrier *Circus cyaneus*

This species breeds in Britain, though its numbers are clearly restricted by cynical, illegal persecution.

Hen is used because it was originally reputed to threaten domestic fowl. The reputation stays with it, because today the bird is frequently the victim of illegal killing, particularly on grouse-moors managed by the shooting industry.

Harrier comes from the verb *to harry*, to pester or pursue. It was first in evidence in Ray and was used primarily for this species. It was only applied later to its congeners.

The specific name *cyaneus* is Latin for 'dark blue', but here it refers to the blue-grey plumage of the adult male.

Northern Harrier *Circus hudsonius*

This American species occurs infrequently in Britain (see also opening note to this section).

Northern is used because it breeds in northern parts of North America.

The specific name *hudsonius* is for Hudson's Bay. (Compare *hudsonicus* for the American Whimbrel.)

* In North America it is often known by the traditional name *Marsh Hawk*, largely because it tends to prefer wetter habitats. It differs from Hen Harrier in that respect and in a number of plumage and structural aspects.

Pallid Harrier *Circus macrourus*

This is a scarce vagrant to Britain.

Pallid is another way of saying 'pale', and the male is the palest of the harriers.

The specific name *macrourus* derives from the Greek *macros* (long) and *ouros* (tail).

* Pallid Harrier has expanded its breeding range westwards and now seems to be more frequent. Werkman (2021) postulates that the end of collective farming in the former USSR has led to a deterioration of its breeding habitats there, and so pushed it westwards into more open arable farming habitats.

Montagu's Harrier *Circus pygargus*

This bird has normally bred in very small numbers in Britain, but seems to be on the brink of ceasing to do so.

Montagu's is named after Colonel George **Montagu** (1751–1815), the British ornithologist who separated it from the Hen Harrier.

The specific name *pygargus* derives from the Greek *puge* (rump) and *argos* (white).

* The name Ash-coloured Falcon was used by Montagu and later modified by Selby to Ash-coloured Harrier, before Yarrell formally dedicated the species to Montagu in 1843. In any case, *Ash-coloured* was also used to refer to male Hen Harriers.

'TRUE' KITES

In times past, the Red Kite was a familiar species in Britain, especially around human habitation where it thrived as a scrounger of carrion and waste. Black Kites migrate into Europe each spring, but seem never to have established themselves in the British Isles, though a few turn up each year. Black Kites themselves were spilt during the past decade, when the fairly static, yellow-billed race was given full species status. The migratory form, seen in Europe, is the nominate race.

Because of their lifestyle, kites do not have large feet or strong talons and only take small live prey, such as small birds and rodents. In spite of that, the Red Kite was an easy target for Victorian persecution of all hook-billed birds. As a result, it was extirpated from Scotland and England, holding out only in an area of Central Wales in small numbers. In the 1980s, a programme of reintroduction of birds from Sweden began in the Chilterns, and that was followed in other areas. Today Red Kites are once again familiar in many parts of Britain, while some predict that Black Kites will colonise naturally with a warming climate.

The former ubiquity of the Red Kite meant that it had more than one dialect name. Kite itself is explained below as an echoic form: the birds are quite vociferous and their whistling call seems to be a communication that travels a long way. An alternative name *Glede* (also *Glead* or *Gled*) began life in Middle English and is related to 'glide', since the bird spends a great deal of time in that flight mode. That name seems to have had a more Northern usage than the word Kite, but was certainly known to both Gilbert White in Hampshire and to Thomas Pennant in Wales. It died out with the last Scottish Gled at the end of the nineteenth century and has not been resurrected with the regeneration of the Kite itself.

Red Kite *Milvus milvus*

The bird is once again familiar in many parts of Britain.

Red describes the rufous tones of the plumage.

Kite comes from the Anglo-Saxon *cýta*, which was originally echoic of the whistling call.

Milvus is simply the Latin for 'kite'.

* The word for the flying toy is taken from the bird.

Black Kite *Milvus migrans*

This is a scarce vagrant to Britain, but breeds widely in mainland Europe.

Black is an exaggeration: the overall ground colour is dark grey-brown.

This species is a summer migrant into much of Continental Europe, hence the name *migrans*.

Yellow-billed Kite *Milvus aegyptius*

As the specific name *aegyptius* implies, this widespread African species is found in Egypt. It was long considered to be the sedentary subspecies of the Black Kite.

Yellow-billed separates this from the darker billed Black Kite.

FISH EAGLES

Perhaps the most confusing vernacular names in this family are those birds until recently classed in the genus *Haliaeetus*. That was split in 2023, removing all but four species into a resurrected genus *Icthyophaga*. That decision was based on recent findings of distinct genetic divisions within this group of birds.

In *Raptors of the World* (2001), Ferguson-Lees and Christie expressed a preference for the use of Fish Eagle to promote uniformity in their names, but local usage has ensured that vernacular names remain far from uniform: Bald Eagle, African Fish Eagle, White-tailed Eagle and White-bellied Sea Eagle are examples of the variations.

Some of these birds are huge, with the White-tailed Eagle on average larger than the Golden Eagle. That fact challenges the sense of 'royalty' of the *Aquila* Eagles. We can carry such considerations too far, of course, as did Benjamin Franklin. He was appalled when the Americans chose the Bald Eagle as their national symbol because 'he is a bird of bad moral character'. He goes on to complain that the unworthy eagle steals food from the 'fishing-hawk'.

The fish-eagles are widespread around the word, though they have no representative in South America. While fish form the core of their diet, some take birds and mammals too, and some scavenge. There are ten extant species, with Steller's and White-tailed among the largest raptors in the world. Four species have been recorded in the Western Palearctic.

Pallas's Fish Eagle *Haliaeetus leucoryphus*

This Asian species has been recorded in Norway, Finland and Poland.

Peter **Pallas** (1741–1811) was a Prussian naturalist and explorer working in Russia.

Note again the use of **Fish Eagle**: this is another species that fishes mainly in bodies of fresh water but, like its African relative, has a broad diet.

Haliaeetus is Greek for 'Osprey' or 'sea-eagle' (note different spelling from that used with Osprey).

The specific name *leucoryphus* means white-headed.

White-tailed Eagle *Haliaeetus albicilla*

Successful reintroductions have brought this sea-eagle back to Scotland and England, where it was once widespread.

White-tailed is self-evident, and that meaning is repeated in *albicilla*.

Eagle is now used alone in the formal name of this species. It has become conventional to omit *Sea* from its former name.

* A traditional English name for this species was *Erne* (from Germanic roots). A number of place names (e.g. Earley, Berkshire) are based on that word, and are part of the evidence that the birds were once widespread in lowland England.

Bald Eagle *Haliaeetus leucocephalus*

This North American species has been recorded in Ireland.

Bald: the bird certainly does not have the sort of baldness seen on a vulture. Instead the word replaced the epithet *White-headed* with a simpler word that originally meant much the same thing – a white patch, such as a blaze on a horse. *Phalacrocorax* (bald raven) is used in a similar way to describe the white head of the (European) Great Cormorant.

The specific name *leucocephalus* means 'white-headed'.

In 2023 the IOC announced a reorganisation of the Fish/Sea Eagle complex. The change does not affect the previous three species, but the following species is redesignated.

African Fish Eagle *Icthyophaga vocifer*

This widespread sub-Saharan **African** species has been recorded, though rarely, in Egypt.

Fish Eagle is used here as recommended by Ferguson-Lees and Christie: their logic against Sea-Eagle was that while some do not fish at sea, they all catch fish. This one fishes primarily in freshwater, but some frequent coastal mangroves and estuaries. However, the diet is not exclusively fish, since they may take birds, reptiles and mammals.

Lesson's genus *Icthyophaga*, meaning 'fish-eating', was resurrected in 2023, but restored to his original spelling, having been long used incorrectly.

The specific name *vocifer* relates to the bird's noisy and frequent calls: it is a vociferous species.

BUZZARDS

In Old World usage, species of the genus *Buteo* are collectively known as buzzards. This is an area where American English usage is very different. Across the Atlantic, *buzzard* is a colloquial word for vulture while members of the genus *Buteo* and their close allies are known as *hawks*.

Rough-legged Buzzard *Buteo lagopus*

This is a scarce winter visitor to Britain from northern Eurasia.

Rough-legged: the legs are heavily feathered.

Buzzard develops via the Old French *busard*, which itself evolved from the Latin *buteo*.

Buteo is a Latin word with echoic roots.

The specific name *lagopus* echoes its use for the Willow Ptarmigan in implying that the heavily feathered legs are like the furry legs of a hare, from the Greek *lagos* (hare) and *pous* (foot).

Long-legged Buzzard *Buteo rufinus*

This is a species of Eastern Europe, the Middle East and North Africa.

The bird's **long-legged** appearance seems to derive from its shorter 'trousers' rather from particularly long legs.

The specific name *rufinus* relates to the overall rufous tone of the plumage and especially to the tail of the adult.

Cape Verde Buzzard *Buteo bannermani*

This endemic island form has now been spilt from Common Buzzard, though this decision may yet be reversed.

The species is entirely confined to the **Cape Verde Islands**.

The specific name commemorates the Scottish ornithologist Dr D.A. **Bannerman** (1886–1979).

Common Buzzard *Buteo buteo*

This is one of the most widespread of all resident birds of prey in Britain.

Common distinguishes this from other species of buzzard.

* The range of the Buzzard in Britain has spread dramatically eastwards in the past 40 years or so. This seems to coincide with better legal protection and seems to be related to the reintroduction of Red Kites, though the species' fortunes are also linked to fluctuating rabbit populations.

* The Common Buzzard complex is widespread and there are still several unresolved issues of speciation surrounding it, as with Cape Verde Buzzard (above).

OWLS

Humans have always given owls a special niche in their cultures, associating them with myth, legend and folklore from early times. When Little Owls moved into the temple of Athene, they became symbols of her cult. Athene was considered to be the epitome of the values that founded the Athenian Republic, and from that grew the completely false idea that owls were wise. An extension of that concept fed into the cult of the Roman goddess, Minerva. Superstitions abounded later in Europe, fed by owls' association with the night, and with churchyard spirits. Our fascination with owls in general is largely because of the humanoid upright stance, round face and forward facing eyes, but largely too to the mystery of species that are rarely seen in broad daylight.

Owls are generally divided into two main families, the Tytonidae being the smaller group, though enlarged in recent years by a number of splits. Although many owls are nocturnal, or at least crepuscular, a few are diurnal. Conventional terminology might exclude owls from the terms 'raptor' or 'bird of prey', but taxonomists have now moved them much closer to the eagles and hawks than previously, while distancing the falcons – so perhaps those terms may now embrace them more comfortably. In fact owls between them have almost equally broad feeding habits, which involve taking live prey, from medium-sized mammals to the meanest of arthropods. More surprisingly, a number of owl species have been observed scavenging on carrion, particularly in times of hardship. The Western Palearctic has a very representative selection.

44. TYTONIDAE: BARN OWLS

The family name is based on *Tyto*.

(Western) Barn Owl *Tyto alba*

This common species fluctuates greatly in numbers.

Western is required in a world context, since the Barn Owl complex has now been divided into a number of species.

Barn Owl: Historically, there were a number of local names, such as *Church Owl* and *White Owl*, but Barn Owl eventually became the formal choice, since many barns (and other farm buildings) served as nesting sites.

Owl has roots in the Latin *ulula*, which in Medieval English became *ule*, before evolving into the form we know now.

Tyto comes from the Greek *tuto* (night owl).

The specific name *alba* echoes the old name *White Owl*. The bird has golden brown uppers and white underparts, but tends to appear mainly white when hunting at dusk.

* Coward listed Dark-breasted Barn Owl, *Tyto alba* **guttata** (streaked), as an irregular migrant from Northern Europe. This is a subspecies that appears fairly regularly and may have bred in the past.

45. STRIGIDAE: Owls

The family name is based on *Strix*.

Boreal (Tengmalm's) Owl *Aegolius funereus*

This is a very rare vagrant to Britain.

Boreal is the formal international name of a widespread species. It is named for the circumpolar Northern coniferous belt, known as the boreal region, which in turn takes its name from Boreas, the Greek god of the north wind. That name seems more suited to the distribution pattern in America and Asia than to that of Europe, where it is also found in montane woodlands as far south as Greece.

The European and nominate form is known generally as **Tengmalm's** Owl, after Peter Tengmalm (1754–1803), a Swedish doctor and naturalist.

Aegolius is rooted in the Greek for 'screech-owl'.

The specific name *funereus* refers to the sombre, funereal grey-black of the mantle, and particularly of the downy juveniles.

Little Owl *Athene noctua*

This European species was not present in Britain until the nineteenth century, when it was deliberately introduced. It subsequently became quite widespread through England and Wales.

In spite of the name '**Little**', it is much larger than the Pygmy Owl and similar in size to Tengmalm's.

Athene recalls the Greek goddess and her temple on the Acropolis. It seems quite possible that this cleft-dwelling species first drew attention to itself by nesting there. That association carried forward into the Roman era, in the cult of Minerva.

The specific name *noctua* reminds us of the nocturnal habits of these owls.

* Both Jobling and Coward indicate that an attempt was made by British ornithologists, in 1915, to overturn Boie's use of *Athene*, because it was too close to *Athena*, a genus of moths. Coward therefore uses the substituted genus *Carine* (a singer of dirges), that Hartert attributes to Yarrell. That was later discarded and *Athene* reinstated. This episode illustrates the value of the international conventions that have been agreed since that episode.

* The variations and range of this species have provoked suggestions of future splits.

(Northern) Hawk-Owl *Surnia ulula*

This is a species of boreal forests, but has been recorded in Britain.

Northern seems to be a redundant word: no other owls have a current formal vernacular name which includes *hawk-owl* and there is no other species in the genus *Surnia*. The word is widely used in North America, but is omitted by the BOU. It was first used by the British naturalist G.R. Gray in 1840, possibly to distinguish this species from the former Philippines hawk-owls, which are now classed in the genus *Ninox* (Boobook owls).

Hawk-Owl is an apt name for a diurnal species that resembles a small hawk.

Surnia is an unexplained word that may or may not be of Greek origin, but was possibly just an invention of the French scientist André Duméril in 1805.

The specific name *ulula* simply uses the Latin for 'owl', perhaps to remove any doubt about whether it was a hawk or an owl.

Eurasian Pygmy Owl *Glaucidium passerinum*

This species has never been recorded in Britain.

Eurasian is needed in a world context.

Pygmy Owl is an obvious name for the smallest species.

Glaucidium pursues that theme, as it is the diminutive form of the Greek *glaux* (owl).

The specific name *passerinum* compares it to a sparrow.

Scops Owl *Otus scops*

This is a very rare vagrant species to Britain.

Scops is a Greek word meaning 'watcher'.

Otus (eared) is used here as a generic name. The use of the term is misleading, because the so-called ears are just mobile feathered tufts, as they are in all 'eared' owls.

Cyprus Scops Owl *Otus cyprius*

This was formerly considered an endemic subspecies of *Otus scops* on the island of **Cyprus**, but was given full species status in 2019.

Pallid Scops Owl *Otus brucei*

This Middle Eastern species winters in southern Israel.

This is a **Pallid** (pale) relative of the previous species.

The specific name *brucei* commemorates an American missionary, the Rev. Henry James **Bruce** (1862–1909).

Long-eared Owl *Asio otus*

This is the most secretive of Britain's owl species.

Long-eared is misleading, because the 'ears' are mobile feathers. The real ears are hidden in the facial disc.

Asio is a name for an eared owl and was created by Brisson in 1760.

The specific name *otus* is the Latin for 'eared owl'.

* American handbooks of the late twentieth century list the Eastern Screech Owl as *Otus asio*, a confusing inversion of the name of this species. That pitfall was removed when the genus *Megascops* was reinstated in 2003.

Short-eared Owl *Asio flammeus*

This species breeds in Scotland, but a good many continental birds boost British numbers in the winter. It is seen more easily than some because it often hunts by day.

Short-eared has far less-obvious 'ears'.

The specific name *flammeus* refers to the flame-orange patch on the outer wing.

* An alternative folk name was the *Woodcock Owl*. Bosworth Smith (1913) explains that the winter owls arrived with the Woodcocks, shared similar habitats and flushed with them when disturbed.

Marsh Owl *Asio capensis*

This African species occurs in Morocco.

Apart from **marsh** habitat, the species frequents grassland and scrub.

The specific name *capensis* refers to the Cape region of South Africa, where the bird also occurs.

Snowy Owl *Bubo scandiacus*

In spite of its wide distribution through Arctic North America and Eurasia, this species is considered monotypic. It is a rare Northern vagrant to Britain, and has bred once in Shetland.

Snowy is an obvious name for a largely white owl from the Arctic.

The specific name *scandiacus* underlines the bird's northern range by reference to Scandinavia.

Eurasian Eagle Owl *Bubo bubo*

This species has occurred in recent decades as a nesting species in remote corners of Britain. It occurs widely throughout Europe.

Eurasian is needed in a world context.

Eagle Owl is a measure of the bird's size and ferocity: it is in fact considerably larger than the Red Kite or Common Buzzard: among British raptors, only the Golden Eagle and White-tailed Eagle are larger.

Bubo is the Latin for 'Eagle Owl'.

* The presence in Britain of this top avian predator is controversial. The population probably descends from escaped or released birds. The spontaneous arrival of birds of Continental origin is considered unlikely, though not impossible: in 1926 Coward described it as 'an occasional visitor' and as 'a rare wanderer from Northern Europe'. While there is prehistoric evidence of the species in Britain, lack of any evidence of its presence at any historical stage works against its acceptance. Meanwhile, the BOU, BTO and RSPB all remain ultra-cautious about its status (see *Owls*, Mike Toms).

Pharaoh Eagle Owl *Bubo ascalaphus*

This is a species of the deserts of North Africa and the Middle East.

Pharaoh is a fairly obvious regal name to give to one of the region's top predators and is in keeping with the symbolism of the *Aquila* eagles.

The specific name *ascalaphus* derives from a Greek legend in which Ascalaphus is transmuted into an owl by a vengeful Persephone.

Brown Fish Owl *Ketupa zeylonensis*

This Eastern species may still exist in Turkey.

Brown distinguishes this form from several others with the fishing habit.

Fish relates to its diet.

Ketupa appears to be rooted in a Malay word for the related Buffy Fish Owl. The genus was created by Lesson in 1831 from a species name previously used by Horsfield.

The specific name *zeylonensis* relates to Ceylon (modern Sri Lanka), one of the places where it is found.

Tawny Owl *Strix aluco*

This is one of the commonest British owls, which is widespread in Europe.

Tawny was introduced by Pennant in 1768 and, like so many of his names, became the standard form. Traditional names included *Brown Owl* and *Ivy Owl*, the latter for the bird's favoured roosting habitat, but Ray had introduced the name *Grey Owl* in 1678. This owl does in fact vary in ground coloration, with brown dominating in Britain, so Pennant's version was a suitable compromise.

Strix is a word derived from the Latin for 'Screech Owl' and was later applied (in Strigidae) to an entire family of owls.

The specific name *aluco* is rooted in two variants of an Italian name, *alocho* and *allocco*.

Maghreb Owl *Strix mauritanica*

This was previously considered to be a subspecies of Tawny Owl, but was split in 2010. It occurs in Mauretania, Morocco, Algeria and Tunisia.

Maghreb derives from an old Arabic word meaning 'the west' and relates traditionally to north-west Africa.

The specific name *mauritanica* refers to the same general region.

Desert Owl *Strix hadorami*

This is a newly designated species, long considered to be a form of Hume's Owl, *S. butleri*. It is found in Egypt and Israel.

The name **Desert** Owl reflects the habitat preferences of this bird.

The specific name *hadorami* was created in 2015 when Desert Owl was split from Hume's as a new species. It commemorates the contemporary Israeli naturalist, Hadoram Shirihai (b.1962).

* The other half of the Hume's complex is now known as the Omani Owl and retains the name *S. butleri.*

* Allan Hume (1829–1912) was a British ornithologist based in India. He named the owl, which was collected by Lt-Col. E.A. Butler of the British Army.

* Confusingly, the entry in the *Collins Bird Guide* (third edition, 2022) has it as Hume's Owl, *Strix hadorami.*

Great Grey Owl *Strix nebulosa*

This is a species of boreal forests of both Eurasia and North America.

Great is deceptive, because the bird's true size is considerably exaggerated by a thick and dense layer of plumage that protects it against extreme cold. This has evolved to suit a sedentary hunting style in which the bird sits and watches for long periods.

Grey is underlined by the specific name *nebulosa* (cloudy).

* The American form is nominate. It was first described in 1772 by Forster, shortly before he replaced Joseph Banks as the naturalist on Cook's second voyage. The Eurasian subspecies is *lapponica* (from Lapland), which was described by Carl Thunberg twenty-six years later.

Ural Owl *Strix uralensis*

This is another species of boreal forests.

Ural and *uralensis* are based on the Ural Mountains of Russia.

46. COLIIDAE: Mousebirds

The family name is based on *Colius.*

Blue-naped Mousebird *Urocolius macrourus*

This sub-Saharan species is found in northern Mauritania and was first recorded in Algeria in 2018.

Blue-naped describes a splash of blue on the nape of a generally grey bird.

Mousebird describes the fact that these small birds scurry along branches like mice.

Urocolius is a 'colius with a tail'. *Colius* itself seems slightly puzzling in a superficial resemblance to the Greek for 'Jackdaw', but Jobling cites an alternative suggestion that it is from the Greek *koleos* for 'scabbard' or 'sheath', in the shape of the tail.

The specific name *macrourus* adds to the picture of the bird by describing its large tail.

Hoopoe to Parrots

47. UPUPIDAE: Hoopoes

The family name is based on *Upupa*.

The Hoopoe has a strange cultural heritage. It was revered in Ancient Egypt and was celebrated in Middle Eastern culture as the messenger between King Solomon and the Queen of Sheba. In Ancient Greece it was chosen by the dramatist Aristophanes to represent kingship and tyranny, the crest representing the crown and the long strong bill the power. That was the image that filtered through into later European cultures via the Roman Ovid. Even so, its French reputation for having a filthy nest gave a less romantic aspect to its story.

Today the strays that turn up in Britain are seen as true exotics, much admired for their beauty. Most of these occur as a result of 'overshooting', when they are carried further north on spring migration than intended. Most do not linger, though a pair apparently bred successfully in Leicestershire in 2023 (*Birdguides*).

(Eurasian) Hoopoe *Upupa epops*

This is a regular vagrant to Britain.

Eurasian serves to remind us that there are other hoopoe species.

All three names are echoic, recalling the bird's call, respectively in English, Latin and Greek.

Hoopoe comes into English via the French *huppe*, which derives from the Latin **upupa**, which is in turn based on the Greek *epops*.

* Lockwood records a whole range of local English variants for this bird including *Hoop*, *Houp*, *Whoopoo* and *Hoopoop*, none of which seems too surprising.

* *Dung Bird* was recorded by Charleton in 1668. That name is in stark contrast to the Hoopoe's association with any aspect of kingly pride, treachery or exotic beauty. It is rooted in the defence mechanism of the nestlings, which spray would-be predators with their faeces. It is unclear where Charleton gleaned the name, but given the rarity of breeding attempts in Britain, it seems highly probable that he borrowed the idea from the French synonyms *coq merdeux* (dung bird) and *coq puant* (stinking bird). The idea catches the imagination and is of ornithological interest, but the scarcity of the bird meant that it did not endure as a valid English name.

48. CORACIIDAE: Rollers

The family name is based on *Coracias*.

European Roller *Coracias garrulus*

This migrant species breeds in parts of Europe, but is a scarce vagrant to Britain.

European reminds us that there are other members of the family.

Roller: The traditional view is that the name derives from the tilting display flights with which the birds defend their territory. Lockwood suggests that the word appears to originate in Gessner (1555) and is based on a Strasbourg-German dialect form. It was adopted by Ray in 1678 and was subsequently used in English. However, the roots are apparently echoic, based on the bird's repetitive nervous call, so the association of the word with the display seems to be coincidental. It is no coincidence that the first appearance of the word Roller in 1663 relates to Willughby and Ray's visit to Strasbourg, where they made the acquaintance of a local figure, Leonhard Baldner, whose anecdotes and experiences made a significant impression on their ornithological lives (Birkhead, 2018).

Coracias is a variant on *corax* (Raven), a hangover from the time when this bird was seen as a relative of the Jay.

The specific name *garrulus* underlines that mistaken link to the noisy Jay, *Garrulus glandarius*, though it appears that this bird is noisy enough in its own right.

Indian Roller *Coracias benghalensis*

This Asian species has been recorded in Syria and Iraq.

Indian is self-evident, while *benghalensis* refers to the Bengal region.

Abyssinian Roller *Coracias abyssinicus*

This sub-Saharan species has been recorded in the Canaries.

Abyssinian and *abyssinicus* relate to a historic name for modern Ethiopia.

Broad-billed Roller *Eurystomus glaucurus*

This sub-Saharan species has been recorded in Israel.

Broad-billed is self-evident.

Eurystomus reinforces the English name with the meaning 'wide-mouthed'.

In the specific name *glaucurus* we learn that the bird has a blue-grey tail.

49. ALCEDIDINAE: Kingfishers

The family name is based on *Alcedo*.

White-throated Kingfisher *Halcyon smyrnensis*

This Middle-Eastern species is found in Israel.

White-throated is self-evident.

For **Kingfisher** see Common Kingfisher below.

Halcyon is a genus derived from the name of Alcyone, a figure in Greek legend. She and her lover Ceyx called each other 'Hera' and 'Zeus': as a result they were punished by the gods for their blasphemy by being turned into kingfishers.

The specific name, *smyrnensis*, recalls the port of Smyrna (now in modern Turkey).

Grey-headed Kingfisher *Halcyon leucocephala*

This African and Arabian species has been recorded in the Cape Verde Islands.

Grey-headed becomes 'white-headed' in the specific name *leucocephala*. It is a very pale grey, of course.

For **Kingfisher** see Common Kingfisher below.

Collared Kingfisher *Todiramphus chloris*

Small populations of this mainly South Asian species live in the Arabian Peninsula. It has been recorded in Egypt.

Collared refers to a broad white neck band.

For **Kingfisher** see Common Kingfisher below

The genus *Todiramphus* likens the stout bill to that of the todies (a family of tiny woodpecker-like birds).

The specific name *chloris* means green, though the Arabian forms are more turquoise that their Eastern congeners.

(Common) Kingfisher *Alcedo atthis*

This is a widespread native species.

Common is only used in an international context.

Kingfisher seems appropriately regal for this jewel of a bird. Lockwood points out that the name *fiscere* was first seen in around the year 1000, but it appears that the regal touch was added later in imitation of a French name *roi-pêcheur*. The origins of that are probably linked to the Arthurian legend of the Fisher King. Pennant lent weight to the name *Kingfisher* in 1768 and that outlasted an alternative form, *King's Fisher*.

Alcedo is Latin for 'kingfisher'.

The specific name *atthis* is a link to a Greek legend in which Perseus quarrels with a resplendent Indian prince (Atthis) and slays him.

Belted Kingfisher *Megaceryle alcyon*

This American vagrant occurs infrequently in Britain.

Belted describes the broad, dark pectoral band of this large kingfisher.

The scientists have found ways of making a number of names around the legend of Alcyone and Ceyx. In *Megaceryle*, Ceyx is fashioned into *ceryle*, with the prefix Mega- underlining the fact that it is a large species.

Yet another variant of Alcyone occurs in the specific name *alcyon*.

Pied Kingfisher *Ceryle rudis*

This species is found in Egypt and Israel.

Pied is an obvious name for a black-and-white species.

Ceryle uses a simple form, based in Ceyx's name.

The specific name *rudis* appears to mean 'wild' or 'rude', but it seems that Hasselqvist confused the Latin *hispida* (kingfisher) with *hispidus* (rough) and left yet another enigma (see also the story of the Cattle Egret, *Bubulcus ibis*). In fairness to Hasselqvist himself, his premature death on his way home from Africa meant that he was unable to explain, discuss or modify his notes and specimens.

50. MEROPIDAE: BEE-EATERS

The family name is based on *Merops*.

European Bee-Eater *Merops apiaster*

Bee-eaters are rare in Britain, but they do breed occasionally.

European distinguishes this form from a number of others.

Bee-eater is self-evident, though they do take other insects. The name originated as a direct translation into English of the specific Latin name *apiaster*.

Merops is also the Greek for 'bee-eater'.

Blue-cheeked Bee-eater *Merops persicus*

This African and Middle Eastern species is a rare vagrant to Britain.

Blue-cheeked describes a subtle feature of this mainly green bird.

The specific name *persicus* is 'from Persia' (modern Iran).

White-throated Bee-eater *Merops albicollis*

This African species has been recorded in the western Sahara, Algeria and Israel.

White-throated is self-evident and rather more accurate than *albicollis* (white-collared) since the white does not extend behind the ear.

> The former Green Bee-eater complex was split into three species in 2021, two of which occur in the Western Palearctic. The Asian form now bears the original name *M. orientalis*.

African Green Bee-eater *Merops viridissimus*

This species is recorded in Egypt.

Green is echoed in *viridissimus*, 'the greenest'. This form has green underparts and face.

Arabian Green Bee-eater *Merops cyanophrys*

This species is recorded in Israel.

The specific name *cyanophrys* refers to the vivid blue eyebrows – Greek *kuanos* (blue) and *ophrus* (eyebrow). The undersides are also blue-toned, in contrast to *viridissimus* above.

51. PICIDAE: WOODPECKERS

The family name is based on *Picus*.

Woodpecker is a simple way of saying what the bird does. That concept is echoed in the numerous Greek and Latin variants that occur in the generic names of this family. The word is first seen in English in 1530, in the writing of John Palsgrave, who was a tutor to Henry Fitzroy, the illegitimate eldest son of Henry VIII. The word was one of a number of regional variants on the same theme (e.g. wood hacker, wood knacker, wood tapper), but this form prevailed after it was formally adopted by Pennant in the late eighteenth century. It was originally associated with the Green Woodpecker, but widened its scope to include Britain's two 'spotted' woodpecker species and, eventually, the vernacular name of a whole family. There are, of course, examples of other vernacular names below, including the Wryneck and two American forms, Sapsucker and Flicker. In each of those cases, it seems that other perhaps more interesting characteristics caught the eye.

(Eurasian) Wryneck *Jynx torquilla*

This is a scarce, mainly passage species in Britain.

Eurasian is used in an international context, as there is another in Africa.

Wryneck uses an archaic word meaning 'twisted' to convey the bird's defensive ploy of twisting its neck to imitate a snake as a threat bluff. (A New Zealand plover is known as the Wrybill for the sideways twist to its beak.)

Jynx has Greek roots in a cry uttered during a superstitious ritual that involved the bird being used to divine the course of love.

The specific name *torquilla* repeats the idea of wry (twisted). The word relates in other contexts to the twisted neck ornament (*torc*) and to the mechanic's *torque wrench*.

* Like the Hoopoe, the Wryneck earned its own legends by its impressive imitation of a snake, a performance that evolved to scare off predators. This eerie and un-bird-like behaviour led to its association with witchcraft, myth and legend. It appears in legends concerning Zeus and Aphrodite, and in the story of the unlucky nymph *Iynx*, whose name became associated with the bird, and with the word 'jinx' as a symbol of bad luck.

Yellow-bellied Sapsucker *Sphyrapicus varius*

This American species has been recorded in Britain, Ireland and Iceland. Significantly, it is a seasonal migrant within North America and therefore susceptible to being carried by autumn storms.

Yellow-bellied is self-evident, though it is only a subtle feature of a boldly marked bird. The name was one of Catesby's creations.

Sapsucker is the name given to several American woodpeckers that open a flow of sugary sap in a tree as a source of food, and also eat the insects that accumulate around the sap.

Sphyrapicus literally means 'hammer-pecker', *sphura* being the Greek equivalent of the Latin *martius* (see Black Woodpecker).

The specific name *varius* (varied) seems to be used for the bird's extravagant patterns of patches and streaks in black, white and red.

Eurasian Three-toed Woodpecker *Picoides tridactylus*

This is a species of boreal forests, which has not been recorded in Britain.

Eurasian distinguishes this species from the American one, from which it has now been split.

Three-toed describes a curious feature of this genus: while most other woodpeckers have four toes, two forwards and two backwards, these birds are lacking one of the backwards-facing toes.

That idea is reinforced by the specific *tridactylus* (with three digits).

Picoides simply states that this genus resembles *Picus* (see below).

Middle Spotted Woodpecker *Dendrocoptes medius*

This species is found in much of Europe but not in Britain.

Middle places this between *Great* and *Lesser*, both of which it resembles in a general way, and that is underlined by *medius*.

Dendrocoptes describes birds of this genus as 'tree-hammerers'.

African Grey Woodpecker *Dendropicos goertae*

This sub-Saharan species has been recorded in Mauretania and Algeria.

The English names are self-evident.

Dendropicos brings together trees and woodpeckers.

Buffon noted the local Senegalese name *Goërtan*, which is used in the specific form *goertae*.

Lesser Spotted Woodpecker *Dryobates minor (comminutus)*

This is fast becoming an endangered species in Britain, which is made even more urgent by the fact that Britain has its own subspecies.

Lesser was eventually favoured over Pennant's *Lest* [*sic*] to describe the smaller of the British 'spotted' woodpeckers, and that is supported by *minor*.

Spotted originally distinguished the two British black-and-white woodpeckers from the Green and applies because each has patches of white on a black upper plumage. In this case the 'spots' are jagged stripes.

Dryobates this time tells us of a 'woodland walker'. This was a recent move from the genus *Dendrocopos*.

The subspecific name *comminutus* means 'separated' or 'diminished'. The name was probably intended to show that it was separated from the Continental form, but sadly it now seems to have drifted closer to the second value.

Syrian Woodpecker *Dendrocopus syriacus*

This species ranges in Eastern Europe and the Middle East.

Syrian and *syriacus* are not exclusive: it is found in a number of countries in the Middle East and in Eastern Europe.

Dendrocopus is another variant on the 'wood-striking' theme.

Great Spotted Woodpecker *Dendrocopus major*

This is a common species.

Great is the larger of the two 'spotted' woodpeckers found in Britain, and that idea is reinforced by the specific name *major*.

Spotted: in this case the 'spots' are large patches and small arrows.

White-backed Woodpecker *Dendrocopus leucotos*

This is a widespread form, mainly found in Eastern Europe. It is unknown in Britain.

White-backed describes a feature that is not present in other European 'spotted' woodpeckers, and is in fact absent in the subspecies found in the south-east of its range.

The specific name *leucotos* is confusing and appears to be an accidental truncation of *leuconotos* (*leucos*, white, and *notos*, backed). Johann Bechstein did apparently attempt to correct this later but under the rules of the Linnaean system the older usage prevails. Other such examples exist, an example being *atricilla* for *atricapilla* (Laughing Gull).

Northern Flicker *Colaptes auratus*

This North American species has been recorded in Denmark.

Northern separates this species from half a dozen others of the same genus found further south in the Americas.

Flicker describes the effect of reflective primary and undertail feathers that catch the sun and flicker like a flames when the bird moves.

Colaptes derives from a Greek word meaning 'chiseller'.

The specific name *auratus* refers to the golden colour of the reflective plumage of the Eastern form, which is sometimes tagged 'golden-shafted' to distinguish it from the 'red-shafted', Western form.

* Audubon called it the *Golden-winged Woodpecker*. One of its more confusing colloquial names arising from that idea is the form *Yellowhammer*: in that usage the 'hammer' is a quite literal rendering of its action, and under that name it is the state bird of Alabama. The name *Ant Woodpecker* reflected a similar ground-feeding habit to the Green Woodpecker.

Black Woodpecker *Dryocopus martius*

This large Continental species has never crossed the Channel.

Black does not take into account the scarlet cap.

Dryocopus means 'tree-beating' – one of many contrived names in this family on the same theme.

The specific name *martius* emphasises the 'tree-beating' idea with a word related to the Latin *martulus* (hammer). The very relationship of those words to the Roman god Mars tends to emphasise the size and might of this species.

European Green Woodpecker *Picus viridis*

This is a moderately common species.

Green is the dominant colour and occurs again in the specific name as *viridis*.

Picus is a Latin name for 'woodpecker', almost certainly related to the concept of pecking, but this word is also linked to another extravagant legend, this time a Roman one, telling that King Picus of Latium rejected the advances of the witch Circe and was transmuted into a woodpecker as her angry vengeance.

The specific name *viridis* simply means 'green'.

* A traditional word for the bird's call is a 'yaffle'. From that grew a series of southern dialect names, of which *Yaffingale* is best known, a word that Lockwood feels is imbued with a touch of humour. A Nightingale it is not! Children and parents of the 1970s and '80s may remember Professor Yaffle, a wooden woodpecker and a bit of a know-all, who featured in *Bagpuss* on children's TV.

* Lockwood's treasure-chest offers a range of alternative forms. Among these curiosities are several variants of *Hickwall* and a touch of folklore in *Rain Bird*. *Hey-hoe* seems to have derived from *Hew-hole*. The word *Woodspite* was used as an alternative by Willughby, and that appears to have roots in Wood Speight, which seems to have come from a proper name. The Green Woodpecker is not a bird to ignore, and the many colourful names suggest that it was much better known in our more bucolic past.

Iberian Green Woodpecker *Picus sharpei*

A 2011 report recommended the split from *P. viridus* to make this a full species.

Iberian is mainly, but not exclusively, limited to the Iberian Peninsula, since the range bleeds into parts of southern France.

The specific name *sharpei* is a dedication to the British zoologist R.B. **Sharpe** (1847–1909), who noted the differences of this form.

Levaillant's Woodpecker *Picus vaillantii*

This is the North African counterpart of the Green Woodpecker.

Levaillant's is dedicated to François **Levaillant**, an eminent French explorer and naturalist. Curiously he started out in life as plain Vaillant, but modified his name in about 1784, almost certainly as a social affectation. The simpler form of the name is recorded in the specific name, *vaillantii*.

Grey-headed Woodpecker *Picus canus*

This Continental species is another that has never crossed to Britain.

Grey-headed appears to limit the extent of the grey, though the specific name *canus* (grey) implies that the whole bird is grey. In fact the mantle is green and the head is a darker shade of grey than the undersides.

52. FALCONIDAE: FALCONS

The family name is based on *Falco*.

The current position of the falcons in the taxonomic sequencing of birds is the most radical change of recent times. After many centuries of assuming that falcons were close relatives of the harriers, hawks and eagles (the Accipitridae), scientists have now placed them closer to the parrots, with owls now deemed to be more closely related to the Accipitridae.

Lesser Kestrel *Falco naumanni*

This migratory falcon breeds in Southern Europe but rarely visits Britain.

Lesser is used because this is the smaller of two European kestrels.

Kestrel: see below.

The dedication is to German artist and naturalist Johann Friedrich **Naumann** (1780–1857).

* Naumann's Thrush commemorates his father, Johann Andreas.

Common Kestrel *Falco tinnunculus*

Common is self-evident for a very widespread species.

Kestrel began life as an English name for this specific form and later was applied generically to a number of similar small falcons. It evolved from the French *crécerelle*, because its courtship call resembles a rattle (*crécelle*).

The specific name *tinnunculus* also refers to the squeaky calls, but this time from a Latin root. The form first appears in Gessner in 1555 and was widely used before being formalised by Linnaeus.

* The 'rattle' analogy makes much better sense in the context of a largely rural medieval society, where wooden rattles served as agricultural tools to scare away crop pests. Such rattles later served as police alarms and to warn of gas attacks in war-time. Until the 1970s, they were regularly used by football fans to vent their enthusiasm.

* One of the most widespread and durable of alternative names for this species is *Windhover*, which clearly reflects the bird's distinctive habit of holding station on the wing while facing into a stiff breeze. In that way it has a platform for searching open ground beneath for prey. That name is perpetuated in a famous 1877 poem by Gerard Manley-Hopkins. The name has other variants that suggest it derives from a Germanic form that was in use before Kestrel became the norm.

American Kestrel *Falco sparverius*

This small American species has been recorded in Britain and elsewhere in Europe. It is a partial migrant, with the Northern population moving south to winter.

American is needed to distinguish this from the other kestrels.

The specific name *sparverius* tells us that it is a 'sparrow hawk': Catesby called it a *Little Hawk*, another illustration of the much looser historical usage of terminology for birds of prey.

Red-footed Falcon *Falco vespertinus*

This migratory species is an irregular vagrant to Britain: it breeds mainly in Eastern Europe and near-Asia.

Red-footed is apt, because the feet of the adult male are vividly scarlet.

Falcon and *Falco* are from the Latin *falx*, *falcis* (sickle) and via the French *faucon*. Its earliest English form dates from 1250 as *faucun*. The image relates to the sickle-shaped wings of many falcons.

The specific name *vespertinus* refers to a bird of crepuscular habits (compare *vespers*, an evening church service).

Amur Falcon *Falco amurensis*

There have been a handful of records of this Far-Eastern species in Western Europe, including Britain.

Amur and *amurensis* relate to the Amurland region of Siberia: this species is the Far Eastern equivalent of the Red-footed Falcon and was formerly treated as a subspecies of that.

Eleonora's Falcon *Falco eleonorae*

This Mediterranean species has rarely been recorded in Britain.

Eleonora of Arborea was a fourteenth-century Sardinian warrior princess, whose legal protection of falcons and hawks on the island was unprecedented.

* This falcon coincides its breeding with the heaviest late-summer passage of hirundines, which are the principal food source for its young.

Sooty Falcon *Falco concolor*

This species replaces Eleonora's in North-east Africa and is unknown in Britain.

Sooty describes a bird that is mainly charcoal grey as an adult.

The specific name *concolor* means 'plain'.

(Eurasian) Hobby *Falco subbuteo*

This is a regular summer migrant to Britain.

Eurasian is used in the context of other hobby species, for example *Oriental*.

Hobby comes from the Old French verb *hober*, to jump about, and is a reflection of the bird's agility on the wing.

The specific name *subbuteo* means 'near to the buzzards', a rather illogical attachment that seems to stem from Aristotle's time (see also the introduction to Raptors and Birds of Prey).

* A curiosity linked to this bird is the fact that the name of the table-football game Subbuteo was adopted when the inventor was refused permission to patent the name Hobby.

Merlin *Falco columbarius aesalon*

As currently recognised, the Merlin is distributed through both Eurasia and North America. It is a resident species in Britain, though there is an element of altitudinal migration from high country in summer to lowlands in winter.

Merlin is a unique name for the smallest European falcon. The word had Germanic roots related to the Franco-German *smeril* and Norse *smyrill*, for a small hawk. It then evolved via French as *esmeril*, then as *merillon* to *merlin*. In its current form it is one of very few single-word traditional vernacular names.

The specific name, ***columbarius***, relates to doves as prey – though this is not generally true of European Merlins. However, the American form is the nominate one, and doves do appear to be on its menu.

* Currently the simple name Merlin applies to one widespread species, but a future split between North American and Eurasian forms has been suggested on the basis of fossil evidence and the fact that there has been no gene flow between the two regions for at least a million years.

* The subspecies breeding in Western Europe is *aesalon*, Greek for 'hawk', and a name linked historically with the Merlin. The Icelandic race *subaesalon* may also be found in Britain in winter.

* In medieval falconry, this small falcon was considered to be a ladies' bird and was flown against prey such as larks.

* The wizard of Arthurian legend may share the name in its modern spelling, but the root of that name lies in Welsh language and folklore.

Saker Falcon *Falco cherrug*

This species of the Eastern Mediterranean and eastwards is not seen as a wild bird in Britain.

Saker has origins in the Arabic *saqr* and means *falcon*.

The specific name ***cherrug*** is from even further east, since it is of Sindhi origins and originally relates to the female.

Lanner Falcon *Falco biarmicus*

Lanner is a Mediterranean species and is not recorded as a wild species in Britain.

Lanner derives from the Old French name *L'anier*, which means 'duck-hunter'. It is rooted in the Latin *anas* (duck) from which, in Old French, the word *ane* meant 'female duck'. The suffix *-ier* has a general sense of being '*for the purpose of*'.

The specific name ***biarmicus*** is a double error. First, Linnaeus attached the name to the Bearded Tit or Reedling, where it implied, wrongly, that the bird came from a White Sea region called Bjarmaland. Temminck later thought that the word referred to the moustachial stripes of the Reedling and attached the word to the moustached Lanner Falcon. Rules governing the Linnaean system normally protect the earliest-used names, regardless of such errors.

Gyrfalcon *Falco rusticolus*

This Arctic species is rare in Britain.

Gyrfalcon is usually used as a single word. The word *gyr* has roots in the Old Norse for a spear and evolved via *geirfalki* to become the Medieval English *gerfauk*. The word's path into modern usage is bumpy owing to the mistaken assumption, by Aldrovandi in 1603, that the word had to do with *gyro* and circling in the sky, hence *Gyrofalco*. Pennant anglicised that, and Gyrfalcon (or alternatively Gyr Falcon) became the standard form.

The specific name, *rusticolus* (country-dweller), seems a mean epithet for this free-spirited bird of the Arctic wilderness.

* Both Hartert (1912) and Coward (1926) distinguish three forms: two subspecies are tagged as *Iceland Falcon* and *Greenland Falcon*. Such discrimination related to the colour and was a tradition stemming from falconry. The 'Greenland' form tends to be whiter. Today the species is treated as monotypic and the colour variations as morphs.

Peregrine Falcon *Falco peregrinus*

This is a widespread resident in Britain.

Peregrine was first used as *peregrinus* by the cleric Albertus Magnus, who captured one in 1250 on its first migration flight, or 'peregrination'.

* The words pilgrimage and pilgrim come from similar routes, leading to one of the bird's alternative names in the late Middle Ages being the *Pilgrim Hawk*, which seems appropriate, given that Albertus was eventually canonised.

* The fate of this name is grammatically curious: having started life as an adjective describing a form of falcon, the word is now mainly used as a noun – *the* Peregrine.

* **Barbary Falcon**, long treated as a full species, is now considered to be one of the eighteen subspecies of Peregrine.

PARROTS

In Medieval England the word *Popinjay* was used as the general name for these birds, that being based on a French import, which itself has roots in Spanish and Arabic. Fashions change and the word *Parrot* came into English in the mid-Tudor period. That is another French derivation, which follows the tradition of using personal names for birds. In this case the name was *Pierrot*, a diminutive of Pierre.

The Western Palearctic has no native parrot species, though these birds are favourites with bird fanciers and the cage-bird trade. As a result, some non-native escapes survive or even thrive in the wild. The survival of exotic parakeets in the temperate climates of Britain surprises many people, but Ring-necked Parakeets, particularly the Indian subspecies, are much tougher than they look, having evolved in warmer latitudes but at high altitudes. Some tropical and subtropical species tend to manage well in warmer parts of Europe and the Middle East. While these birds add colour to the local avifauna, some species can become problematic: Ring-necked and Monk Parakeets tend to proliferate most readily and seem to cause the greatest nuisance.

The representative selection below is perhaps arbitrary and could probably include several more species.

It is interesting to note that parrot enthusiasts and the cage-bird trade may use different names from those used in ornithology. One example, Blue-crowned Parakeet, also known as the Blue-headed Conure, is included below.

Taxonomically, the world's parrot species are now divided into four families. New Zealand parrots and cockatoos are not involved here, but the two remaining families are represented below.

53. PSITTACIDAE:
African & New World Parrots

The family name is based on *Psittacus*.

Monk Parakeet *Myiopsitta monachus*

A small escaped population existed for a while in the British Isles, but was eliminated before it became a nuisance. This South American escape still thrives in a number of European countries, where its bulky communal nests can cause real problems, especially when they involve such structures as power installations.

Monk describes the hooded effect of its plumage – and that is reinforced by *monachus* (hooded).

Parakeet means 'small parrot' and derives from the French *perroquet*, itself rooted in the proper name *Pierrot* (see above).

Myopsitta derives from the Greek *mus*, *muos* (mouse) and *psitta* (parrot) to describe a small and rather drab parrot, which is grey-green above and grey below.

* This species is often called the *Quaker Parrot* by the cage-bird community, which name, like *Monk*, appears to be derived from the hood. That name seems to be used frequently in discussions of the bird's situation as an established escape.

Nanday Parakeet *Aratinga nenday*

This South American species has an established feral population in Israel.

Nanday and the specific form *nenday* began life respectively in the South American native languages of Tupi and Guarani. In Tupi *Nandáia* means 'noisy talker'.

Aratinga also stems from Tupi and means 'bright parrot'.

Blue-crowned Parakeet *Thectocercus acuticaudatus*

This South American species has small local breeding colonies in Kent and in Spain, which derive from escaped cage-birds. It is known in aviculture as the **Blue-headed Conure**.

Blue-crowned/Blue-headed are self-evident.

Conure derives from a Greek root meaning 'cone-tailed'.

Thectocercus and *acuticaudatus* both say, in Greek and then in Latin, that the bird has a 'sharp tail'. That description relates to its tapering shape, as in Conure.

Red-masked Parakeet *Psittacara erythrogenys*

This South American species has a thriving feral population in and around Valencia in Spain. It has expanded rapidly over the last decade or two and was admitted to the Spanish lists in 2019.

Red-masked is a distinctive feature of an otherwise green bird.

Psittacara derives from the Greek *psitta* (parrot) and *kara* (head).

The specific name *erythrogenys* means 'red-cheeked'.

54. PSITTACULIADAE: OLD WORLD PARROTS

The family name is based on *Psittacula*.

Rose-ringed Parakeet *Psittacula krameri* (INTERNATIONAL NAME)
Ring-necked Parakeet *Psittacula krameri* (BRITISH NAME)

This escaped Asian/African species is naturalised and widespread in the south of England and elsewhere in Europe.

Ring-necked describes a black collar, though an alternative name is **Rose-Ringed** for the male's addition of pink in the breeding season.

Wilhelm **Kramer** (1724–1765) was a German naturalist.

* There is an element of urban legend surrounding the origins of the birds in Britain, in which there may be some truth, but what is certain is that this very popular cage-bird escaped in sufficient numbers to colonise south-west London and then expand from there.

Alexandrine Parakeet *Psittacula eupatria*

This Asian species has self-sustaining populations in several European countries.

Alexandrine relates to the fact that the bird was first brought from India by Alexander the Great.

The specific name *eupatria* reinforces the royal connection in relating to the Greek for a 'noble daughter'.

Fischer's Lovebird *Agapornis fischeri*

There are well-established feral populations of this East African species in south-west France and in Portugal.

Gustav **Fischer** (1848–1886) was a German explorer in Africa.

Lovebird relates to the affectionate behaviour of this very sociable species, while *Agapornis* derives from the Greek for the same idea.

Tyrants to Larks

55. TYRANNIDAE:
TYRANT FLYCATCHERS

The family name is based on *Tyrannus*.

This is a large American family that generally parallels the behaviour of the Old World flycatchers in feeding style: the birds tend to share the upright posture on the branch and make rapid sorties to feed. Some have crown stripes and some are very feisty defendants of their territories, hence the addition of 'tyrant' to the general name, which distinguishes them from the unrelated Old World species.

Some of the North American species are highly migratory and may get caught in North Atlantic storms. Consequently a few of this family have been recorded as vagrants to Britain and other parts of the Western Palearctic.

Eastern Phoebe *Sayornis phoebe*

This species has been recorded once in England.

Eastern distinguishes this species from the *Western* Phoebe: here the terms relate to an American context.

Phoebe is an alternative name for the goddess Diana.

Bonaparte originally named the Say's Phoebe after Thomas Say (1787–1834), an American naturalist, and from that came the generic name *Sayornis* (Say's bird).

Eastern Wood Peewee *Contopus virens*

This species has been recorded twice in the Azores.

Eastern pertains because this bird also has a Western counterpart, again in an American context.

Wood implies a habitat preference.

Peewee is in imitation of the bird's call. Curiously, the same name is applied colloquially to an unrelated Australian species for similar reasons (see also *Peewit* for Lapwing).

Contopus: Choate points out that *kontos* is the Greek for short and that this bird does indeed have small feet, as the name implies.

The specific name *virens* means green (though that is very subdued in this case).

Yellow-bellied Flycatcher *Empidonax flaviventris*

This species was recorded in Scotland in 2020.

It is not easy in the field to spot the **yellow-belly** of this shadowy bird, but that separates it from similar species. That name is repeated in the Latin version *flaviventris*.

Empidonax tells us that these birds have 'mastery of gnats' (Greek *empis*, gnat, and *anax*, lord).

Acadian Flycatcher *Empidonax virescens*

This species has been recorded in England and Iceland.

Acadian relates to a seventeenth- and eighteenth-century French settlement in the Canadian Maritimes that was subsequently overrun by the British and later reinstated as a refuge. Today, several communities in New Brunswick and Nova Scotia pride themselves on maintaining a French-based cultural history. *Acadie* was a version of Arcadia, the central region of Ancient Greece that was seen as a safe refuge. During the original troubles, a great many Acadians fled to Louisiana, where their name developed into the local form *Cajun*.

The specific name *virescens* tells us that this is a 'greenish' bird.

Alder Flycatcher *Empidonax alnorum*

This species has been recorded in England, Iceland and Norway.

Alder and the Latin form *alnorum* (as in the genus *Alnus*) tell us that this species is found in riverside habitats where alders dominate.

Least Flycatcher *Empidonax minimus*

This species has been recorded in Iceland.

Least and *minimus* underline the fact that this is the smallest of a small genus.

Eastern Kingbird *Tyrannus tyrannus*

This species has been recorded in Ireland and Scotland.

Eastern separates this from other kingbirds, again in an American context.

The **Kingbird** is so-called because of its crown stripe and imperious character.

Tyrannus is for a bird that defends its territorial rights with vigour and determination. When Benjamin Franklin was discussing the Bald Eagle, he wrote: 'Besides he is a rank coward; the little kingbird, not bigger than a sparrow attacks him boldly and drives him out of the district.'

Western Kingbird *Tyrannus verticalis*

A vagrant was recorded in Flores in the Azores in 2018.

Western underlines that this is a bird of the western parts of North America.

The specific name *verticalis* links to the Latin *vertex* for the top of the head, and therefore means 'crowned' (thus Kingbird).

Fork-tailed Flycatcher *Tyrannus savana*

This Central and South American species has been recorded in North America, but was a surprising record in Spain in 2002, where it arrived after an adverse weather event. It appears to be a unique example of vagrancy to Europe of a species normally resident in Latin America.

Fork-tailed describes a dramatic tail structure that is considerably longer than the bird's body.

The specific name *savana* was first applied by Buffon when he found the species in flooded grasslands.

56. MALACONOTIDAE: Bushshrikes

The family name is based on the genus *Malaconotis*, which means soft-backed.

This family was previously considered to be part of the shrike family. It is now understood that the similarity is due to convergent evolution.

Black-crowned Tchagra *Tchagra senegalus*

This is a North African species.

Black-crowned is self-evident.

Tchagra appears to be an onomatopoeic name.

The specific name *senegalus* is another variant on a word relating to Senegal.

57. VIREONIDAE: Vireos

The family name is based on *Vireo*.

Vireos are an American family with some characteristics of warblers. Again, some are highly migratory. Three species have been recorded as vagrants to Britain and two more have been recorded in the Azores.

Red-eyed Vireo *Vireo olivaceus*

This is a fairly regular North American vagrant to Britain and has been recorded as far as Poland, Italy and Malta.

The curious name **Red-eyed**, which is literal, contrasts this with a 'White-eyed' species.

Vireo is taken from a Greek word for a green bird, possibly originally a Greenfinch or the female Golden Oriole.

The specific name *olivaceus* continues the Vireo theme with a word for a shade of green.

Philadelphia Vireo *Vireo philadelphicus*

This species has been recorded several times in Britain.

Philadelphia and *philadelphicus* are both to do with the American city of that name, near which the type specimen was collected.

Warbling Vireo *Vireo gilvus*

This species has been recorded in the Azores.

Warbling records the fact that this bird has a strong song.

The specific name *gilvus* tells us that this species has some pale yellow (the underparts), though that applies only to the Eastern form.

Yellow-throated Vireo *Vireo flavifrons*

There has only been one record of this species in Britain and another in Germany.

Yellow-throated is an understated description, since the yellow also extends over the breast.

The specific name *flavifrons* describes a different area of yellow – the frons (forehead). In this case there is a narrow line of yellow above the bill that extends back to a yellow eye-ring.

White-eyed Vireo *Vireo griseus*

There have been several records of this species in the Azores.

White-eyed points to a feature that distinguishes this species, though the specific name *griseus* (grey) presents a relatively dull image of a bird that has a lot of yellow (in that respect it is almost as unjust as the name *Grey Wagtail*).

58. ORIOLIDAE: Old World Orioles

The family name is based on *Oriolus*.

Most of the true orioles are found in the East, with several in Australia. American orioles are lookalikes that are not related in any way: those are icterids. The Eurasian species, which gives us the word, is the only representative in the Western Palearctic.

(Eurasian) Golden Oriole *Oriolus oriolus*

This is a rare species in Britain, where it has bred.

The word **Eurasian** serves to remind us that there are Indian and African Golden Orioles, but the word is not generally needed.

Golden describes a richly yellow male (but omits the black that sets off the yellow so well).

Oriole and *oriolus* are rooted in the Latin *aureolus* – from *aurum* (gold). The suggestion that the bird's fluting call is at the root of the name was made as early as the thirteenth century by Albertus Magnus (see also Peregrine Falcon), but that is a happy coincidence.

59. DICRURIDAE: Drongos

The family name is based on *Dicrurus*.

This family originated in Southeast Asia and spread into Africa and Australia.

Ashy Drongo *Dicrurus leucophaeus*

This is an Asian species that has been recorded in both Israel and Kuwait.

The name **Ashy** applies more aptly in the eastern parts of the range, where the bird is paler. In the west it is blue-black.

Drongo originated as a Malagasy word, which was adopted into English in the nineteenth century.

Dicrurus means 'fork-tailed'.

The specific name *leucophaeus* combines the Greek for 'white' with 'dusky' to echo the concept of the vernacular name.

* In spite of the Australian use of the word *Drongo* as a mild insult (which is apparently to do with a failed racehorse of that name), some members of this family exhibit unusual intelligence.

Black Drongo *Dicrurus macrocercus*

This Asian species has been recorded in Kuwait.

The specific name *macrocercus* means 'long-tailed'.

60. LANIIDAE: Shrikes

The family name is based on *Lanius*.

Shrike possibly originated as a name for the Mistle Thrush (referencing 'shriek', for its somewhat shrill song) and was definitively applied to this family by Pennant in 1768.

The genus *Lanius* applies to all of this family. It is the Latin for 'butcher', which is a reminder that an alternative English folk name, for particularly the Red-backed Shrike, was Butcherbird, due to its habit of skewering a larder of uneaten prey on thorns and spines. (In international ornithology that name now applies formally to an unrelated family of Australian birds.)

Issues of speciation and naming within this family have long been unsettled. Since 2018 there has been some reorganisation of particularly the larger 'grey' shrikes: the American/Eastern Siberian species now owns the name Northern Shrike, while the Eurasian form is the Great Grey Shrike. The word Southern has now been dropped in favour of Iberian Grey Shrike. One or two other names have also been changed. In 2023 the whole family was moved within the taxonomic sequencing and its internal sequencing was also revised.

Great Grey Shrike *Lanius excubitor*

This is a scarce winter visitor to Britain from Northern Europe.

Great and **Grey** are both obvious. To Pennant it was the *Great Shrike*, but Yarrell modified that to the present form in 1843.

The specific name **excubitor** (sentinel) refers to the shrike's habit of perching on a vantage point to spy out potential prey.

Iberian Grey Shrike *Lanius meridionalis*

Iberian is not exclusive, as this species is also found outside Iberia, along the south coast of France.

The specific name *meridionalis* is a reminder that this species was long called the *Southern Grey Shrike* until a name change in 2017.

* The replacement of the former name *Southern* removes any potential for geographical confusion, since the Northern Shrike is primarily a North American species.

Northern Shrike *Lanius borealis*

The American/Siberian species, which was formerly seen as conspecific with Great Grey, but was split in 2017, has been recorded in Finland and Norway.

Northern distinguishes this species in America from the Loggerhead Shrike, the more southerly species on that side of the Atlantic.

The specific name *borealis* underlines the English name.

Masked Shrike *Lanius nubicus*

There are just four British records of this Eastern Mediterranean species.

Masked describes a broad black eye-stripe.

The specific name *nubicus* relates to the Nile region of Nubia where it appears in winter.

Lesser Grey Shrike *Lanius minor*

This is a species of south-east Europe, which has been recorded regularly as a vagrant to Britain.

The vernacular names are self-evident: it is considerably smaller than the other grey European shrikes.

The specific name *minor* underlines that last fact.

Woodchat Shrike *Lanius senator*

This Mediterranean species turns up fairly regularly in Britain.

The name **Woodchat** was discovered posthumously in Ray's notes and was unexplained, but was adopted by Pennant for this species. It was literally a spare name. All else is speculative.

The specific name *senator* appears to compare the red cap and nape of the bird with the stripe of the Roman senator's toga.

Bay-backed Shrike *Lanius vittatus*

This small South-east Asian species has been recorded in Kuwait and Iran.

Bay-backed describes the warm brown colour of the mantle.

The specific name *vittatus* means 'banded' and describes the broad, wrap-around, black facial mask.

Isabelline Shrike *Lanius isabellinus* (INTERNATIONAL NAME)
Daurian Shrike (BOU NAME)

This is a rare but regular vagrant to Britain from Mongolia and Western China.

Isabelline and *isabellinus* describe a faded tawny colour. It is thought to originate with the fifteenth-century Queen Isabella of Spain, who vowed not to change her under-linen until the Moors had been defeated.

BOU prefers to use the name **Daurian Shrike**, which derives from a region of Siberia.

Red-backed Shrike *Lanius collurio*

This was once a relatively common breeding species in Britain, but now breeds rarely. It is a fairly regular passage migrant.

Red-backed tells that the male has rufous-brown mantle plumage, which stands out in contrast to a blue-grey nape and cap.

The specific name *collurio* appeared in both Aristotle and Hesychius and was assumed to be this species.

Red-tailed Shrike *Lanius phoenicuroides* (INTERNATIONAL NAME)
Turkestan Shrike (BOU NAME)

This is a rare but regular vagrant to Britain from Central Asia.

Red-tailed is a self-evident name, but the Isabelline Shrike also has a red tail.

The specific name ***phoenicuroides*** tells us that it also resembles *L. phoenicurus* (crimson-tailed), which was a former scientific name of the Brown Shrike.

Turkestan is an area of central Asia. Again, this name is preferred by the BOU.

Brown Shrike *Lanius cristatus*

This Siberian species has been recorded as a rare vagrant to Britain.

Brown relates to the fairly uniformly brown upper parts of the bird.

The rufous-brown of the cap is brighter, which may explain why the bird has the specific name ***cristatus*** (crested).

Long-tailed Shrike *Lanius schach*

This Asian species has been recorded twice in Britain and only very rarely elsewhere in Western Europe.

Long-tailed is self-explanatory.

The specific name ***schach*** is onomatopoeic, rather like the *chack* of the Jackdaw.

61. CORVIDAE:
CROWS AND THEIR ALLIES

The family name is based on *Corvus*.

The crow family is well represented in Britain by eight resident species. A few other species are found in other parts of Europe, in North Africa, and in the Middle East.

Members of the Crow family are generally difficult to ignore: they are all relatively large, often appear to be numerous, and are sometimes showy and noisy. The birds themselves have long known how to exploit human activity, so are willing, if cautious, neighbours. As a result they, perhaps more than any other group of birds, have long been judged by human values, and their habits have become part of folklore and legend. Ravens in particular are recorded in many cultures with a mix of fear, awe and admiration. Likewise, Jays, Jackdaws and Magpies epitomise human flaws and become involved in tales of trickery, envy and greed. That tradition is marked by linking such birds to human names. It has to be said that the habits of some members of the family are appalling to human eyes, and that fact is reflected in such collective names as 'a murder of crows', but the cultural and physical pressures on some of this family are today fuelled not only by self-interest but by a long-held belief that humans are somehow 'in charge'.

Siberian Jay *Perisoreus infaustus*

This boreal species is found in Scandinavia and eastwards, hence the name **Siberian**.

Jay: see next species.

The generic name *Perisoreus* has unclear origins. Bonaparte was responsible for replacing Linnaeus's use of *Corvus* with the current genus, but his reasons for the use of this word are unclear. The simplest possibility seems to lie in a Greek word for 'hoarding' (which the birds do with winter food stores). Some 50 years later, Coues suggested an alternative Latin root linked to augury, but that seems a little contrived and he may well have been swayed by the bird's folk-reputation as an ill omen.

On the other hand, the specific name *infaustus* (unlucky) certainly seems to have been selected by Linnaeus for that reason and may have influenced Coues.

(Eurasian) Jay *Garrulus glandarius*

It is the only Jay in Britain, but there are other species around the world.

Jay has roots in the medieval version of the French name *geai*, which in turn links in Latin to the name Gaius. There are also probable links to the Old French roots of the English *popinjay*, which was originally used for a parrot, but was loosely applied to other bright species.

Garrulus describes a noisy, chattering bird.

The specific name *glandarius* invokes the bird's favourite food, the acorn.

* *Jackdaw* and *Magpie*, in this same section, show similar links to personal names. The fact that the tradition traces back to Roman culture is not surprising, given that corvids were kept as pets in Ancient Rome, and Greece too: Mynot (2018) quotes a number

of examples, including a tale by Plutarch of a Roman barber's pet Jay with prodigious talents for mimicry.

Iberian Magpie *Cyanopica cooki*

This species is limited to **Iberia**.

Cyanopica means 'blue magpie'.

The specific name *cooki* commemorates Samuel Cook (1787–1856), a British collector/naturalist.

* It is still often called *Azure-winged Magpie* from which it was formally split in the early 2010s. The related species is found only in the Far East. It was long thought that the Iberian form had been introduced by returning mariners, but fossils found in Gibraltar were 44,000 years old, disproving that assumption. Subsequent genetic studies showed the European form to be distinct.

(Eurasian) Magpie *Pica pica*

This is one of the most familiar of British birds, which is found all over Europe.

The word **Magpie** started life in English as *Pie*, which seems appropriate for a pied bird. This was originally an Old French word, which had evolved from the Latin *pica*, an image suggesting that a white bird was 'daubed with tar' (see also *Sea Pie* for the Oystercatcher). In the late Middle Ages the female name *Mag* (one of the many variants of Margaret) became attached in English, imported from a French tradition, which is illustrated in an extant rhyme telling the story of *Margot la Pie*. That accounts for Shakespeare's much misunderstood reference to 'magot pies and choughs and rooks' (Macbeth, Act V, Scene iv). The use of personal names for birds is surprisingly frequent: see also Jay and Jackdaw in this section.

Maghreb Magpie *Pica mauritanica*

This species is endemic to north-west Africa from Morocco to Tunis.

Maghreb is an old Arab name for north-west Africa.

The specific name *mauritanica* refers to Mauretania, another word historically used for the north-west of Africa.

(Spotted) Nutcracker *Nucifraga caryocatactes*

This Eurasian species makes very rare incursions into Britain in small numbers during irruptions.

The word **Spotted** separates this one from a North American relative but is of little interest, other than as a descriptive word.

The word **Nutcracker** is a translation of Turner's Latin, *Nucifraga*, which started life as the German *Nussbrecher*.

That is capped by the Greek-rooted specific name *caryocatactes*, which says the same thing. That was in fact the work of Gessner, and was also based on the German word.

(Red-billed) Chough *Pyrrhocorax pyrrhocorax*

This bird is a scarce resident on some western coasts of Britain.

The word **Red-billed** is used internationally, but only the name **Chough** is needed in Britain. The history of that word is complicated, since it seems to have evolved as a regional name for the Jackdaw and was based on the *chack* call. A related crow species found in Cornwall became known as the *Cornish* Chough, a name that was used by Turner in 1544. Later, as Jackdaw became the dominant name, the Cornish bird retained the name Chough. Later, when that name was also attached generically to its yellow-billed relative (next species) it was realised that *Cornish* was too narrow in usage for a bird that was found in many countries. The replacement name *Red-billed* distinguishes it from the yellow-billed Alpine Chough.

The Generic and specific names *Pyrrhocorax* tell us that this is a 'fire-crow', because of its red bill and legs.

Alpine Chough *Pyrrhocorax graculus*

This species has never been recorded in Britain. It is found no closer than the Alps and Pyrenees.

Alpine implies that it is found at relatively high altitudes, though not just in the Alps.

The specific name *graculus* is a Latin word that became attached to the Jackdaw (as evidenced in Merrett in 1666), though it seems to have had a less specific usage originally. That same word was later compressed into the modern American name *grackle*.

(Western) Jackdaw *Coloeus monedula*

It comes as a surprise that this familiar name may need some qualification, but the Daurian Jackdaw exists further east.

Jackdaw seems to be an amalgam of two onomatopoeic elements. The *chack* call gives the first element and the *daw* part is probably a variant of *caw*. The adaptation to *Jack* illustrates the convention of giving familiar birds personal names, and one in this case that tends to be associated with dubious character ('Jack the Lad' or the Jack or Knave in the card pack). The Jackdaw's reputation was already sullied in Ancient Greece (see *monedula*, below).

Coloeus is a Latinised version of the Greek *koloios* for 'Jackdaw'.

The specific name *monedula* (money-eater) is based on the Greek legend of the greedy Arne of Thrace, who betrayed her nation for money and was turned into a Jackdaw as punishment.

* Probably the best known literary tribute to the Jackdaw is the whimsical nineteenth-century poem by Ingoldsby, 'The Jackdaw of Rheims'.

Daurian Jackdaw *Coloeus dauuricus*

This Siberian species has been noted in various countries of Northern Europe, but not in Britain.

Daurian derives from the Dauria region of eastern Russia (and has nothing to do with the *daw* element of the vernacular name).

The spelling of *dauuricus* may be modified at some time: it is currently used in the simpler form elsewhere.

House Crow *Corvus splendens*

This South Asian species seems to be a fairly regular ship-borne vagrant, which has established itself in Egypt, has visited some parts of Europe and may have bred in the Netherlands.

It survives by being an efficient scavenger around humans, hence **House** Crow.

Corvus is the Latin for 'Raven'.

The specific name *splendens* means 'shining'.

* This bird is now subject to strict new control laws in Europe. An incipient population at the Hook of Holland, a busy shipping port, was removed by 2019.

Rook *Corvus frugilegus*

This is a common and familiar species.

Rook is an Anglo-Saxon word that probably has its roots in onomatopoeia.

The specific name *frugilegus* suggests that this a 'crop-gatherer', but that is a distortion of the true story, since Rooks actually eat a higher proportion of non-plant material, such as soil creatures.

Carrion Crow *Corvus corone*

This is another well-known species in Britain and Western Europe.

The association with **Carrion** (dead creatures) perhaps does not flatter, but carrion-eaters perform a useful service in removing potential sources of infection.

Crow, like Rook, is almost certainly rooted in the call.

The specific name *corone* is rooted in a Greek word for 'Crow'.

* These birds have often been treated as conspecific with the next species (see note below).

Hooded Crow *Corvus cornix*

This species replaces the Carrion Crow in northern parts of the British Isles and in much of Europe.

Hooded describes the grey and black plumage.

The specific name *cornix* is the Latin for 'Crow'.

* This species has, in the past, been lumped with the Carrion Crow. Current analysis suggests that the two are separated by the slimmest of genetic differences. The evidence that they rarely interbreed relates on the one hand to those genes that produce colour, and on the other to those that control the bird's perception of colour.

Pied Crow *Corvus albus*

This sub-Saharan African species has been recorded as far north as Morocco.

Pied implies that this species is white and black. It is otherwise patterned similarly to the Hooded Crow.

The specific name *alba* emphasises the white.

(Northern) Raven *Corvus corax*

Northern serves to distinguish this species on the international stage, but is not used generally. In any case it is a somewhat dubious name for a species found as far south as North Africa and Central America.

Raven is another deep-rooted Germanic name: its Old English form *hræfn* probably had echoic origins.

The specific name *corax* is the Latin for 'Raven', another echoic rendition.

* The Raven, a large and intelligent species, is widely recorded in folklore and legends in many cultures, admired for its intelligence, but feared for its strength. It still has a powerful symbolic representation today at the Tower of London, where legend says that if the Ravens leave the Tower, the kingdom will fall. A ravenmaster oversees the resident birds.

Brown-necked Raven *Corvus ruficollis*

This is a North African species.

Brown-necked and the Latin equivalent *ruficollis* describe a bronze sheen on the neck feathers.

Fan-tailed Raven *Corvus rhipidurus*

This is a species of the Arabian Peninsula.

Fan-tailed and *rhipidurus* mean the same thing (Greek *rhipis*, fan, and *ourus*, tailed).

62. BOMBYCILLIDAE: THE WAXWINGS

The family name is based on *Bombycilla*.

In Europe just Waxwing normally suffices, because there is only the one species, but the Americans have two, and one of those is the same species as the Eurasian bird. Since the Cedar Waxing occasionally makes its way to Europe, we sometimes need to distinguish them.

(Bohemian) Waxwing *Bombycilla garrulus*

This species is an irregular winter visitor, arriving during 'irruptions'. These are erratic and spontaneous migratory movements that are provoked by events such as food shortages and extreme weather.

Bohemian is used in the sense of 'wandering' or 'nomadic' because the name was once a French synonym for their Romany population, who were reputed to have arrived via Bohemia.

Waxwing comes from the little wax-like red blobs on the wings.

Bombycilla derives from *bombyx* (silkworm) and *cilla* (tail), an idea originally rooted in the German word *Seidenschwanz* (silk-tail) for the species.

The specific name **garrulus** is evidence of Linnaeus's mistaken assumption that the Waxwing was related to the Jay *Garrulus glandarius* (see above). This association explains the origin of the archaic name *Waxen Chatterer*, which seems odd for a fairly quiet bird.

Cedar Waxwing *Bombycilla cedrorum*

This is an American species that has been recorded twice in Britain.

The berry of the cedar tree is a main food among the variety of fruits eaten, hence the names **Cedar** Waxwing and *cedrorum* (of the cedar).

63. HYPOCOLIIDAE: Hypocolius

The family name is based on *Hypocolius*.

(Grey) Hypocolius *Hypocolius ampelinus*

This is a Middle Eastern species.

The mystery is why the word **Grey** remains at all: this is a unique species in a unique genus.

The word *Hypocolius* suggests that the bird 'resembles a Mousebird' (*Colius*). Its exact taxonomy has always been an enigma. That hesitation accounts for the fact that it is also likened to the waxwings in the specific name *ampelinus*.

* *Ampelinus* is a Greek name found in Aristophanes and was originally used by Linnaeus as the generic name of the Waxwing and its supposed allies. It seems that his classification in this area was confusing, so Vieillot later created a new genus, *Bombycilla*, for the Waxwing, and that is now preferred.

64. PARIDAE: Tits and Chickadees

The family name is based on *Parus*.

Parus was originally a Latin generalisation for birds of the tit family. It was applied by Linnaeus as a generic name for the whole family. Recent DNA analysis has caused the family to be subdivided into 11 genera, with *Parus* now restricted to just four species.

In Britain the accepted vernacular form for this family is now 'Tit', which first appeared alongside Titmouse in 1703. For much of the nineteenth century either form was acceptable, but Victorian prudery seems to have caused the shorter form to be off-limits for some decades until about 1930: the older and longer forms 'Titmouse/Titmice' appeared in Coward's 1926 list as *Willow-Titmouse*, and so on (complete with hyphen). Those forms slowly faded from fashion, but were still seen regularly until at least the 1950s.

The truth is that neither form is strictly accurate. For an explanation of that we must go back to Medieval English to find the root word *Titmose*. In this case *tit* was 'little' and *mose* was the bird. It was only later that *mose* was aligned (or confused) with 'mouse' and gained its irregular plural. Therefore 'tit' is only half the story, an adjective bereft of a noun, while 'titmouse' is something of a muddled idea. Be that as it may, the Americans still use titmouse in full for some species and a locally coined echoic word 'chickadee' for others. In short, any coyness about the use of 'tit' has nothing to do with the true roots of the word in this context.

Coal Tit *Periparus ater*

A common resident in Britain.

Coal tells of the black head and slightly sootier look of this tiny species, which is reinforced by the specific name *ater* (black).

Periparus tells us that there is a relationship to the *Parus* genus.

(European) Crested Tit *Lophophanes cristatus*

This species is limited in Britain to the Caledonian forests, but has more catholic habits in mainland Europe.

European is needed only when looking at a wider range of species.

Crested and *cristatus* mean the same.

The generic name *Lophophanes* repeats the idea in a Greek form, which means 'showing a crest'.

Sombre Tit *Poecile lugubris*

This is a species of the Eastern Mediterranean and Middle East.

Sombre and *lugubris* both underline the greyness of this species.

Grey-headed chickadee *Poecile cinctus*
(Siberian Tit)

This Palearctic species occurs from Scandinavia and across Siberia. For some reason, the North American name has now become the international usage.

It was originally collected in Siberia and was known for many years as the **Siberian Tit**, and that is indeed the range of the nominate form. It does not appear on the British list, but that name is still preferred in the *Collins Bird Guide* (third edition, 2022).

Grey-headed is self-evident, though the grey is distinctly brown-tinged.

Chickadee is echoic of the call of the Black-capped Chickadee. The word came into use in about 1834, but it is unclear where or how it originated.

* The specific name *cinctus* means 'banded'. Buffon described it as 'white-banded' for its broad white cheek.

Marsh Tit *Poecile palustris*

This is a widespread species, though it is patchy in distribution.

Marsh is a misnomer, since this species prefers damp, deciduous woodland. The mistake is perpetuated in *palustris*, which also implies a marshy habitat.

Poecile was one of those names retrieved, as *poikilis*, from Greek ornithology and representing a small bird.

Caspian Tit *Poecile hyrcanus*

This is a species of the coastal forests of the southern **Caspian** Sea.

That geographical theme is echoed in *hyrcanus*: to the Ancient Greeks, the people of that region were the Hyrcani.

Willow Tit *Poecile montanus*

This is a species in considerable decline in many parts of Britain.

Willow is not an unreasonable tree to link to this species, which is found in damp, lowland habitats.

On the other hand, the specific name *montanus* tells that this species may also be found at altitude and may also frequent conifers.

African Blue Tit *Cyanistes teneriffae*

The African form is found in Morocco and the Canaries.

Cyanistes: see below.

The specific name *teneriffae* relates to the island of Tenerife.

(Eurasian) Blue Tit *Cyanistes caeruleus*

This is one of the most familiar species in Britain.

The fact that it has several relatives gives the occasional need for **Eurasian**.

The **Blue** Tit has a considerable amount of blue (though it sports plenty of yellow too).

Cyanistes is dark blue in Greek, while *caeruleus* says it again in Latin.

Azure Tit *Cyanistes cyanus*

This Eastern European species has a subtly different blue from the previous species, as the words **Azure** and *cyanus* imply.

Great Tit *Parus major*

This is a common and widespread species.

Great is a suitable name for the largest of the family and that is supported by the specific name *major*.

Parus: Today this once-overarching genus of about 50 species has been reduced by the 'taxonomic industry' to just four species, a story that parallels the fate of the warbler genus *Sylvia* (see later).

65. REMIZIDAE: Penduline Tits

The family name is based on *Remiz*.

Eurasian Penduline Tit *Remiz pendulinus*

This species is a scarce visitor to Britain from the Continent, where it is widespread.

Eurasian: there are a surprising number of species of penduline tit, of which this is the most westerly species.

Tit here is a name of convenience as it is very tit-like, but is not a member of the Paridae.

Penduline and *pendulinus* refer to the hanging nest.

Remiz is the Polish name for the species, a reflection of its easterly bias within Europe.

66. PANURIDAE: Reedling

The family name is based on *Panurus*.

There is only the one species in this family.

Bearded Reedling *Panurus biarmicus* (INTERNATIONAL NAME)
Bearded Tit (BRITISH USAGE)

This is a scarce and local species in Britain.

Reedling is a name imposed by the IOC in the interests of underlining the uniqueness of this species. The word was originally an East Anglian name used without qualification and is a unique name in itself for a small bird living in reeds. The diminutive ending *-ling* is also found in such names as Starling and Sanderling to indicate smallness. (Starling was originally the young Stare.)

Bearded was originally a way of separating this from the other 'tit' species (though the 'beard' is more of a 'moustache'). It originated in the French *mésange barbue* and was translated into *Parus barbatus* in 1760, before being anglicised in 1771 as *Bearded Titmouse*.

Panurus was created to convey the sense of 'long-tailed', but seemingly with a nod to *Parus*.

Before the adoption of the name Bearded, Linnaeus had already contributed the idea of *biarmicus*, in the mistaken belief that the bird was found in the Bjarmaland area of Russia. That error was later compounded by Temminck, who took it to mean 'bearded' and used it with that intention in the naming of the Lanner Falcon, *Falco biarmicus*.

67. ALAUDIDAE: LARKS

The family name is based on *Alauda*.

The 'lark' (which is primarily the Skylark in the English tradition) inspires some of the most beautiful poetic and musical imagery in British culture. Its appeal lies in the panache of its exuberant and trilling song which is delivered on the wing as the male rises so high as to become invisible. It personifies sweetness, exuberance, aspiration and towering love – and of course metaphorically links earth to heaven in the Christian tradition.

The family is very much under-represented in Britain, with just Skylark and Woodlark as breeding species, plus the odd regular visitors. Within the Western Palearctic as a whole there are many more species and some of those have reached Britain.

In Western Europe, lark species become more abundant in Mediterranean regions. In her monograph on the Montagu's Harrier, Elvira Werkman recounts the unexpected discovery that abundant lark species, many of those also on migration, provide the main diet for Montagu's Harriers during their spring-time stopover in Morocco.

Greater Hoopoe Lark *Alaemon alaudipes*

This is a North African species.

Greater also has a *Lesser* counterpart.

This lark has pied wings and a flight style reminiscent of a **Hoopoe**.

Alaemon is from the Greek for 'a wanderer'.

The specific name *alaudipes* tells us that it is 'lark-footed'.

Thick-billed Lark *Ramphocoris clotbey*

This is a North African species.

Thick-billed is self-evident.

Ramphocoris is composed of the Greek *rhamphos* (bill), and the Modern Latin *coryx* (lark).

The specific name *clotbey* is a real curio, because it unites the Ottoman honorific *Bey* with the name of an esteemed French physician, Antoine-Barthélmy Clot (1778–1858), who was awarded the title for his services to Egyptian medicine under the Viceroy, Muhammad Ali Pasha.

Desert Lark *Ammomanes deserti*

This is a North African species.

Both **Desert** and *deserti* tell of a habitat.

Ammomanes reinforces that image by stating that these birds have 'a passion/mania for sand'.

Bar-tailed Lark *Ammomanes cinctura*

This is a North African species.

Bar-tailed is reinforced by the specific form *cinctura*, which also tells of a 'banded tail'.

Black-crowned Sparrow-Lark *Eremopterix nigriceps*

This species has bred in Israel.

Black-crowned is applicable to the male, and that is reinforced by the specific name *nigriceps* (black-headed).

Sparrow-Lark refers to the sparrow-like shape and 'jizz' of this species.

Eremopterix is composed of *erēmos* (desert) and *pterix* (wing, i.e. bird).

Chestnut-headed Sparrow-Lark *Eremopterix signatus*

This East African species has been recorded in Israel.

Chestnut-headed is self-evident.

The specific name *signatus* suggests that the bird stands out as 'well-marked': the bright chestnut head-patches are offset by clean white.

Kordofan Lark *Mirafra cordofanica*

This is as sub-Saharan species whose presence in the northern part of Mauretania is disputed.

The name **Kordofan** comes from a former province of Sudan.

That is echoed in the specific name *cordofanica*.

The word *Mirafra* offers more of a challenge, since its origins have been lost since it was first used by Horsfield for *Mirafra javanica*, Jobling favours the idea that it originated in some unidentified local use in Southeast Asia. An alternative idea suggests that the *-afra* element refers to Africa.

Woodlark *Lullula arborea*

This is a partial migrant in Britain, but also breeds regularly in southern heathlands.

Woodlark was probably in use as a single word before Skylark came into use. Together with the specific form, *arborea*, it is clear that this species is more associated with trees.

Lullula was contrived from *Lulu*, an eighteenth-century onomatopoeia based on its lovely fluting notes. The name was created by Buffon.

White-winged Lark *Alauda leucoptera*

This Asian species has been recorded in Britain.

White-winged and *leucoptera* say the same thing.

Alauda: see Eurasian Skylark.

Raso Lark *Alauda razae*

This is a species limited to the Cape Verde Islands.

The bird is found only on the uninhabited **Raso Island**.

Alauda: see Eurasian Skylark.

The specific name *razae* Latinises the name of the island.

Oriental Skylark *Alauda gulgula*

This bird has been noted in Israel.

Oriental indicates an eastern range.

Alauda: see Eurasian Skylark.

The specific name *gulgula* may relate to the song (via *gula* throat), but may also have links to *Gulgul*, a name of the Indian Mynah.

Eurasian Skylark *Alauda arvensis*

This is the 'common' lark in Britain and its root name radiates into those of the entire family.

Eurasian now separates this from the Oriental form.

Skylark, as a specific name, dates only from 1678, when Ray modelled it on Gessner's *Himmellerch*. Previously the word **lark** served alone. That is a deeply rooted Germanic form.

Alauda was the original Latin name, but the Roman Pliny left a comment to the effect that the Latin name itself was rooted in a Celtic word meaning 'a great songstress'.

The specific name *arvensis* describes this bird as being 'of the fields'.

Thekla's Lark *Galerida theklae*

This species is found in Iberia and North Africa.

A recent adjustment changed this name by the addition of the possessive. This reminds us that the name was dedicated as a memorial to **Thekla** Brehm (1833–1857) by her father Christian Brehm and brother Alfred, who were both German naturalists. It seems probable that Alfred collected a specimen during a prolonged trip to Spain in 1856 and sent a specimen to his more sedentary ornithologist-father.

Galerida is from the Latin *galerita* for 'a lark with a crest'.

* Alfred's brother Reinhold, travelled with him and settled in Spain to later name the Spanish Imperial Eagle, which is recorded under his father's name.

Crested Lark *Galerida cristata*

This species has been recorded in Britain.

Crested and *cristata* say the same thing. The long crest is visible even when not erected.

Maghreb Lark *Galerida macrorhyncha*

This is a species of the **Maghreb** region of north-west Africa (see also Maghreb Magpie).

The specific name *macrorhyncha* refers to the bird's long bill.

Horned Lark *Eremophila alpestris* (INTERNATIONAL NAME)
Shore Lark (BRITISH NAME)

This is a scarce but regular winter visitor to Britain. Most orginate from Europe, but the American form has also been recorded.

Horned refers to the pattern of horn-like black feathers worn by the male in breeding plumage.

Shore derives from the fact that these birds are traditionally found on beaches and coastal dunes when wintering in Britain.

Another word containing *erēmos*, *Eremophila*, tells us that this is a 'desert lover'.

The specific name *alpestris* tells us that the species is found at 'high altitude'.

The names of this bird contain three apparently contradictory habitat indicators, but the total picture is of a widespread bird found in remote areas.

* There are in fact over 40 subspecies, with over half of those in the Americas. A loose convention separates the New World forms as the *Horned Lark* and the Eurasian forms as *Shore Lark*. It would not be surprising to see some future radical reorganisation of this complex and the resultant formalisation of *Shore Lark*.

Temminck's Lark *Eremophila bilopha*

This is a North African and Arabian species, closely related to Horned Lark.

C.J. **Temminck** (1778–1858) was a Dutch ornithologist.

The specific name *bilopha* means 'double-crested' (see *Horned* above).

Hume's Short-toed Lark *Calandrella acutirostris*

This Central Asian species has been recorded in Israel.

Allan **Hume** (1829–1912) was a British ornithologist working in India.

For **Short-toed**, see next species.

The specific name *acutirostris* means 'sharp-billed'.

(Greater) Short-toed Lark *Calandrella brachydactyla*

This is a regular vagrant to Britain.

Greater originally provided a simple formal distinction from the *Lesser Short-toed Lark*, but the latter species has now been split and renamed (see below).

Short-toed, and the specific name *brachydactyla*, describe the same feature.

Calandrella is a diminutive form of *Calandra* (see below).

Bimaculated Lark *Melanocorypha bimaculata*

This is the Eastern equivalent of Calandra Lark.

Bimaculated and *bimaculata* refer to the twin dark spots on the sides of the breast. Confusingly, these are even more pronounced in the Calandra Lark.

Melanocorypha is an enigmatic name: it uses a combination of the Greek *melas* (black) with *koruphos*, a name used by Hesychius for an unidentified bird, but here confused with *korudos* (lark).

Calandra Lark *Melanocorypha calandra*

This species is found in south-west and south-east Europe.

Calandra is from the Greek *kalandros* (lark).

Black Lark *Melanocorypha yeltoniensis*

This Asian species has been recorded in Britain.

Black is self-evident.

The specific name *yeltoniensis* refers to Lake Yelton in Astrakhan, Central Asia.

Dupont's Lark *Chersophilus duponti*

This species is found in Spain and North Africa.

L-P. **Dupont** (1795–1828) was a French naturalist and collector.

Chersophilus means that this is a 'lover of bare land'.

Dunn's Lark *Eremalauda dunni*

This is North African species that is found in the western Sahara, Morocco and Algeria. The species was split in 2022, with the Middle-Eastern form becoming the Arabian Lark (see next).

Dunn's and *dunni* commemorate Captain H.N. **Dunn** (1864–1952), a British hunter and collector.

Eremalauda tells us that this is a 'desert lark' (Greek *erēmos*, desert, plus *alauda*).

Arabian Lark *Eremalauda eremodites*

This was split from Dunn's Lark in 2022.

The name **Arabian** is a little narrow, since the bird's range extends into Syria, Israel, Lebanon and Jordan.

The specific name *eremodites* means 'desert-diver'. Many lark species conclude their song flight with a swift drop to the ground.

Mediterranean Short-toed Lark *Alaudala rufescens*

This species has been recorded in Britain.

Mediterranean now refers to the European form of what was the *Lesser Short-toed Lark* complex, which was split into several species in 2020. The name *Lesser* no longer applies to any of this genus.

Alaudala is the diminutive form of Alauda.

The specific name *rufescens* is appropriate to the red-brown cap of this form.

Turkestan Short-toed Lark *Alaudala heinei*

Until 2020 this was part of the *Lesser Short-toed* complex (see previous species).

The species is not limited to **Turkestan**: there is a breeding population in Ukraine and there have been occasional records from other European countries, including Belgium.

The specific name *heini* commemorates Ferdinand Heine (1809–1894), a German ornithologist and a prolific collector.

* The proliferation of 'short-toed' larks raises the question of whether the names would be any less efficient if 'short-toed' were dropped altogether to give Hume's Lark, Mediterranean Lark, Turkestan Lark and so on. In that case *Calandrella brachydactyla* could revert to its original form as the simple Short-toed Lark, with no further need of the word Greater.

Bulbuls to
Silviid Babblers

68. PYCONOTIDAE: Bulbuls

The family name is based on *Pyconotus*.

This large family of about 160 species is found across Africa and Asia. It was first introduced by the English zoologist, G.R. Gray, who considered it to be a subfamily of the thrushes. Modern taxonomy places these birds closer to the larks and hirundines.

Three species occur naturally within the region and two more occur as introduced feral species. Lees and Gilroy (2021) describe them as largely sedentary, so they are little known in Western Europe, though it is predicted that Common Bulbul may soon colonise Spain from North Africa.

Red-whiskered Bulbul *Pyconotus jocosus*

This oriental species is represented in Spain, where an expanding feral population has been recorded near Valencia. It was added to the Spanish list in 2019.

Red-whiskered refers to the red-stripe behind the ear.

Bulbul originally meant 'nightingale' in Persian but evolved to be used for a different family of birds here.

Pyconotus means 'thick-backed', 'compact'.

The specific name *jocosus* implies merry or jovial – as apparently befits a bewhiskered figure.

Red-vented Bulbul *Pyconotus cafer*

This Asian species has been introduced to several other regions where it lives as a successful feral species. It is found in Kuwait and the Canary Islands, for example. It is a notable fruit-pest.

Red-vented is for the red undertail of an otherwise dark, white-rumped bird.

The specific name *cafer* appears to be erroneous, in that the word is of Arabic origin and is associated with South Africa, with which this species has no connection. The error was originally Brisson's, but was perpetuated by Linnaeus.

White-Spectacled Bulbul *Pyconotus xanthopygos*

This species is found in the Eastern Mediterranean and Egypt.

White-spectacled describes a wide white eye-ring.

The specific name *xanthopygos* means 'yellow-rumped', though the yellow is on the vent rather than the rump.

White-eared Bulbul *Pyconotus leucotis*

This is a Middle-Eastern species.

White-eared describes the white patch on the ear-coverts and is restated with *leucotis*.

Common Bulbul *Pyconotus barbatus*

This is a North African species.

Common is self-evident.

The specific name *barbatus* means bearded, though the bearded effect is created by the darker shading on the throat rather than by any longer plumage.

69. HIRUNDINIDAE:
Swallows and Martins

The family name is based on *Hirundo*.

Three members of this family normally breed in Britain, where the names swallow and martin became the established usage. Today those names are collective terms used for many more of their relatives. The two words represent the two main roots of Modern English, *swallow* from its Germanic, Anglo-Saxon lineage, and *martin* from the Latin-based Old French language, which was introduced after the Norman Conquest in 1066.

Southern Europe has several more species, some of which show up as vagrants to Britain, as do a few American species.

Banded Martin *Neophedina cincta*

This sub-Saharan species was recorded for the first time in Egypt in 2019.

Banded describes the broad V-shaped chest band and is reinforced by the specific name *cincta*, which has the same meaning.

For **Martin** see next species.

Neophedina adds *neo* (new) to the genus *Phedina*. The latter was concocted by Bonaparte, meaning brown swallow (from Greek *phaios*, brown, and Italian *rondine*, swallow).

Sand Martin *Riparia riparia*

This is a common summer visitor.

Sand tells of the bird's preference for sandy cliffs for nesting. It was a relatively late-coming usage, first in evidence in Charleton in 1668.

Martin arrived on the scene much later than Swallow. It was imported into English from Old French. Its French origins, according to Buffon, link to the proper name and possibly to *Saint* Martin. This is very much in the tradition of attaching proper names to familiar species. The word was later formally applied to the two smaller British species of swallow, *House* and *Sand* Martins. (In Modern French, forms of the word are connected with swifts and kingfishers, but no longer with swallows of any sort.)

Riparia is the Latin for 'a river-bank dweller'.

* Lockwood points out that the Old English, tenth-century, name for the bird was *staep-swealwe* and that '*bank*' replaced the *staep* element in Middle English to become *Bank Swallow*. That was long used in England until Sand Martin was formally adopted in the eighteenth century. The older name then travelled to North America with English settlers and has been preserved as the preferred name there. It is clearly much closer to the Latin form.

* In spite of the wide separation of the two populations, the species is considered to be monotypic.

Brown-throated Martin *Riparia paludicola*

This species is found in western Morocco.

Brown-throated describes a feature that contrasts with the white throat of the similar Sand Martin.

The specific name *paludicola* tells us that this is a marsh-dweller.

Pale Martin *Riparia diluta*

This is an Asian species, which has been recorded in Kuwait.

This is a **pale** relative of the Sand Martin.

The specific name *diluta* suggests that the colour is diluted.

Grey-throated Martin *Riparia chinensis*

This Asian species has been recorded in Kuwait and Israel.

Grey-throated is self-evident.

The specific name *chinensis* is used very loosely to mean 'eastern': while it is found in south-west China, the main range is between Pakistan and Southeast Asia.

Tree Swallow *Tachycineta bicolor*

This American species has been recorded as a vagrant to Britain.

Tree Swallow relates to the tree-holes in which it naturally nests.

For **Swallow** see Barn Swallow (below).

Tachycineta suggests that birds of this genus are 'fast-moving'.

The specific name *bicolor* describes a bird with two very distinct colour zones: grey-blue above and clean white below.

Purple Martin *Progne subis*

This American species has been recorded as a vagrant to Britain.

Purple is self-evident.

Progne is a name for a Swallow, which derives from a Greek legend in which Tereus ravishes his sister-in-law, Philomena, and provokes a dreadful vengeance from his wife, Procne (or Progne). As the gods settle the chaos, the main protagonists are transmuted into birds. Procne becomes a swallow.

The specific name, *subis*, is a curious one, since it relates to a legendary bird of great courage. Native Americans traditionally hung up hollow gourds to encourage these birds into their villages: the name *subis* implies that they chase off predators, but at a more mundane level they also eat nuisance flies. Today, Purple Martin numbers are sustained by nest boxes provided in gardens and public places.

* Because of their site loyalty, these were the first birds in the Americas to be tracked by simple geolocators, which could be collected and analysed on their return to the nest site.

Eurasian Crag Martin *Ptyonoprogne rupestris*

This species has been recorded in Britain.

Eurasian reminds us that there are other Crag Martin species.

Crag links this species to the rocky habitat where it nests and that is reinforced by the specific name *rupestris* (rock-dweller).

Ptyonoprogne describes a 'fan-tailed swallow' (Greek *ptuon*, fan, and *progne*, swallow; see Purple Martin above).

> The next two entries are complicated: Pale Crag Martin has been split from Rock Martin and is treated as the more northerly form. In the *Collins Bird Guide* (third edition, 2022) the distinction is not acknowledged, while some online references show Pale Crag Martin as a Western Palearctic species, but now exclude Rock Martin.

Pale Crag Martin *Ptyonoprogne obsoleta*

This form is found widely from Morocco to Arabia.

Unsurprisingly, it is a notably **pale** version of the Eurasian Crag Martin.

The specific name *obsoleta* means 'plain' with a hint of 'worn out'.

Rock Martin *Ptyonoprogne fuligula*

See note above.

Rock, like Crag, is for habitat.

The specific name *fuligula* means sooty-throated, which is truer of the southern form, but is inappropriate to the disputed Pale Crag Martin.

Barn Swallow *Hirundo rustica*

This is a common species in the British summer, but has declined in many areas.

The use of the prefix **Barn** is more necessary in America, which has to distinguish this from a number of other species, but the word was originally used in England, alongside names such as *Chimney Swallow*, to underline the attachment of the species to the shelter provided by buildings.

Swallow is a word of Germanic origins, meaning 'cleft stick' (after the forked tail). It was already in evidence in Old English as *swealwe* and evolved from that into the modern form.

Hirundo is Latin for 'swallow'.

The specific name *rustica* implies that it always had rural preferences when compared with the House Martin.

* In this case the American form is a different subspecies from that found in Britain. There are eight subspecies in all, three of which occur regularly in the Western Palearctic.

Ethiopian Swallow *Hirundo aethiopica*

This African species has been recorded in Israel.

Ethiopian and *aethiopica* mean the same thing. However, the bird has a much wider range through more than 20 other African countries.

(Western) House Martin *Delichon urbicum*

This is a common summer visitor, though it is declining in some areas.

The word **Western** is attached internationally to acknowledge the fact that three other species of House Martin are now recognised.

House and *urbicum* (town-dwelling) contrast this species with the Barn Swallow, because they more readily nest on dwellings, which serve as an alternative to rock faces. The attachment of the word *House* is first seen in Gilbert White in 1767. He recorded the behaviour of the birds in his home village of Selborne, in Hampshire in what was, in effect, the first modern monograph study of a bird species.

* In 1822, Boie wanted to underline the differences between the Barn Swallow and the House Martin, so he created an anagram of the Greek-based *Chelidon* (then preferred for the swallows). From that, *Delichon* was born. Ironically *Chelidon* was eventually replaced by the Latin *Hirundo*, while *Delichon* remains. Similar anagrams are found in *Dacelo* (from *Alcedo*) in the kingfisher family and in *Zapornia* (from *Porzana*) in the crakes.

Asian House Martin *Delichon dasypus*

As the name **Asian** implies, this species has a far more easterly range than the previous species, yet it has been recorded in Israel.

The specific name *dasypus* suggests that the bird has 'hairy feet': the legs are, of course, heavily feathered (as are those of *D. urbicum*).

Red-rumped Swallow *Cecropis daurica*

This is a relatively frequent vagrant to Britain.

Red-rumped is self-evident.

Cecropis relates to Cecrops, the mythical founder of Athens, half man, half creature, whose story embodies some of the basic values that underpinned classical Athenian life.

The specific name *daurica* (this time in a simplified spelling) refers to the region of Dauria, to the east of Lake Baikal in Russia. The Finnish-Swede Erik Laxmann obtained a specimen during a 1769 expedition to Siberia.

American Cliff Swallow *Petrochelidon pyrrhonota*

This species has been recorded as a vagrant to Britain.

American distinguishes this from other cliff swallow species.

Cliff echoes the names Crag and Rock above in indicating a nesting habitat (though they do now use buildings as cliff substitutes).

Petrochelidon combines *petros* (rock) with *Chelidon* (swallow).

The specific name *pyrrhonota* tells that the bird is 'flame backed'.

Preuss's Cliff Swallow *Petrochelidon preussi*

This is a sub-Saharan species that was recorded on Cape Verde Islands in 2020.

Paul **Preuss** (1861–1926) was a German botanist and collector in West Africa and elsewhere.

Streak-throated Swallow *Petrochelidon fluvicola*

This South Asian species has been recorded in Turkey, Kuwait and Egypt.

Streak-throated is an obvious description of a bird otherwise known as the *Indian Cliff Swallow*.

The specific name *fluvicola* means 'river-dweller'.

WARBLER: A CHANGE OF VALUE

Warbler was used as a collective name by Pennant in 1773. Its roots can be found in Norman French, in the verb *werbler* (to sing or play melodiously). Pennant's intention was to create a word that had a value similar to woodpecker or thrush – that is, as a collective vernacular name for a group of birds considered to be closely related. The concept pertained for many years in an Old World context, but in recent years the taxonomy of that group of birds has been reviewed completely. Old World warblers have become a series of divided and rather disparate families. Consequently, the word 'warbler' now has a much looser meaning, so it is more conventional to qualify the various forms with words such as 'bush', 'leaf' and so on.

70. CETTIIDAE: Bush Warblers etc.

The family name is based on *Cettia*.

Cetti's Warbler *Cettia cetti*

This non-migratory species colonised Britain in the latter half of the twentieth century.

Francesco **Cetti** (1726–1788) was an Italian mathematician and zoologist who observed the species in Sicily. The triple use of his name here is unusual. What is more, the birds sometimes seem to call his name.

71. SCOTOCERCIDAE: SCRUB WARBLER

This is a relatively new family name, based on *Scotacerca*.

According to the IOC there is just one species, **Streaked Scrub Warbler** *Scotacerca inquieta*, but that is flagged for a future split. Research is ongoing, but there appears to be compelling evidence for this and even further splits. The *Collins Bird Guide* (third edition, 2022) treats the split as having been completed and that is the pattern followed here, since wider acceptance seems probable.

Levant Scrub Warbler *Scotacerca inquieta*

The range of this species extends from Egypt into Arabia.

Levant echoes Levant Sparrowhawk in relating to countries of the Eastern Mediterranean. It originates as a figurative reference to the rising sun (from the Latin *levare*, to rise or lift).

Scrub describes an uncultivated habitat with generally sparse, low vegetation.

Warbler suggests that it is a passable songster or is otherwise warbler-like. (For warbler, see Wood Warbler below.)

Scotocerca indicates that the bird has a dark tail (Greek *skotos*, dark, and *kerkos*, tail).

The specific name *inquieta* describes the bird as 'restless'.

Saharan Scrub Warbler *Scotocerca saharae*

As the names imply, this species ranges across the northern fringes of the **Sahara**. It was previously treated as a subspecies of *inquieta*.

72. AEGITHALIDAE: BUSH TITS

The family name is based on *Aegithalos*.

This is a small, widespread, mainly Asian family, with just one representative in America (the Bushtit) and one in Europe.

Long-tailed Tit *Aegithalos caudatus*

This is a common species in Britain.

Long-tailed is self-evident.

Tit is a traditional usage that is applied to several families of small birds (see Paridae).

Aegithalos is a Greek name for 'tit', which was used by Aristotle.

The specific name *caudatus* means 'tailed'. The length is assumed.

* This familiar species had a variety of attractive local folk names that described assorted characteristics, examples being *Kitty Longtail* for shape, *Creak Mouse* for sound, *Bum Barrel* for nest. (*Mouse*, of course, relates to *Titmouse*, *mose* being the Old English name used for this and other tit species – see Paridae.)

73. PHYLLOSCOPIDAE: Leaf Warblers

The family name is based on *Phylloscopus*.

Leaf warblers form a large family of over 80 species, many of which are highly migratory. As a result, a number of vagrants from the east swell the British list and provide immense challenges for birdwatchers. As their collective name implies, they are usually found in foliage, but are also very active and acrobatic. In birdwatcher jargon, these are often known as 'phylloscs'.

Wood Warbler *Phylloscopus sibilatrix*

This is a summer migrant to Britain, with a bias towards damp western woodlands.

Wood relates to the woodland habitat preferred by this bird.

Warbler was initially a collective name adopted by Pennant in 1773. Its roots can be found in Norman French, in the verb *werbler* (to sing or play melodiously).

Phylloscopus means 'leaf-searching', which relates to the feeding style of these birds, which gather insects, bugs and so on as they glean through the foliage.

The specific name *sibilatrix* means 'whistler'. Gilbert White referred to 'a sibilous grasshopper-like noise' that precedes the rattling part of the song, which has been likened to a small coin spinning on a tin-lid.

Western Bonelli's Warbler *Phylloscopus bonelli*

This species has been recorded in Britain.

It ranges through the **Western** Mediterranean.

Franco **Bonelli** (1784–1830) was an Italian ornithologist and entomologist. He is also commemorated in the name of the eagle.

Eastern Bonelli's Warbler *Phylloscopus orientalis*

This species has been recorded in Britain.

It ranges through the **Eastern** Mediterranean, a fact that is supported by the specific name *orientalis*.

Hume's Leaf Warbler *Phylloscopus humei*

This Central Asian species has been recorded in Britain.

Hume's and *humei* both commemorate Allan **Hume** (1829–1912), a British ornithologist based in India.

Yellow-browed Warbler *Phylloscopus inornatus*

This Siberian species is an increasingly regular passage visitor to Britain.

Yellow-browed describes a variable supercilium stripe.

The specific name *inornatus* rather unjustly describes the bird as 'plain': as well as the eye-stripe, it has a yellowish double wing-bar.

Pallas's Leaf Warbler *Phylloscopus proregulus*

This Siberian species is recorded fairly regularly in Britain.

Peter **Pallas** (1741–1811) was a Prussian zoologist who spent most of his life in Russia.

The specific name *proregulus* draws attention to its similarity to the 'crests' of the genus *Regulus*: it is similar in shape, size and behaviour, and has a yellow crown-stripe.

Radde's Warbler *Phylloscopus schwarzi*

This Siberian species is a regular but scarce visitor to Britain during migration.

The names commemorate both the finder, the German naturalist Gustave **Radde** (1831–1903), and the leader of his expedition, the German astronomer Ludwig **Schwarz** (1824–1894). It is unusual in being one of very few names where two individuals are recorded.

Sulphur-bellied Warbler *Phylloscopus griseolus*

This species was recorded in Britain in 2019. It normally breeds in Central Asia and winters in the Indian subcontinent.

Sulphur-bellied describes a bird that wears a mid-yellow wash.

The rest of the description is furnished by *griseolus*, which means grey, but is used in a diminutive form to emphasise the smallness of this bird.

Dusky Warbler *Phylloscopus fuscatus*

This Siberian species is another regular but scarce visitor to Britain during migration.

Dusky and *fuscatus* both describe this as a 'darkish' bird.

Plain Leaf Warbler *Phylloscopus neglectus*

This Central Asian species has been recorded in Sweden.

Plain is self-evident.

The specific name *neglectus* suggests that it is easily 'neglected' or 'overlooked'.

Willow Warbler *Phylloscopus trochilus*

This is a fairly common summer visitor.

Willow and birch are often favoured habitat trees.

The specific name *trochilus* is derived from a Greek word for 'wren' and was used to acknowledge the fact that until the end of the eighteenth century this was often called the Willow *Wren*.

* Wood Warbler, Willow Warbler and Chiffchaff were not distinguished properly before Gilbert White separated them in the late eighteenth century. They were often collectively known as *Lesser Pettychaps*, Garden Warbler being the *Greater Pettychaps*. Today an unidentified sighting might be reported as a 'willow-chaff'.

The name Chiffchaff was first applied to the form known in Britain, the Common Chiffchaff. Some closely related species are now named separately, with other qualifying names necessitating the addition of Common to the original local folk name. The name Chiffchaff was originally perpetuated by Gilbert White as a way to distinguish it from two very close allies, which had long been known under a single umbrella name (see Willow Warbler, above). In short, 'chiffchaffs' collectively are not a separate group within the leaf-warbler family and might readily go by plain 'warbler' names, like so many of their congeners. It seems possible that further studies of the current subspecies of Common Chiffchaff might yet see more species similarly named.

Mountain Chiffchaff *Phylloscopus sindianus lorenzii*

The vernacular name **Mountain** related initially to the Caucasian subspecies, while the nominate form, *sindianus*, was named for the Sindh Province (in modern Pakistan).

For **Chiffchaff** see below. The subspecies name *lorenzii* commemorates the German zoologist Theodore K. Lorenz (1842–1909).

Canary Islands Chiffchaff *Phylloscopus canariensis*

This sedentary species is endemic to the **Canary Islands**.

For **Chiffchaff** see below.

Common Chiffchaff *Phylloscopus collybita*

This is one of our commonest warblers.

The word **Common** is now used increasingly, as some forms have been split and others can be identified by plumage and voice.

The name **Chiffchaff** was popularised for this species by Gilbert White. It was probably a local Hampshire name. It normally has a regular two-note song that seems to say that name quite clearly (though the same concept is rendered slightly differently in some other European languages, e.g. *Zilpzalp* in German and *Siffsaff* in Welsh). In other related species the song may be somewhat similar in parts or in general, but with some distinct differences.

A local Norman name, *compteur d'argent*, likened the two-note song to the sound of chinking coins when counting money. That idea was used by Vieillot to form the specific name *collybita*, which is based on the Latin for 'money changer'.

Scandinavian Chiffchaff *Phylloscopus collybita abietinus*

Many of the Chiffchaffs overwintering in Britain are of the Scandinavian race, *abietinus*, a name derived from the Latin for 'fir tree'.

Siberian Chiffchaff *Phylloscopus collybita tristis*

This Eastern form is often identified by plumage and voice among other Chiffchaffs during a British winter, and is reported under the name **Siberian**. While it is currently treated as a subspecies of Common Chiffchaff, it may eventually be split.

The subspecific name *tristis* (sad) is for the grey tones of its plumage.

Iberian Chiffchaff *Phylloscopus ibericus*

This species has been recorded in Britain, usually identified by its different song.

This one has full species status and is normally found in the **Iberian Peninsula** (hence the specific names) and in North Africa.

Eastern Crowned Warbler *Phylloscopus coronatus*

This Far Eastern species has been recorded in Britain.

Eastern because there is also a Western form.

Crowned and *coronatus* say the same thing. The 'crown' pattern is a pale stripe running through a darker olive-grey cap that is clearly demarked by pale yellow supercilium stripes.

Green Warbler *Phylloscopus nitidus*

This Caucasian species has been recorded in Britain.

Green is very obvious, but there is always a trap in the similar name Green*ish* Warbler (below), which is another species.

The specific name, *nitidus*, means 'shining. The upper parts are in fact a brighter green than the Greenish Warbler.

Two-barred Warbler *Phylloscopus plumbeitarsus*

This is a rare visitor to Britain. It breeds in Mongolia, Manchuria and southern Siberia.

The worrying feature about the splits of recent years is the proliferation of extra descriptive words, an example being (Eastern) Bonelli's. This species was long seen as a subspecies of Greenish Warbler. In this case, the IOC avoided both the clumsiness and confusion of Two-Barred Greenish Warbler (which had been widely used previously for the then subspecies) by omitting *Greenish* and settling for **Two-barred** Warbler – and that is perfectly adequate. Beware confusion with *Barred Warbler* (see Sylviidae below). Here the 'bars' are wing-bars, while the Barred Warbler has a barred breast.

The specific name *plumbeitarsus* tells us that the bird has 'lead-grey legs'.

Greenish Warbler *Phylloscopus trochiloides*

This Eastern European species is a rare visitor to Britain.

Greenish indicates a grey-green shade.

The specific name *trochiloides* cross-references to the name of the Willow Warbler (*trochilus*) to mean 'Willow Warbler-like'. The name *trochilus*, of course, already likens the Willow Warbler to a Wren.

Pale-legged Leaf Warbler *Phylloscopus tenellipes*

This Manchurian Chinese species has been recorded in Britain, possibly more than once.

Pale-legged is self-evident, though the specific name *tenellipes* refers to the legs as 'delicate'.

The addition of **Leaf** to this and a few other names in this family relates to the genus *Phylloscopus*, but it seems unnecessary.

Kamchatka Leaf Warbler *Phylloscopus examinandus*

This Far-Eastern species was split from the Arctic Warbler in 2011 and was recorded and confirmed in Finland in 2021.

Kamchatka is the name of the massive peninsula that stretches into the North Pacific off Siberia.

The specific name *examinandus* means thoughtful or careful, but the reason for its use here is unclear.

Arctic Warbler *Phylloscopus borealis*

This is another regular but rare visitor to Britain.

Arctic and *borealis* (northern) together underline the very northerly range of this bird.

74. ACROCEPHALIDAE:
REED WARBLERS AND ALLIES

The family name is based on *Acrocephalus*.

Basra Reed Warbler *Acrocephalus griseldis*

This migratory species that breeds along the rivers Tigris and Euphrates has recently spread into Israel and maybe Syria: it was first recorded breeding in Israel in 2007.

Basra is the name of the main port town at the confluence of the rivers.

Acrocephalus is a curious name because it seems that its creators (the Naumanns, father and son) wanted to create a word intended to describe the attenuated head (*cephalus*) of many of this family. It seems that they misused the Greek *akros* (topmost) when they intended something like *acutus* (sharp).

The specific form *griseldis* seems to be a unique variant of *griseus* (grey). It was concocted by the German K.J. Gustav Hartlaub in 1891.

* This species was particularly threatened during the 1990s when Saddam Hussein ordered the drainage, for economic and political reasons, of the extensive marshes that was its main habitat. By the time of his downfall in 2003, only 10% of the marshes remained. These have been in part restored since then, but the bird is still considered endangered.

Cape Verde Warbler *Acrocephalus brevipennis*

As the name implies this species is endemic to some of the **Cape Verde** islands.

The specific name *brevipennis* means 'short-winged'. Island species often evolve shorter wings or even become flightless.

Great Reed Warbler *Acrocephalus arundinaceus*

This is a scarce vagrant to Britain from mainland Europe.

Great tells of its size, which Linnaeus compared to a thrush when he originally named the bird *Turdus arundinaceus*.

The specific name *arundinaceus* still survives, associating this species with yet more reeds (Latin *arundo, arundinis,* reed).

Oriental Reed Warbler *Acrocephalus orientalis*

This Far-Eastern species has been recorded in Israel.

Oriental and *orientalis* are self-evident.

Clamorous Reed Warbler *Acrocephalus stentoreus*

This largely Asian species is also found on the Nile.

Clamorous means that it is noisy, and *stentoreus* underlines that fact.

* The scope of this species has been greatly reduced during the past decades as a result of splits that separated the Australasian forms.

Moustached Warbler *Acrocephalus melanopogon*

This is a species of the Mediterranean and Middle East, not recorded in Britain.

Moustached tells of a black moustachial stripe, though *melanopogon* is a gross exaggeration, meaning 'black bearded'.

Aquatic Warbler *Acrocephalus paludicola*

This Eastern European species is a scarce vagrant to Britain.

The marshy habitats of this bird are reflected in the name **Aquatic** (water loving) and in the specific name *paludicola*, which means 'marsh-dweller'.

Sedge Warbler *Acrocephalus schoenobaenus*

This is a common summer migrant.

Sedge indicates the use of a slightly different habitat from the Reed Warbler: it is often found in scrubby water margins, though the two often overlap.

The specific name *schoenobaenus* (Greek *skhoinos*, reed) was constructed around a Swedish name, *Savstigare* (a reed-climber).

* In Greek legend Schoenus was one of the siblings of Anthus (see pipits), Acanthis (redpolls), and Erodias (herons) who were all turned into birds following a contretemps with a herd of hungry horses.

Paddyfield Warbler *Acrocephalus agricola*

This is a rare vagrant to Britain from central and southern Asia.

Paddyfield takes its name from the Malay *padi* (rice), which is usually grown in wet fields.

The specific name *agricola* embraces the field but ignores the crop.

Blyth's Reed Warbler *Acrocephalus dumetorum*

This is a scarce vagrant to Britain from Northern Europe.

This species commemorates the British zoologist Edward **Blyth** (1810–1873).

The specific name *dumetorum* tells that the bird prefers thickets (though not necessarily in wet environments).

Common Reed Warbler *Acrocephalus scirpaceus*

This is a common summer migrant.

Common has replaced *Eurasian* following the 2022 lumping of the African and Eurasian species.

Reed is an obvious habitat word, with the species often favouring *phragmites*.

The specific name *scirpaceus* (Latin *scirpus*, reed or rush) tells of the bird's favoured habitat.

Marsh Warbler *Acrocephalus palustris*

This is a scarce summer migrant to Britain, breeding in very small numbers.

Marsh and *palustris* mean the same, though lush, rank, damp herbage is the bird's actual preference.

Thick-billed Warbler *Arundinax aedon*

This Siberian-breeding species migrates to India, but has been recorded in Britain.

Thick-billed is obvious.

The exact status of this species has been long been contentious: prior to 2018 it was classified variously as *Iduna* or *Acrocephalus*, but has now been moved into the resurrected genus ***Arundinax***, which means 'master of the reeds' (Latin *arundo, arundinis*, reed, and Greek *anax*, master).

The specific name *aedon* means 'nightingale', and is the name of yet another of those tragic Greek figures suffering transmutation. (That legend seems to be a variant on the story of Philomela – see Song Thrush – as there are many parallels.) The fact that this bird seems to sing with some virtuosity is significant.

Booted Warbler *Iduna caligata*

This Eastern European species has been recorded in Britain as a scarce passage visitor.

Booted and *caligata* mean the same, referring to the fact that the toes are darker than the 'legs' (see Glossary).

∗ The word *caligata* is evocative of the nickname of the infamous third Roman emperor, Gaius. When, as a child, he appeared in front of his father's troops in full battle gear, the legionaries dubbed him *Caligula* (Little Boots), under which name he is best remembered for his deranged and tragic behaviour while emperor.

Sykes's Warbler *Iduna rama*

This Central Asian species has been recorded in Britain.

Col. W.H. **Sykes** (1790–1872) was an English ornithologist working in India.

The roots of the genus *Iduna* puzzle Jobling, who offers three unconvinced explanations. The possibility of a phonetic link to *aedonis* (nightingale) seems the most reasonable.

The specific name *rama* is from the Indian god Rama, a manifestation of the god Vishnu.

The vernacular naming of the next two species is not universally accepted. *Collins Bird Guide* (third edition, 2022) maintains the original names, but notes the alternatives in parenthesis. The IOC versions of these names seem to complicate matters needlessly. (Here the original names are in parenthesis.)

Eastern Olivaceous Warbler *Iduna pallida* (Olivaceous Warbler)

This Eastern Mediterranean species has not been recorded in Britain.

Eastern was appended to the original name, **Olivaceous** (of an olive-green colour), when it was decided to underline the closeness of this and the next species.

The specific name *pallida* (pale) describes the somewhat 'washed-out' greenish-grey of the mantle.

Western Olivaceous Warbler *Iduna opaca* (Isabelline Warbler)

This Mediterranean species has been recorded in Britain.

Western Olivaceous is a newly coined name to link this to the previous species. For many years it had been known as the **Isabelline** Warbler, largely for its off-white under-plumage. The radical change seems to be an odd and inelegant complication.

The specific name *opaca* means dark: the term is relevant to the browner bird of the two species.

* *Isabelline* links to a legend that Queen Isabella I of Spain vowed not to change her under-garments until the Moors had been expelled from Spain. The word gained currency in the English language, via French, as a word describing a pale grey-fawn colour and was later applied to the names of several pale birds, another example being the *Isabelline Wheatear*.

Upcher's Warbler *Hippolais languida*

This Middle-Eastern species is found in Israel and Turkey.

H.M. **Upcher** (1839–1921) was an English naturalist who was very active in Norfolk. The bird was dedicated to him by his friend H.B. Tristram (see *Tristram's Warbler* below).

The genus *Hippolais* seems to have been appropriated by Conrad von Baldenstein from a different sort of bird: the supposed origin is *hupolais*, a bird mentioned by Aristotle, which may have been a Wheatear.

The specific name *languida* means weak. The reason for that name is not obvious: the song is 'energetic', so the epithet may refer to the fact that the bird glides before landing, giving a rather lazy impression.

Olive-tree Warbler *Hippolais olivetorum*

This Eastern Mediterranean species has rarely been recorded in Britain.

In spite of its name it is not exclusive to **Olive trees**, but that concept is reinforced in *olivetorum*.

Melodious Warbler *Hippolais polyglotta*

This species breeds widely in Spain, Italy and France, but is recorded as a regular passage species in Britain.

Melodious tells of its pleasant voice, while *polyglotta* suggests that it sings 'in many tongues', in other words that it is imitative in its repertoire.

Icterine Warbler *Hippolais icterina*

This more northerly species is recorded as a regular passage species in Britain.

Icterine and *icterina* together suggest that the bird is 'jaundice-yellow', though that is limited to a yellowish wash on the undersides.

75. LOCUSTELLIDAE:
GRASS BIRDS AND THEIR ALLIES

The family name is based on *Locustella*.

Gray's Grasshopper Warbler *Helopsaltes fasciolatus*

This eastern Siberian species has been recorded in France and Denmark.

This species is named after George R. **Gray** (1808–1872), an English ornithologist at the British Museum. He was younger brother of John Edward Gray. He described this species from a specimen collected by Alfred Russell Wallace.

Grasshopper: see Common Grasshopper Warbler below.

The genus name *Helopsaltes* is a new one, created in 2018. It combines the Greek *helops* (marsh) with *psaltis* (chanter).

The specific name *fasciolatus* means 'banded': there is only faint banding on the breast.

* The IOC is currently considering shortening the name to *Gray's Warbler*. That may be logical and desirable, but is not a precedent easily carried forward to its congener, below, since there is also a leaf warbler named after Pallas.

Pallas's Grasshopper Warbler *Helopsaltes certhiola*

This is a rare vagrant to Britain. It normally breeds in Siberia, Mongolia and other parts of eastern Asia.

Peter **Pallas** (1741–1811) was a Prussian naturalist-explorer.

Grasshopper: see Common Grasshopper Warbler below.

The specific name *certhiola* suggests that it is like a small treecreeper (*Certhia*).

Lanceolated Warbler *Locustella lanceolata*

This species has been recorded as a vagrant to Britain. Its nearest breeding area is in Russia.

Lanceolated and *lanceolata* refer to the spear-shaped spots on the plumage of this species.

River Warbler *Locustella fluviatilis*

This species breeds in Eastern Europe, but is only a rare vagrant to Britain.

River refers to the bird's preference for riverside habitats.

The specific name *fluviatilis* reinforces the riverside image.

Savi's Warbler *Locustella luscinioides*

This is a scarce summer visitor to Britain and has bred here rarely.

It is named for the Italian ornithologist, Paolo **Savi** (1798–1871).

The specific name *luscinioides* implies that the bird resembles a nightingale (*luscinia*). That is a mystery: the monotonous, reeling song certainly does not lend itself to that comparison, while the physical resemblance is tenuous. The only other explanation is that, like the Nightingale, this is a nocturnal singer.

Common Grasshopper Warbler *Locustella naevia*

This is a regular summer migrant.

Common is rarely used: this is a difficult bird in any case.

Grasshopper derives from the bird's sibilant song, which resembles the sound made by the insect.

Locustella relates to Grasshopper in a form meaning 'a small locust'.

The specific name *naevia* refers to spotting, which in this case is on the undertail plumage.

76. CISTICOLIDAE:
Cisticolas and their allies

The family name is based on *Cisticola*.

The cistocolas are mostly found in Africa. They are generally small, brown birds that are sometimes so similar in appearance that only their voices, locations or habits can separate them. As a consequence, a list of their English names reads like a parody of serious ornithology.

Only one species of Cisticola is found within the Western Palearctic and that arguably has the most unlikely name of all.

The other representatives of the family have recently increased from one to three, with the 2021 split of Graceful and Delicate Prinias, and the 2011 discovery that Cricket Warblers are now breeding in the region.

Zitting Cisticola *Cisticola juncidis*
(Fan-tailed Warbler)

This is a scarce vagrant to Britain.

The name **Cisticola** literally means 'rock-rose dweller'. It was given to this species by Temminck and became the name for the whole genus.

Zitting makes better sense when seen alongside other names for members of this family: Rattling, Tinkling, Wailing and Churring cisticolas all make some sort of noise – and this one 'zits'. It is hard to treat a name like that seriously. Little wonder that the BOU adheres to the simple old name of **Fan-tailed Warbler**: it is fan-tailed and looks generally like a warbler, but such vastly different names are confusing.

The specific name *juncidis* means 'of the rushes'. *Juncus*, a species of rush (see also Dark-eyed Junco) is a characteristic plant of its Coto Doñana habitats. Although it is a mainly grassland species in the southern part of its range, the bird has adapted to other habitats during its northwards expansion.

Graceful Prinia *Prinia gracilis*

This is a Middle Eastern species, found from Syria down through the Eastern Mediterranean and into the Nile Valley and Red Sea. It formerly included the new species, Delicate Prinia (see below).

Graceful is an apt word for a delicate-looking bird, and that is reinforced by the specific name *gracilis*, which means the same.

Prinia is a Javanese name for a local relative of this species.

Delicate Prinia *Prinia lepida*

This Middle-Eastern species was split from Graceful Prinia in 2021. It is found in Turkey and Syria and further east.

Delicate seems to be a carefully selected near-synonym of Graceful.

The specific name *lepida* has the sense of 'charming' or 'elegant'.

Cricket Warbler *Spiloptila clamans*

This sub-Saharan species seems to be spreading its range: it was first recorded as breeding in Morocco in 2012 and bred in Algeria in 2022.

The name **Cricket** derives from its cricket-like song.

Warbler is used here very loosely because it is warbler-like in appearance.

Spiloptila combines the Greek *spilos* (spot) with *ptilon* (feather).

The specific name *clamans* (shouting) reflects its loud, piercing calls.

77. SYLVIIDAE: Sylviid Babblers

The family name is based on *Sylvia*.

This family has been much altered in recent years as a result of genetic studies. Most radical has been the removal of most species from the genus *Sylvia* (once the umbrella name for most Old World warblers) into the new genus *Curruca*, named for the Lesser Whitethroat, which is treated as the type species for the genus.

Several new species have been designated, and there have been some subtle changes to spellings.

(Eurasian) Blackcap *Sylvia atricapilla*

This is a common summer migrant and also a winter visitor.

Eurasian distinguishes this species from the *Bush* Blackcap internationally, but is otherwise unused.

Blackcap was once widely used in England to name other black-capped species, such as Reed Bunting and Marsh Tit, but was fixed on this species by Pennant in the late eighteenth century, and was subsequently sometimes used as the *Blackcap Warbler*.

Sylvia, which relates to woodland, was originally created as an overarching genus for most the Old World warblers, but that has slowly been eroded in importance by the many modern subdivisions of the original warbler family.

The specific name *atricapilla* suggests that the species has 'black hair'.

* Traditionally, the British breeding population of Blackcap travels to Southern Europe and North Africa in winter. During the late twentieth century an increasing number of wintering birds were seen in Britain. Research shows that these are mostly individuals from Central Europe, attracted by a combination of the milder maritime climate and widespread garden feeders. The two populations still seem to behave separately.

Garden Warbler *Sylvia borin*

This is a common summer visitor, though it is fairly shy.

Garden is something of a misnomer, since the species generally frequents marginal scrub. Bewick first used it in 1832, when he erroneously anglicised Gmelin's *Sylvia hortensis*, which was an Orphean Warbler. Gould and Yarrell both cemented that usage.

The specific name *borin* appears to be another mistake, since the word is based on the Latin *bos* (ox), which has nothing whatever to do with this species. The problem was that the bird was named by Pieter Boddaert from a museum specimen, with no context to describe its behaviour, so he may have confused it with such as the Yellow Wagtail, which does feed around cattle.

* A rather unkind joke among birders is that the 'g' has been omitted from *borin*: the bird is very plain, as befits a skulker, but it does sing beautifully.

* This rather dull vernacular name replaced an older one, *Greater Pettychaps*, which was current until the end of the eighteenth century. It meant the larger of two 'small fellows', which has a certain appeal. The smaller form was one of three *Phylloscopus* warblers: Willow Warbler, Wood Warbler or Chiffchaff.

The genus *Curruca* derives from the Latin name for an unidentified small bird. It was originally used by Linnaeus as the specific name for the Lesser Whitethroat. Later that species was preferred as the type species for the genus Sylvia because of some ambiguity in Linnaeus's record of the Common Whitethroat. That priority accounts for the choice of Curruca here under the revised taxonomy of the family, which was adopted by the IOC in 2020.

Barred Warbler *Curruca nisoria*

This is a scarce passage migrant.

The male in particular has **barred** undersides.

Curruca: see Lesser Whitethroat below.

The specific name *nisoria* is an indirect reference to the barring, because it refers to *Accipiter nisus*, the Sparrowhawk, which is barred on the breast.

The next three entries are unresolved business in terms of taxonomy. The Lesser Whitethroat complex shows variations and divisions, with some birds wintering in Africa and others in South Asia. Currently, three relevant 'species' are listed in the Western Palearctic by some authorities: Lesser Whitethroat, Desert Whitethroat and Hume's Whitethroat. However, it is generally acknowledged that the exact taxonomies of these are currently unresolved. The IOC is now proposing to treat them as one species.

Lesser Whitethroat *Curruca curruca*

This is a common summer migrant to Britain and much of Europe from East Africa.

This is the **Lesser** of the two original Whitethroat species and was so designated by Gilbert White.

Whitethroat is a simple descriptive folk name that was in use in England long before bird names were formalised.

Curruca was in use long before Linnaeus adopted it as the specific name of the Lesser Whitethroat. It was used by the second-century Roman Juvenal and later by Gessner (1555) and Aldrovandi (1599). Ray (1678), in writing about the Dunnock, explained that *curruca* linked to the concept of a bird cheated by a Cuckoo (as in 'cuckolded'). Given that Lesser Whitethroat is not strongly in evidence as a regular host of the Common Cuckoo, it seems quite likely that Linnaeus's choice of *curruca* for the species was to do with the word's similarity to an obsolete Swedish name *kruka* (see Jobling).

Desert Whitethroat *Curruca (curruca) minula*

This Central Asian form has been recorded several times in Britain. It normally winters in the Indian subcontinent.

Desert tells of habitat preferences, though these might be more accurately described as 'drier lowlands'.

The (sub)-specific name *minula* means very small.

* An alternative common name, **Small Whitethroat**, echoes the specific name.

Hume's Whitethroat *Curruca (curruca) althaea*

This larger Asian form has been recorded in Finland, the Netherlands and Egypt. It overlaps the range of Desert Whitethroat, but prefers to breed at higher altitudes quite separately from *C. minula*.

Allan **Hume** (1829–1912) was a British ornithologist based in India.

The (sub)specific name recalls, in Greek mythology, **Althaea**, the tragic wife of King Oënus of Calydon.

Arabian Warbler *Curruca leucomelaena*

This species is found on the **Arabian** Peninsula, in Egypt, and in Israel.

The specific name *leucomelaena* means 'black and white'.

Western Orphean Warbler *Curruca hortensis*

This species has been recorded in Britain.

This **Western** Mediterranean form was long seen as conspecific with the next species. The new names illustrate the use of 'Western' and 'Eastern' in a specifically Mediterranean context in a number of recent splits.

Orphean is an accolade to a skilled songster, since Orpheus was a renowned poet and musician in Greek legend, the son of Calliope, the chief Muse.

The specific name *hortensis* is to do with gardens.

Eastern Orphean Warbler *Curruca crassirostris*

See above species. This is a species of the **Eastern** Mediterranean, which has been recorded in Britain.

The specific name *crassirostris* means 'thick-billed'.

Asian Desert Warbler *Curruca nana*

This migratory species has been recorded in Britain.

The **Asian** Desert Warbler ranges from the Middle East eastwards.

In spite of the name **Desert**, it prefers semi-desert and dry steppe.

The specific name *nana* means 'dwarf': it is Chiffchaff-sized.

African Desert Warbler *Curruca deserti*

This is a sedentary North **African** species.

Like the Asian form, it prefers semi-desert, so the specific name *deserti* is a little exaggerated.

Tristram's Warbler *Curruca deserticola*

This is a partially migratory North African species.

H.B. **Tristram** (1822–1906) was an English theologian, traveller and naturalist.

The specific name *deserticola* means 'desert-dweller'.

Menetries's Warbler *Curruca mystacea*

This is a Middle Eastern species.

Édouard **Ménétries** (1802–1861) was a French zoologist.

The specific name *mystacea* means moustached (from the Greek *mustax*).

* This and the following name conveniently illustrate the decision by the IOC 'to use no accents, except umlauts, for certain proper names of people'.

Rüppell's Warbler *Curruca ruppeli*

This Eastern Mediterranean species has been recorded in Britain.

Eduard **Rüppell** (1794–1828) was a German naturalist and explorer.

The specific name is now modified to *ruppeli* under the rule of precedence, because that was the misspelling used when the bird was first named.

Cyprus Warbler *Curruca melanothorax*

This species is limited to **Cyprus** in the breeding season.

The specific name *melanothorax* speaks of the 'dark breast', which is grey and heavily spotted.

Sardinian Warbler *Curruca melanocephala*

This widespread Mediterranean species has been recorded in Britain.

Sardinia is merely a type location.

The specific name *melanocephala* tells us that this is a 'black-headed' species.

In 2020, the IOC accepted a further split of the former **Subalpine Warbler**. Moltoni's had been separated from it previously. Since that was the nominate form, it took with it the specific name *subalpina*. As a result of the splits, there are now three separate species: Moltoni's, Western Subalpine and Eastern Subalpine, all of which have been recorded as vagrants in Britain.

Western Subalpine Warbler *Curruca iberiae*

Western reflects its distribution in the Mediterranean (North Africa, from Tunisia to Morocco) and in Europe (Iberia, southern France and extreme north-west Italy).

Subalpine refers to its habitat at the base of mountains.

The specific name *iberiae* means Iberian.

Moltoni's Warbler *Curruca subalpina*

It is found in north-central Italy, Corsica, Sardinia, and the Balearics.

This form was named for the Italian naturalist, Edgardo **Moltoni** (1896–1980).

As the former nominate form of the original Subalpine Warbler, this species now carries the name *subalpina*.

Eastern Subalpine Warbler *Curruca cantillans*

The distribution is not entirely **Eastern**, since it is found in parts of Italy, as well as in the Balkans, Greece and western Turkey.

The specific name *cantillans* means 'warbling'.

(Common) Whitethroat *Curruca communis*

This is a well-known common summer migrant that breeds in Britain.

Common is rarely used, because it is far easier to see than the shy Lesser Whitethroat and is far better known.

This was the original **Whitethroat**, a simple descriptive name, first in written evidence in 1676. However, Lockwood speculates that, like other simple descriptive names, such as *Redbreast*, it was probably in currency as early as the fourteenth century.

The specific name *communis* means 'common'.

Spectacled Warbler *Curruca conspicillata*

This largely Mediterranean species has been recorded in Britain.

Spectacled reflects the fact that the male has a conspicuous white eye ring.

The specific name *conspicillata* also means 'spectacled'.

Dartford Warbler *Curruca undata*

This is a precarious resident in Britain, at the northern extremity of its range and very susceptible to prolonged cold spells.

Latham received specimens of this species from a correspondent in **Dartford**, Kent, and shared them with Pennant, who fixed the name.

The specific name *undata* describes the diagnostic speckles on the throat.

* Kentish Plover specimens came to Latham from the same source.

Marmora's Warbler *Curruca sarda*

This Mediterranean species has been recorded in Britain.

It commemorates General A.F. La **Marmora** (1789–1863), an Italian soldier and naturalist who spent much of his life studying the topography and wildlife of Sardinia.

The specific name *sarda* refers to a Sardinian woman.

Balearic Warbler *Curruca balearica*

This was long treated as a subspecies of Marmora's but now has full-species status.

Balearic and *balearica* refer to the Balearic Islands where it is resident.

Parrotbills to Treecreepers

78. PARADOXORNITHIDAE:
PARROTBILLS AND ALLIES

The family name is based on the genus *Paradoxornis* (an extraordinary bird).

A number of families and species in this spectrum have puzzled the scientists for some time and there has been some reshuffling: this new family was created in 2019. It would not be represented here were it not for two oriental species that have established small feral breeding communities in several European countries: these originate from escapes or from deliberate releases.

Parrotbills closely resemble Bearded Reedlings or Long-tailed Tits in shape and were once classed with them, but the similarity is now considered to be due to convergent evolution rather than to any relationship.

There is some doubt about the exact identification of the birds in each feral population, but both species appear to be involved. In any case, some hybridisation occurs in their native region of Southeast Asia and China.

Ashy-throated Parrotbill *Suthora alphonsiana*

Ashy-throated indicates a grey throat, though the grey is more extensive.

Parrotbill describes the short, deep bill.

Suthora is a Nepalese name for a local parrotbill species (Ashy-eared), which was first used as a generic name by Brian H. Hodgson in 1837.

The specific name *alphonsiana* commemorates **Alphonse** Milne-Edwards (1835–1900), a French zoologist, whose father is commemorated in the petrel *Calonectris edwardsii*.

Vinous-throated Parrotbill *Suthora webbiana*

Vinous-throated suggests a strong red-wine colour, but the colour is a pinkish shade of brown, altogether warmer than the Ashy-throated.

The specific name *webbiana* commemorates the British botanist, Philip Barker **Webb** (1793–1852).

79. LEIOTHRICHIDAE:
LAUGHINGTHRUSHES AND ALLIES

The family name is based on *Leiothrix*.

The birds in this family were previously classed in Timalidae, but in 2018 a new arrangement was proposed. At the same time the three babblers below were moved from *Turdoides* into the genus *Argya*.

Other than the feral populations of the first species, this mainly tropical family is not normally represented in Europe. The four *Argya* species below all live in adjacent regions of North Africa or the Middle East.

Red-billed Leiothrix *Leiothrix lutea*

This is a native of South Asia, but feral populations now exist in Italy, France, Spain and Portugal. They were originally imported as cage-birds for their colour and song, an alternative popular name being *Pekin Nightingale*.

Red-billed is a tiny feature of a Robin-sized bird.

Leiothrix literally means 'smooth-haired'.

The specific name *lutea* means saffron-yellow: that is for a striking throat-patch. (Latham called it the *Luteous Flycatcher*.)

Fulvous Babbler *Argya fulva*

This is a North African species.

Fulvous means 'tawny red-brown', and that is repeated in the Latin *fulva*.

Babbler means 'talker': these are gregarious birds that communicate freely.

The genus *Argya* reinforces the meaning of Babbler. It derives from the Latin *argutus* (noisy).

Afghan Babbler *Argya huttoni*

This species is also found in Kuwait and Iran.

The specific name *huttoni* refers to Captain T. **Hutton** (1807–1874), who served a short tour with the British Army in Afghanistan.

Arabian Babbler *Argya squamiceps*

This bird is indeed found in the **Arabian** Peninsula, but also in Israeli deserts.

The specific name *squamiceps* implies that the bird is 'scaly headed', though that is an illusion created by streaked plumage.

Iraq Babbler *Argya altirostris*

This species is found in Turkey and Syria as well as **Iraq**.

The specific name *altirostris* describes the bird as 'high-billed'. The bill is vertically deep at the base.

80. REGULIDAE: CRESTS AND KINGLETS

The family name is based on *Regulus*.

Ruby-crowned Kinglet *Corthylio calendula*

A vagrant of this American species was added to the British list in 2020.

Ruby-crowned is part of a 'royal' theme running through this family. The crown is a rich red.

The Americans have long preferred the name **Kinglet** for this family: it simply anglicises *Regulus*.

Molecular studies have shown this bird to have evolved differently from the genus *Regulus*. The revised genus, adopted by IOC in 2021, resurrects a disused name, *Corthylio*, which derives from *korthilos*, a small wren-like bird mentioned by the Greek Hesychius.

The specific name *calendula* means glowing, but is also the botanical name of the marigold plant, which has golden and orange flowers.

(Common) Firecrest *Regulus ignicapilla*

This species has proliferated as a resident in the south of England since the late twentieth century.

Common is only used to distinguish the bird from its Madeiran cousin, therefore is rarely needed.

Firecrest parallels the name Goldcrest, but tells us that this species has a flame-coloured crest.

Regulus is the diminutive of the Latin *rex, regis* (king), therefore 'prince' or 'kinglet'.

The 'firecrest' concept is repeated in the specific name *ignicapilla*, which literally means 'burning hair'.

Madeiran Firecrest *Regulus madeirensis*

This island species is endemic to **Madeira**, as both names suggest.

Its separation from the above species was accepted by most authorities in 2003.

Goldcrest *Regulus regulus*

This a common breeding resident, boosted by winter visitors.

Goldcrest is an obvious name for a bird with a golden-yellow crown-stripe. It is ultimately a shorthand version of the older names for this bird, which included the *Golden crown'd Wren* and the *Golden-crested Wren*, the latter evolving into the modern form. The wren element in this context was for an even smaller bird: the Goldcrest is the smallest British native species. That link was echoed in another local name *Golden Cutty* (*Cut* or *Cutty* represented the upright tail of the Wren). In 1843, the name was still so unsettled that Yarrell, often a final arbiter in these matters, hedged his bets by recording it as a *Little Golden-crested Regulus or Kinglet*, but it was a simple *Goldcrest* for Coward by 1926.

* One of the most curious names for this little bird is *Woodcock Pilot*, which stems from a disbelief that so many could appear in autumn on North Sea coasts. Their occurrence coincided approximately with the migration of Woodcock, so bemused locals once assumed that the tiny birds had hitched a lift on the sturdier species. In fact they arrive slightly before the Woodcock, a point made by Lockwood, but that in itself is hardly less wonderful. Although they are somewhat the smallest, they are often accompanied by tiny *Phylloscopus* warblers, such as Pallas's, Radde's, Dusky and Yellow-browed, many of which originate much further east. We tend to underestimate the amazing durability of migrating birds in general. The truth is often stranger than the fictions of folklore.

81. TROGLODYTIDAE: Wrens

The family name is based on *Troglodytes*.

There are almost 90 species of wren, but only one of these is found outside the Americas. Ironically it was the European form of this species that gave the entire family its common English name. The word Wren normally suffices in isolation in Britain.

The ancestor of the Eurasian Wren is thought to have crossed from North America some 4.3 million years ago at the Pacific end of Eurasia. There are now almost 30 subspecies ranging throughout the region. The main British form is *T. t. indigenus*, but five more forms are found in offshore islands, the most famous of which is the St Kilda Wren, *T. t. hirtensis* (Hirta being the main island of the archipelago).

There was a period around the turn of the twenty-first century when the Eurasian Wren was designated as a form of the American species, the Winter Wren, and for some time that name was used formally, though it was largely ignored by British birdwatchers. Matters have moved on since then: further research not only returned the European form to its independence, but also split the American species into Winter Wren and Pacific Wren.

(Eurasian) Wren *Troglodytes troglodytes*

This is a common resident, with several subspecies recorded, even within the British Isles.

Wren has Germanic roots and evolved via the Old English form *wraenna*, through *wrenne* to the modern spelling. The word's origins are related to the distinctive upright tail posture, which is also featured in the folk name *Cutty*.

Troglodytes refers to its 'cave-dwelling' habits, since it builds a round nest with a side entrance that is often placed in a crevice or hole at ground level.

* The Wren is deeply associated in British folklore with the Robin. The similarity of two small and common, ground-level species gave rise to the idea that they were partnered as *Jenny Wren* and *Robin Redbreast*. 'Jenny Wren' has endured until modern times, but the Robin's forename was so popular that it became the formal name for the species and many have forgotten his surname.

* An old Scottish form, *Wrannock*, paralleled the Medieval English words *Ruddock*, *Tinnock* and *Dunnock*, the *-ock* element being a diminutive. *Ruddock* was later replaced by *Redbreast* and then *Robin*; *Tinnock* (Blue Tit) faded from use; but *Dunnock* retains its heritage.

* Prior to the modern formalisation of bird names, the word *wren* was often used in a generic way for other small songbirds, such as *Willow Wren* and *Reed Wren*.

82. SITTIDAE: Nuthatches

The family name is based on *Sitta*.

The word **Nuthatch** evolved around the British species, now known as the European Nuthatch. It is rooted in older words such as *nuthache*, *nuthak* and *nuthagge*, all of which indicate a link to the word 'hack'. This is because the bird feeds by wedging a nut or seed in order to hack it open with its bill. It follows that the 'hatch' element has nothing to do with hatching in its modern usage. A similar dialect name *Nutjobber* was a corruption of nut-jabber.

Names such as *Mud-dauber* and *Mud-stopper* referred to its habit of reducing the entrance to oversized tree nest holes with mud, which hardens to provide a predator barrier. The Rock Nuthatches use a similar method to adjust rock crevices to a suitable size.

Sitta derives from a Greek word *sittē*, which was used by Aristotle to refer to a small woodpecker-like bird.

Algerian Nuthatch *Sitta ledanti*

This is an endangered endemic population, first discovered in **Algeria** as recently as 1975.

The finder was J-P. **Ledant**, a Belgian botanist.

Krüper's Nuthatch *Sitta krueperi*

This species is found in parts of Greece, Turkey and the Caucasus.

The names commemorate T.J. **Krüper** (1829–1921), a German naturalist.

Corsican Nuthatch *Sitta whiteheadi*

This species is endemic to the island of **Corsica**.

The specific name commemorates the English naturalist John **Whitehead** (1860–1899).

Red-breasted Nuthatch *Sitta canadensis*

This North American species has been recorded in Britain.

Red-breasted is more meaningful in North America, where there is also a *White-breasted* Nuthatch.

The specific name *canadensis* refers to Canada, though it is also found widely in the United States.

Western Rock Nuthatch *Sitta neumayer*

This species is found in Greece and Turkey.

Western is relevant to its range, which is generally east of the Adriatic, but west of the range of the next species.

Rock indicates that the birds inhabit rocky terrain and nest in rocky crevices.

The specific name is a dedication to Franz **Neumayer** (1791–1842), an Austrian botanist and collector.

Eastern Rock Nuthatch *Sitta tephronota*

This is a more easterly form, found in eastern Turkey and eastwards.

The specific name *tephronota* uses the Greek *tephros* (ash coloured) and *nōtos* (backed) to describe the pale grey upper-parts of the bird.

* The use of *Western* and *Eastern* in these last two names is a good example of the weakness of those words as qualifiers. Compare their use in *Western* Sandpiper and *Eastern* Cattle Egret, for instance.

(Eurasian) Nuthatch *Sitta europaea (caesia)*

This is a common species of mixed and deciduous woodland with some older trees.

Eurasian is needed in an international context.

The specific name *europaea* is obvious in its meaning but it has rarely been used (see also Nightjar), probably because much of the early scientific naming of birds was Eurocentric. Curiously, *Sitta europaea* was used by Linnaeus in the *Systema Naturae* without reference to any other nuthatch species, so one wonders why that specific name was chosen.

The nominate subspecies, found in Northern and Eastern Europe, is a white-bellied form.

The form found in Britain and Western Europe is *S. e. caesia* (blue-grey, for the colour of the mantle). However, the outstanding contrast with the nominate species lies with its bright orange-buff undersides.

* In fact this is a very widespread species, with over 20 subspecies spread through Europe and Asia, so it is strange to see apparently contradictory combinations such as *Sitta europaea asiatica* or *Sitta europaea formosana* (Taiwan).

83. TICHODROMIDAE: WALLCREEPER

The family name is based on *Tichodroma*.

Wallcreeper *Tichodroma muraria*

This is a species of high mountains of Southern Europe. It has been recorded occasionally in Britain, mainly in the south.

The **Wallcreeper** exhibits similar agility on vertical rock surfaces to that of the treecreepers on trunks and branches of trees.

Tichodroma repeats the vernacular name in a Greek-based form that means a 'wall-runner'.

The specific name *muraria* underlines this in Latin by reminding us that it belongs on walls.

* The strangest feature of the names given to this unique bird is that none mention its dramatic crimson, black and white wings.

84. CERTHIIDAE: Treecreepers

The family name is based on *Certhia*.

Eurasian Treecreeper *Certhia familiaris*

This is a widespread species, as implied by the name **Eurasian**, which is not normally required in Britain.

Treecreeper is a good description of the quiet, mouse-like movements made as it runs up trunks and under branches in a search for food items. The usage as a single unhyphenated word is very modern (see note below).

Certhia began life at the time of Aristotle as the Greek *kerthios*, possibly for this species.

The specific name *familiaris* seems to have a very literal use here, since Linnaeus recorded the use of tame treecreepers to keep houses insect-free.

* The name *Creeper* was the original form, the words Tree and Wall being added to distinguish them, though not formally until 1883. Coward's 1926 form was indexed as *Creeper, British tree*, as opposed to *Southern European* for the next species. It was still indexed under *Creeper* in James Fisher's 1966 *Shell Bird Book*. A period of hyphenation followed, as *Tree-creeper*, which morphed into the single word of modern usage.

Short-toed Treecreeper *Certhia brachydactyla*

This species rarely crosses the Channel, in spite of being fairly widespread on mainland Europe and present in the nearest parts of France.

Short-toed and *brachydactyla* mean the same thing.

* It is the hind claw that is relatively short: other field markings are extremely subtle.

Mockingbirds
to Chats etc.

85. MIMIDAE: Catbirds, Mockingbirds and Thrashers

The family name is based on *Mimus*.

This is an American family, now represented on the British list by three moderately common species.

Grey Catbird *Dumetella carolinensis*

This species has been recorded several times in a number of European countries.

The word **Grey** separates this from the Black Catbird of Central America.

Catbird relates to one of the calls made by this species. It is not related to the catbirds of Australia.

Dumetella relates to the thickets frequented (Latin *dumetum*).

The specific name *carolinensis* relates to the Carolina Colonies, named after Charles II and dedicated to Charles I. A number of early names relate to this relatively small area of the south-east of the continent that contained the fledgling British colony.

* It seems a pity that Vieillot's evocative name *felivox* (cat-voice) has been lost somewhere along the line: the rule of precedence favours Linnaeus's choice.

* In international names, the standard British spelling of *grey* is the norm, though American field guides use the local form *gray*.

Northern Mockingbird *Mimus polyglottos*

There are just two British records and one in the Netherlands.

Northern separates this from birds such as the Tropical Mockingbird. It is the only mockingbird in North America, so the qualifier is not usually needed there.

Mockingbirds are imitators.

Mimus repeats that concept: they mimic the songs of other species. The result is generally very attractive.

The mimicry is revisited in the specific name *polyglottos* – a speaker of many languages.

Brown Thrasher *Toxostoma rufum*

There is just one British record of this, in the winter of 1966–7.

The **Brown** is of a distinctly rufous shade, hence the specific name *rufum*.

Thrasher relates to a number of English dialect words for thrush, such as *thresher*. This one survived transplantation into the New World. In a general way this species is quite thrush-like, with a heavily marked breast, though some of its congeners are less thrush-like.

Toxostoma underlines the most un-thrush-like aspect of its appearance: the bill is distinctly down-curved. The word literally means 'curved mouth'.

86. STURNIDAE: Starlings

The family name is based on *Sturnus*.

Starlings and mynas are closely related. They differ primarily in the fact that the word 'starling' is of English origin, while 'myna' is rooted in Hindi.

These are often intelligent and gregarious species, with highly evolved communication skills and a capacity for imitation. The Common Starling's song may not be flamboyant, but in the wild it is full of imitations of other species. Mozart famously kept a pet Starling for three years, fascinated by its vocal offerings, and marked its passing with a composition.

A number of other species occur in the Western Palearctic, some as vagrants. However, several members of the family are kept as cage-birds. That in part has led to them being transported and introduced into non-native areas, though nostalgia and pest control have motivated some releases. Common Starlings now abound in North America and Australia, while the Common Myna and others have feral populations in several areas of the Western Palearctic. In general such introductions are regretted, as these are intelligent and thrusting species that often outcompete local native species.

Crested Myna *Acridotheres cristatellus*

There are feral populations in Spain and Portugal. Its native range is China and Southeast Asia.

Crested: the crest is a frontal tuft.

Myna was adapted from Hindi form *mainā*, which has Sanskrit roots.

Acridotheres means 'a locust hunter'.

The specific name *cristatellus* is a diminutive form of *cristatus* (crested), as befits a mere tuft.

∗ An alternative name is the *Chinese Starling*.

Bank Myna *Acridotheres ginginianus*

This South Asian species is established in Kuwait.

Bank comes from its habit of nesting in earth banks.

The specific name *ginginianus* comes from Gingi, an alternative name for the Coromandel Coast of India.

Common Myna *Acridotheres tristis*

This is by far the most widespread and best-known of the mynas. Its native range is South and Southeast Asia. It has an established feral foothold in several Middle-Eastern countries, as well as in Spain and Portugal and in other parts of the world too. It is considered one of the most invasive species in the world because it adapts well to city environments and outcompetes many native species.

Common is self-evident.

The specific name *tristis* (sad) seems to be more about its dull grey-brown plumage than its harsh song.

Vinous-breasted Starling *Acridotheres burmannicus*

This Southeast Asian species has a well-established feral population in Israel.

Vinous-breasted describes a warm purple-brown shade.

The specific name *burmannicus* refers to Burma (modern Myanmar), one of the countries in which the bird is native.

Daurian Starling *Agropsar sturninus*

This Far-Eastern species has only rarely been recorded as a vagrant to Europe. The first was noted in Shetland in 1985.

Daurian refers to a region of Siberian and occurs in the name of the Daurian Jackdaw, for example.

Agropsar derives from the Greek *agrops* (field) and *psar* (starling).

The specific name *sturninus* sometimes means 'speckled', but here it means 'starling-like' (see *Sturnus* below).

Rosy Starling *Pastor roseus*
(Rose-coloured Starling)

This largely Asian species appears annually in Britain, though in very small numbers.

The IOC currently prefers the simple form **Rosy**, which is far more succinct than some of their preferences elsewhere. **Rose-coloured** says the same thing. The bird in breeding plumage is certainly rosy (a fact underlined by the specific name *roseus*) though it has plenty of black too. The pink is duskier in winter and non-existent in the pale grey juvenile form. All three plumages occur in Britain.

Pastor (shepherd) is the currently preferred genus, which is unique to the species.

* This species seems to have undergone changes in both its vernacular and scientific names at fairly regular intervals: in Coward's 1926 list it was the Rose-coloured Pastor.

* In China, in the 1980s, it was realised that this species gave better control of locusts and created fewer side-issues than did the use of pesticides. For that reason large numbers of these birds were bred and released to boost the population.

Common Starling *Sturnus vulgaris*

The **Common** Starling is certainly well known, but the qualifying word is only used when other species of starling are concerned.

In spite of its current familiarity, the word **Starling** is something of a curio: until the late eighteenth century the adult was known as a *Stare* (cf. the Greek *psar*) while the diminutive suffix *-ling* was attached only to young birds, which are visibly different from their parents for much of the summer. During the nineteenth century, the word Starling slowly became the preferred name for the species.

Sturnus is the Latin for 'Starling'.

The specific name *vulgaris* was intended and used in its original sense of *common*.

* The old form *Stare* took on some strange variants: from *Sheepstare* (the birds feed among sheep) came such names as *Shepster*, *Chepster* and even *Ship Starling* (Lockwood).

* It is highly fashionable for commentators and reporters to draw attention to Starling 'murmurations', which term generally refers to their swirling pre-roost gatherings. Indeed, a number of roost sites hold special events to witness the spectacle. The word *murmuration* appeared as a collective term in the Book of St Albans in 1486, when the Abbess Juliana Berners listed a number of such terms, which included 'a cherme (charm) of goldfinches', 'an unkyndeness of ravens', 'a covey of pertryches' (partridges), 'a falle of woodcocks' and '**a murmuracion of stares**'. Murmuration has historically been used in a general way to describe the sound of quiet conversation, such as of a congregation. The word became popular in the twentieth century after Mary Webb used it in her 1924 best-selling novel, *Precious Bane*, to describe the sound made by waking Starlings, and that would seem to be a more appropriate image than that of the swirling masses. Today it is probably the best known of such traditional collective terms.

Spotless Starling *Sturnus unicolor*

This species occurs only in the Western Mediterranean, where it appears to have evolved while separated by an Ice Age from the main population of *Sturnus vulgaris*.

Spotless applies to this species in all seasons: such small pale spots as there are in winter are quite insignificant.

The specific name *unicolor* also underlines the fact that the summer plumage seems plainer: it has less iridescence.

Tristram's Starling *Onychognathus tristramii*

This species is found in the Arabian Peninsula.

H.B. **Tristram** (1822–1906) was an English theologian, traveller and naturalist (see also Tristram's Warbler).

Onychognathus is composed from the Greek words for 'claw, nail or talon' and for 'jaw'. The name seems not to make a lot of sense, though it perhaps conjures up the image of a noisy, competitive feeding flock.

87. TURDIDAE: Thrushes

The family name is based on *Turdus*.

Members of the thrush family are well represented in the Western Palearctic. There are six regular 'British' species: three residents, two winter visitors and one summer migrant.

A number more species stray, regularly or irregularly, from either North America or from the East: seven American species have been registered in Europe, and unsurprisingly all of those have been recorded in Britain. Two of those are single records and relate to species that are not highly migratory, whereas the other five are more frequent, probably because they are more vulnerable to adverse weather during their migration. Seven more species have reached Britain from the east, and one further species, Tickell's Thrush, has been recorded just once in the Western Palearctic, in Germany.

Two common species in Britain, Song Thrush and Mistle Thrush, had many regional name variants that Lockwood discusses in full: they represent a whole museum of curiosities, but most seem to have similar Germanic roots, the French-based *Mavis* being the odd-name out. In the end, *Thrush* was preferred over Thrasher, Thrissel and Throstle, for example, to become the standard word after 1770. The alternative names that competed longest with Thrush were *Throstle* and *Mavis*. Both are seen frequently, particularly in poetic use during the nineteenth century and, in that context, into the twentieth century. Throstle is clearly related to the Modern German *Drossel* and was preferred by Pennant in 1768, but, unusually, he did not prevail in that choice. *Mavis* was first in evidence in about 1400 and has Old French roots in *mauvis/malvis* of which the origins are unclear and disputed. It still exists formally in Modern French as *grive mauvis* and Modern Spanish as *malvis*, both for the Redwing.

Varied Thrush *Ixoreus naevius*

This is a very rare North American vagrant, which was recorded in Britain in 1982 and Iceland in 2004.

Varied is a multicoloured thrush, with zones of orange, grey, black and white.

Ixoreus is based on *ixos* (mistletoe), though the connection is spurious and results from an error made by Bonaparte when he modified Gmelin's designation.

The specific name *naevius* suggests that the bird is 'spotted'. The bird has none of the dark breast spots usually associated with the word thrush, but instead the grey-olive edgings of the orange underside plumage give the illusion of puffy spots. Gmelin is accredited with the formal description of *Turdus naevius* in 1789, which he based on Latham's description of specimens of 'Spotted Thrush' from the collection of Joseph Banks.

Wood Thrush *Hylocichla mustelina*

This is a North American vagrant recorded only once in Iceland (1967) and once in Britain (1987).

As the name **Wood** implies, this is a woodland species. The vernacular name is reinforced by the unique generic name **Hylocichla**, which is derived from the Greek *hulē* (woodland/forest) and *kikhlē* (thrush).

The specific name *mustelina* tells of a bird that is (coloured) like a weasel (i.e. chestnut and white).

Swainson's Thrush *Catharus ustulatus*

This is a rare North American vagrant to Britain.

William **Swainson** (1789–1855) was an English naturalist. He never visited North America, but the dedication was made by his contemporary, the English explorer/naturalist Thomas Nuttall. Both Audubon and Bonaparte made similar gestures (warbler and hawk). In later life, however, Swainson became a controversial figure. The AOS announced in 2023 that all eponyms in vernacular names of American species will eventually be replaced.

The genus *Catharus* appears to originate in the fact that Bonaparte so admired the immaculate plumage of the first-named of this genus that he based the word on the Greek *katharos*, signifying 'pure'.

The bird is variable in plumage, but one form is smoky grey, leading to the specific name **ustulatus** (burnt).

Hermit Thrush *Catharus guttatus*

This is a rare North American vagrant to Britain.

Hermit Thrush was apparently so-called for its solitary habits, particularly in winter.

The specific name **guttatus** refers to the heavy droplet-shaped spots on the breast.

Grey-cheeked Thrush *Catharus minimus*

This is a rare North American vagrant to Britain.

Grey-cheeked Thrush is that and a bit more, with very grey tones overall.

The specific name **minimus** tells us that it is the smallest of the North American thrushes.

Veery *Catharus fuscescens*

This is a rare North American vagrant to Britain.

The name **Veery** is echoic. Sibley describes the call as *veer, verre* or *veeyer*.

The specific name *fuscescens* (dark) describes the greyish-brown upper-parts.

White's Thrush *Zoothera aurea*

This is a rare Eastern vagrant to Britain.

White's Thrush is the one name that commemorates Gilbert White (1720–1793), a Hampshire 'parson-naturalist' and an important English pioneer of the study of nature in the field.

While the name *Zoothera* means 'an animal hunter', it relates more to the bird's ground hunting style than to the pursuit of dramatic prey. It eats mainly earthworms, insects and fruits, but actively flushes out some prey items.

After some considerable upheaval in the taxonomy of this family, White's Thrush is now considered to be of the species *aurea* (golden). (Older references have it as *dauma*, when it was treated as a subspecies of Scaly Thrush.)

Siberian Thrush *Geokichla sibirica*

This is a rare Eastern vagrant to Britain.

Siberian is an obvious enough name, which supports the Latin *sibirica*.

The name *Geokichla* indicates that this is a 'ground thrush' (Greek *geo-*, ground, and *kikhlē*, thrush).

Song Thrush *Turdus philomelos philomelos*
Song Thrush *Turdus philomelos clarkei/hebridensis*

This is a common resident species in the form *clarkei*, but numbers of nominate race arrive from the Continent in winter.

Song is for its distinctive voice.

Turdus is the Latin for 'thrush'.

The specific name *philomelos* derives from the Latin/Greek words for a nightingale – a compliment to the bird's singing skills.

* In 1909 Hartert gave the subspecific name *clarkei* to the form found in most of Britain. This commemorated W.E. Clarke, an English ornithologist. In 1913, Clarke himself identified the form *hebridensis*, which as the name implies is found in the Hebrides and adjacent areas. (There is only one other subspecies and that occurs in Siberia.)

* Coward points out that, but for some confusion in Linnaeus's notes, this species might still have been named *Turdus musicus*. In 1915, BOU applied an ICZN clause designed to avoid confusion. This decision broke the rule of precedence to impose Christian Brehm's name *Turdus philomelos*.

Mistle Thrush *Turdus viscivorus*

This is a relatively common resident species.

Mistle relates to its love of mistletoe berries, which is also reflected in the specific name *viscivorus*, 'an eater of *visci*' (mistletoe berries). The berries are viscous (sticky) and readily stick to the bark of trees where they grow parasitically. They are transferred there when the bird wipes its beak.

* Among the many local traditional names for this species, Storm Cock is probably the best known; it derives from the male's habit of singing in bad weather.

* The shrillness of some of its notes produced such names as *Shriek* and *Shrike*, the latter of which was later borrowed for birds of the *Lanius* genus.

* The bird's aggressiveness in defence of its nest gave rise to a dialect name, the *Norman Thrush*, which appears to be an ironic Anglo-Saxon comment on their bullying feudal lords. The fact that this appears to be a very old usage is supported by another regional equivalent, *Norman Gizer*, in which the second word is corruption of the French *guis* for mistletoe.

Redwing *Turdus iliacus*

This is a 'winter thrush', migrating into Britain from further north.

Redwing describes the inner underwing, but there is also a rufous-red patch on the flank which shows when the bird is perched and is the basis of the specific name *iliacus* (Latin *ile*, flank).

(Common) Blackbird *Turdus merula*

While this is by far our most familiar thrush (see below), the use of **Common** is quite unnecessary unless another species of 'blackbird' is involved (and there are none in Europe).

The form **Blackbird** came into use for this species relatively late. It was first in evidence in 1486 and eventually prevailed over *Ouzel/Ousel* from the eighteenth century.

The specific name *merula* is the Latin for 'Blackbird' (cf. Modern French *merle*).

* The older name for this species was *Ouzel/Ousel*, an Old English form that persisted until the seventeenth century, and is still in use for the Ring Ouzel (below). A dialect form *Woosell* appears in Shakespeare's *A Midsummer Night's Dream*.

* It sometimes comes as a surprise that this species is a thrush, but that relationship is much more obvious in the speckle-breasted young.

Eyebrowed Thrush *Turdus obscurus*

This is a Siberian species that is a rare but regular vagrant to Europe.

Eyebrowed describes a thrush that has a broad white supercilium.

The specific name *obscurus* means 'dark' or 'dusky'.

Tickell's Thrush *Turdus unicolor*

This Himalayan species was recorded once in Germany in 1932.

Captain Samuel **Tickell** (1811–1875) was a British army officer and ornithologist who collected in India.

The specific name *unicolor* describes a bird that has a plain, shaded, grey plumage with just a few throat spots.

Fieldfare *Turdus pilaris*

This is another 'winter thrush', migrating into Britain from further north.

It is also a bird with enigmatic names in both the vernacular and specific forms.

Lockwood challenged any complacency over the apparently straightforward **Fieldfare**. With a name that was in evidence as early as 1100, it was too simple to assume that the birds 'fared over fields'. He argued that the *field* element was the Germanic *feld* (grey), referring to the extensive grey of the upper plumage, and that the *-fare* element likened the call to the grunt of a piglet: names in Welsh and Frisian supported that interpretation.

As for the specific name *pilaris*, Jobling points out that it is now Modern Latin for 'thrush', though it was born out of an error that originally confused two Greek words *trikhas* (thrush) and *thrix, trikhos* (hair) and then transposed that into a Latin form that appears to mean 'depilated'.

Ring Ouzel *Turdus torquatus*

This is a relatively scarce summer migrant to Britain, once widely known as the *Mountain Blackbird* because of its upland breeding habitats. The British population winters in the Atlas Mountains of North Africa.

The old name **Ouzel**, originally used for Blackbird, is retained exclusively for this species.

Ring described the white crescent on the breast, which is also recorded in the specific name *torquatus*. A torc was a breast ornament of twisted metal mostly worn in ancient times.

The next four species are all rare vagrants to Europe from areas of Siberia and all appear on the British list.

Black-throated Thrush *Turdus atrogularis*

Black-throated and *atrogularis* mean the same thing.

Red-throated Thrush *Turdus ruficollis*

Red-throated is paired by *ruficollis* (red-collared). The former is more accurate.

Dusky Thrush *Turdus eunomus*

Dusky seems to be an appropriate name for the bird's mantle plumage, while the specific name *eunomus* means 'orderly', seemingly to describe its generally smart appearance.

Naumann's Thrush *Turdus naumanni*

This species commemorates J.A. **Naumann** (1744–1826), a German farmer, amateur naturalist and author.

Only one *Turdus* thrush has been recorded as a scarce vagrant from North America.

American Robin *Turdus migratorius*

Use of the word **American** is necessary in Britain, but is usually omitted in America.

This species behaves more like a Blackbird than a Robin, but American colonists saw this red-breasted bird as a substitute for a well-loved old world species.

The specific name *migratorius* describes the fact that at least part of the population is migratory.

88. MUSCICAPIDAE: Old World Flycatchers and their allies

The family name is based on *Muscicapa*.

This family has been broadened in recent years and now embraces robins, chats, flycatchers, rock thrushes and wheatears. Future and perhaps radical changes might be expected. These are generally insectivorous species, so many are long-distance migrants that exploit the insect-rich northern latitudes in summer and retreat south in winter.

To some extent, retreat is also necessary for some populations of species that would not be generally considered migratory in Britain. When Robins came down to Ancient Greece in winter and left again in spring, they were replaced by Redstarts arriving for the summer. Aristotle tried to rationalise that by offering 'transmutation' as the explanation.

In the naming of the next two species and their several congeners, there seems to be an ongoing difference over the use of the words Bush and Scrub. The IOU and the other authorities now use the latter, but *Collins Bird Guide* (third edition, 2022) retains Bush.

Black Scrub Robin *Cercotrichas podobe*

This African species has now colonised Israel.

This is a **black** species with some white about the tail.

Scrub refers to marginal habitat, such as transitional areas of woodland regeneration or to areas of scattered bushes and weeds, such as Mediterranean garrigue.

Colonial French gives rise to the specific name *podobe*. Both D'Aubenton and Buffon used the name *Podobé*, which would seem to be a locally corrupted French word: Jobling suggests that the word started life as *peau daubé* (daubed skin). The French meaning of *dauber* is strictly limited to 'daubing with white', so focus appears to be on the white tipped under-tail feathers of an otherwise black plumage.

Rufous-tailed Scrub Robin *Cercotrichas galactotes*

This Mediterranean species has been recorded twice this century in Britain: in Norfolk in 2020 and in Cornwall in 2021. Both seem to have been of the Eastern race (see note below), the form that was also recorded by Coward in 1926.

The vernacular name seems to have been very changeable in recent history, but the IOC has now settled on the current form. The bird is indeed *Rufous-tailed*, and the generous tail is a significant feature.

The word **Robin** is used in a generic sense to describe a 'robin-like' species, though it is not that distant in relationship from the original (Eurasian) Robin.

Cercotrichas means 'thrush-tailed', while the specific name *galactotes* describes the 'milky-white' of the undersides.

* There are striking differences between the paler birds (*C. g. galactotes*) of the Western Mediterranean and the duskier form (*C. g. syriaca*) in the Eastern Mediterranean. In his 1926 list, Coward listed this species among the warblers under the names *Brown-backed Warbler* and *Grey-backed Warbler*, with the note that the latter form had been recorded in Britain four times.

Spotted Flycatcher *Muscicapa striata*

This is a declining summer visitor to Britain.

Spotted is something of a misnomer, since the bird is only lightly streaked.

Flycatcher describes an active manner of taking flying insects by sallying out from favoured perches.

Musicapa is based on the Latin *musca* (fly), and *capere* (to catch).

The specific name *striata* gives a more accurate picture than spotted: it means 'streaked'.

* This bird was once much more numerous in Britain and was often seen around houses, where its familiarity gave rise to such local names as *Wall Bird*, *Wall Chat* and *Wall Robin*, for its habits of nesting in recesses and ledges, or *Beam Bird* for similar reasons. These days, one of the favourite sites of this now-scarce species is a well-vegetated, shady churchyard.

Mediterranean Flycatcher *Muscicapa tyrrhenica*

In spite of evidence of genetic differences, this recent split from Spotted Flycatcher has not been acknowledged by all authorities. It breeds in the Balearics and in Corsica and Sardinia.

Mediterranean is slightly misleading, as it is very local within the Western Mediterranean.

The specific name *tyrrhenica* relates to the area surrounding the Tyrrhenian Sea, in this case specifically Corsica and Sardinia.

Dark-sided Flycatcher *Muscicapa sibirica*

Astonishingly, this Eastern Palearctic species was recorded in Iceland in 2016. It is a known long-distant migrant, but this bird had travelled halfway round the world. It winters in South and Southeast Asia.

Dark-sided is a simple description of a bird that is more streaked than the Asian Brown Flycatcher.

The specific name *sibirica* refers to its Siberian breeding areas.

Asian Brown Flycatcher *Muscicapa dauurica*

This is a very rare vagrant to Britain. Its normal range is broadly similar to the previous species.

Asian and **Brown** are both self-evident.

The specific name *dauurica* associates the bird with the region of Dauria, east of the Lake Baikal. The exact spelling is currently under review: elsewhere *daurica* suffices.

(European) Robin *Erithacus rubecula*

This familiar European species has evolved particularly confiding habits in Britain.

The word **Robin** has a complex history: in Old English, variants on the form *Reddock/Ruddock* were used (in parallel to Dunnock, which survives, and where *-ock* means 'a small bird'). Later the form *Redbreast* was the norm. In medieval England, personal names were often attached to familiar creatures such as *Reynard the Fox*, *Brock the Badger* and *Mag the Pie*. *Jenny Wren* and *Robin Redbreast* became 'characters' in folklore in their own right. Eventually Robin began to prevail in popular usage, but during the eighteenth and nineteenth centuries, and even into the twentieth century, official preferences swayed between Robin and Redbreast (see note below). Only Redbreast appears in Coward's 1926 list. Robin finally dominated after that. Meanwhile, birds elsewhere were given the name 'robin' because they showed some resemblance to the familiar form (see American Robin above).

Erithacus (in Greek as *erithakos*) is a name for an unknown species, which may have been the Robin.

The specific form *rubecula* is a Medieval Latin word for the Robin. Like *Erithacus*, the word includes the colour red.

* In Victorian times, postmen wore red uniform jackets and were nicknamed 'robins'. During that period the fashion for exchanging greetings cards at Christmas – delivered by the 'robins' – led to the bird becoming a symbol of Christmas.

* The founding chairman of the BOU, Alfred Newton, disliked the popular use of Robin and insisted on the formal use of *Redbreast*, which was one reason why the BOU retained the word contrary to popular usage.

* In 2015, writer and broadcaster David Lindo organised an unofficial poll in which the Robin was voted as Britain's national favourite bird, ahead of the Barn Owl and the Blackbird.

There is a current proposal to split two Canary Islands forms of Robin as full species under the names shown below. The recommendation is based on a combination of genetic, plumage and vocal differences. Each new species is endemic to the island after which it is named. Elsewhere in the archipelago, the recognised form is **E. rubecula**.

Tenerife Robin *Erithacus superbus*

The specific name *superbus* is self-evident.

Gran Canaria Robin *Erithacus marionae*

The original subspecies, which is now upgraded, was originally proposed by Dietzen et al. in 2015. The name *marionae* is dedicated to German biologist Marion Steinbüchel (b.1977).

White-throated Robin *Irania gutturalis*

This is a rare vagrant to Britain. It migrates from Africa to breed from Turkey eastwards.

White-throated is self-evident.

Irania indicates that the bird is normally found in and around Iran.

The specific name **gutturalis** underlines the significance of the throat (Latin *guttur*).

Bluethroat *Luscinia svecica*

This is a scarce passage visitor to Britain, even though it is widespread in Europe.

Bluethroat is a simple descriptive name. Birds are often reported as 'red-spotted' (subspecies *svecica*) or 'white-spotted' (subspecies *cyanecula*), the two forms most often seen in Britain. The spots, if present, are at the centre of the blue bib.

Luscinia is derived from the Latin for 'Nightingale'.

Linnaeus named this species, hence the use of the specific name *svecica* (Swedish).

Thrush Nightingale *Luscinia luscinia*

This is a scarce vagrant to Britain during migration. It has a more north-easterly range than the Common Nightingale.

The **Thrush** element appears to derive from the fact that the breast is lightly mottled (see *Sprosser*, below).

Nightingale (see below).

Luscinia luscinia suggests that the form known to Linnaeus was treated as the type species. (Common Nightingale was not distinguished until later (see below) and does not occur as far north as Sweden.) It is now accepted that this is the more northerly of two sister species, which can sometimes hybridise.

* In 1926, Coward lists the alternative name *Sprosser*, which was a name much more familiar in Victorian England. From Classical times nightingales were sought after as valued cage-birds. The trade came to a peak in the nineteenth century, when many thousands were trapped in the wild. These were then traded under the names *Nightingale* and *Sprosser*, the German name (meaning 'freckled') being used for the birds traded from Northern and Eastern Europe. Connoisseurs debated the relative merits of their song. The fashion faded fairly spontaneously in Britain after the First World War, especially after new bird protection laws were passed in 1921, and the name *Sprosser* faded with it.

(Common) Nightingale *Luscinia megarhynchos*

This is an annual, but scarce summer visitor to southern England, which is at the very north of its range.

The word **Common** is rarely used unless the Thrush Nightingale is juxtaposed. It is in fact far from common in Britain.

Nightingale is an old word of Germanic origins, which occurred in Old English as *nihtegale*. It is nominally and literally a 'night singer', though it often also sings in the daytime when attracting a mate.

The specific name *megarhynchos* means 'great-billed', but that seems to be a figurative way of saying that it has a big voice. Christian Brehm eventually described this as a separate species in 1831.

* The Nightingale has long been revered for its thrilling song, the virtuosity of which is enhanced by the added mystery of its often nocturnal delivery. Ever since it was linked by Homer to Penelope's laments for Odysseus, the bird and its song have been part of myth and poetry, a history that is too long to tell here. Isabella Tree (2018) offers a summary of its cultural importance in England in her chapter on the species. Sadly its magic is known to few in Britain today.

* One of the surprising things, however, is that neither nightingale species, in English or in Latin, relates in its name to the Greek legend of Philomela: her very name is the Greek for Nightingale. Philomela was the sister of Procne/Progne and both were transmuted into birds after their tragic tangle with King Tereus (see Ovid's *Metamorphoses*). Philomela became a Nightingale and Procne a Swallow. The French, however give a new twist to

the concept of 'sister species', because *Luscinia megarhynchos* is *le rossignol philomèle*, while *Luscinia luscinia* is *le rossignol progné*.

* Somewhat confusingly, the Song Thrush is *T. philomelos*, which appears to be a tribute to its singing skills.

Siberian Rubythroat *Calliope calliope*

This **Siberian** species is a rare vagrant to Britain.

Rubythroat underlines the rich red throat patch.

Calliope was the chief of the Greek Muses (nine goddesses who inspired the arts) and was renowned for having a beautiful voice.

Siberian Blue Robin *Larvivora cyane*

This Far Eastern species has been recorded as a vagrant to Britain.

There is no mystery in the vernacular names.

This bird was moved in 2014 from *Luscinia* to the resurrected genus *Larvivora* (meaning 'a caterpillar-eater').

The specific name *cyane* means 'cyan blue'.

Rufous-tailed Robin *Larvivora sibilans*

This Far Eastern species has been recorded as a vagrant to Britain.

As with the previous species, this bird was moved in 2014 from *Luscinia* to the resurrected genus *Larvivora* ('a caterpillar-eater').

Rufous-tailed is self-evident.

The specific name *sibilans* relates to its whistling song.

Mugimaki Flycatcher *Ficedula mugimaki*

Sightings of this Eastern species have been claimed in Britain, Italy, Norway and Russia. The British claim remains unproven, partly because of known imports in that period by the cage-bird trade.

Mugimaki is a Japanese name that means 'wheat-sowing', since the birds appear there during autumn migration when that activity occurs. They actually breed in Siberia and China and winter in Southeast Asia, Indonesia and the Philippines, so long-distance vagrancy westwards seems perfectly feasible.

Ficedula: see Pied Flycatcher below.

Taiga Flycatcher *Ficedula albicilla*

This species is a rare vagrant to Britain. It was formerly treated as a subspecies of the next species.

Taiga refers to the region of Eurasian boreal forest.

The specific name *albicilla* tells that the bird is 'white-tailed'.

Red-breasted Flycatcher *Ficedula parva*

This species is a fairly regular vagrant to Britain from its Eastern European breeding areas.

Red-breasted is rather more orange-toned in fact.

The specific name *parva* means 'small'.

Semicollared Flycatcher *Ficedula semitorquata*

This species is found in summer mostly in the Eastern Mediterranean and has not been recorded in Britain.

Semicollared and *semitorquata* both describe a limited collar of white that extends behind the bird's ear.

(European) Pied Flycatcher *Ficedula hypoleuca*

This is a summer visitor to Britain.

European separates this from at least one other 'Pied' Flycatcher.

Pied describes the black and white plumage of the breeding male.

Ficedula means fig-eating: it is a surprising fact that, outside the breeding season, this species eats figs, as well as some other fruit.

The specific name *hypoleuca* describes the bird as 'whitish', which seems odd for a pied form, even in faded grey winter plumage.

Collared Flycatcher *Ficedula albicollis*

This species is a rare vagrant to Britain from its Eastern European breeding areas.

Collared and *albicollis* (white-collared) refer to the broad white band on the neck of the male.

Atlas Pied Flycatcher *Ficedula speculigera*

This is a species of the **Atlas** Mountains in North Africa and has not been recorded in Britain.

The specific name *speculigera* relates to the fact that it bears (Latin *gerere*) a white 'wing mirror' or *speculum* (a term that is used much more in describing ducks).

Red-flanked Bluetail *Tarsiger cyanurus*

This is a scarce vagrant to Britain from the Russian taiga.

All parts of the vernacular name are descriptive and accurate, but it is a typical late-coming 'book name'.

Tarsiger relates to the *tarsus* (lower leg) and the Latin verb *gerere* (to bear), but the relevance is unclear.

The specific name *cyanurus* refers to the blue tail.

Common Redstart *Phoenicurus phoenicurus*

This summer visitor is found in old open woodland and parkland.

The word **Common** is used relatively frequently to distinguish this species from the Black Redstart. Both may coincide on passage.

The name **Redstart** uses the archaic word 'start' (tail) in a name that originally paired with *Redbreast* (compare the French *rouge-queue* and *rouge-gorge* for the same two birds). In 1544 Turner had introduced the form Red Tail, but Redstart was later preferred by both Ray and Pennant. *Start* is an old form with Germanic roots, which also occurred historically in *wagstart*, an early alternative to wagtail and in *whitestart* (wheatear).

Phoenicurus is from the Greek *phoinix* (crimson or red) and *ouros* (tailed).

Daurian Redstart *Phoenicurus auroreus*

This Asian species has been reported in European Russia, though a British record is considered a probable escape.

Daurian refers to a region of Central Asia.

The specific name *auroreus* derives from *aurora* (dawn); the connection is that this is an Eastern species and the sun rises in the east.

Black Redstart *Phoenicurus ochruros*

This is a summer migrant. Small numbers may winter at favoured sites.

The male is dark enough to be called **Black**, though the female is sooty-brown.

In the specific name *ochruros*, Gmelin used a shade of ochre, but that is a very variable substance. In reality the shade is not so different from the Common Redstart.

* A rufous-bellied form known as the 'Eastern Black Redstart' has been recorded in Britain.

Moussier's Redstart *Phoenicurus moussieri*

This small North African species has been recorded in Britain.

Its two names commemorate a French surgeon-naturalist Jean **Moussier** (1795–1850).

Güldenstädt's Redstart *Phoenicurus erythrogastrus*

This is a high-altitude species of the Caucasus and eastwards that has never been recorded in Britain. It is also considerably larger than any of the previous three species.

The name commemorates J.A. **Güldenstädt** (1745–1781), a Latvian-German naturalist and explorer who worked in Russia.

The specific name *erythrogastrus* means 'red-bellied.

Eversmann's Redstart *Phoenicurus erythronotos*

This Central Asian species has been recorded in Russia and Israel.

The name commemorates Alexander E.F. **Eversmann** (1794–1860), a Prussian biologist who worked in Russia. His name also occurs in the specific name of the Yellow-eyed Dove.

The specific name *erythronotos* describes the male's red back.

(Common) Rock Thrush *Monticola saxatilis*

This montane and migratory species of Southern Europe has been recorded in Britain.

Common separates it from other similar species, but is often omitted.

Rock relates to its preference for rocky terrains.

Thrush relates to its similarity to the thrushes, with which it and the Blue Rock Thrush were previously grouped. The move into this family is relatively recent.

Monticola states that it is a 'mountain dweller'.

That is underlined by the specific name *saxatilis*, which speaks of a 'rock-frequenter'.

* The IOC prefers the name *Common*, but the *Collins Bird Guide* (third edition, 2022) has it as *Rufous-tailed*. The bird also has rufous undersides.

Blue Rock Thrush *Monticola solitarius*

This is a Southern European species that is mainly resident. Unlike the previous species, this bird breeds at a range of altitudes and often near human habitation. In spite of being mainly sedentary, it has been recorded in Britain.

Blue certainly fits the blue-grey male, though shadow or silhouette views can make it look almost black at times. The female, on the other hand, is dark brown.

The specific name *solitarius* seems to originate in 1678 from Willughby and Ray, the former providing the name and the latter commenting that the bird sat alone to sing in the morning.

Whinchat *Saxicola rubetra*

This is a long-distance summer migrant that has declined in Britain in recent years.

Whinchat is composed of two elements. The first, *Whin*, is an old dialect word for gorse or furze. It may well have been tagged onto this species in error, since the Whinchat actually tends to prefer damper habitats with grass and bramble. The second is '*chat*', which relates to a sharp 'tek-tek' alarm call.

Saxicola indicates that the bird lives in stony territory, though that seems to apply here only because of its close relationship to the stonechats.

The specific name *rubetra* seems to redress the balance in including *rubus* (bramble).

European Stonechat *Saxicola rubicola*

This is a partial migrant.

In recent times the status of the stonechat complex has undergone a complete overhaul. As a result the British form is now the **European** Stonechat, while other full species have been created.

Historical aspects of the name **Stonechat** are confused with the Wheatear, which was once known as the Stone Chack and is far more inclined to favour rocky habitats.

Saxicola retains the 'stone' theme in a form meaning 'stone-dweller'.

The new specific name is, confusingly, *rubicola* (bramble-dweller), but that was the name of the former subspecies.

* The specific name *torquatus*, which pertained for many years and is still found in older references, has been retained by the African species that was split from the Eurasian forms.

Siberian Stonechat *Saxicola maurus*

This species has been recorded in Britain.

Siberian tells us that this is an eastern species.

The specific name *maurus* derives from Moorish and means black.

Amur Stonechat *Saxicola stejnegeri*

The **Amur** region is in eastern Siberia. This is the formal international name.

The alternative name **Stejneger's** Stonechat is also found in the specific name *stejnegeri*.

It commemorates the Norwegian-American naturalist L.H. **Stejneger** (1851–1943) (see also Scoter.)

Canary Islands Stonechat *Saxicola dacotiae*

This form is endemic to Fuerteventura in the **Canary Islands** and formerly carried the name of that island.

The specific form *dacotiae* relates to *Dacos*, a former Moorish name for that island.

Pied Bush Chat *Saxicola caprata*

This widespread Asian species has been noted in Israel and Cyprus.

The English name is self-evident.

Brisson apparently noted the local name in Luzon, Philippines, as *Maria-capra*, and the specific name *caprata* was derived from that.

Anteater Chat *Myrmecocichla aethiops*

This sub-Saharan species has been recorded in northern Chad.

The name **Anteater** is self-evident.

Myrmecocichla is a compound of two Greek works: *murmēx* (ant) and *kikhlē* (thrush).

The specific name *aethiops* relates to Ethiopia, which is part of its range.

Wheatears of the genus Oenanthe make up a considerable slice of this family and 21 of them are recorded in the Western Palearctic.

As well as the Northern Wheatear, six vagrant wheatear species have been recorded in Britain. Two more species are found within the political boundaries of Europe and a number more are found within the peripheral regions of North Africa or the Middle East.

Wheatear evolved from an Anglo-Saxon form *hwit aers* (white arse), which described the white rump and mostly white tail of the familiar form. A little-known archaic name *Whitestart* conveyed the same idea. Most wheatears share that feature, though there are a few exceptions in the list below.

All wheatears are now classed in the genus **Oenanthe**, which derives from the Greek for 'vine blossom'. It was used by Aristotle for a bird whose migration northwards coincided with that season in Greece.

(Northern) Wheatear *Oenanthe oenanthe*

'Greenland' Wheatear *Oenanthe oenanthe leucorhoa*

This is a summer migrant and a passage species.

The use of **Northern** is required in a world context.

'Greenland' Wheatear, *O. o. leucorhoa* (white rumped) is a subspecies, which migrates through Britain later and is often identifiable from its colour and size, aided by a later time of migration.

Atlas Wheatear *Oenanthe seebohmi*

This was formerly *Seebohm's* Wheatear, but the IOC preferred the **Atlas** Mountains of Morocco as the basis of the new name, when changes were made in 2022.

Henry Seebohm (1832–1895) was an English ornithologist and still appears in the specific name *seebohmi*.

Isabelline Wheatear *Oenanthe isabellina*

This Middle-Eastern species is sometimes recorded as a vagrant to Britain.

Isabelline and *isabellina* describe a faded tawny colour. The word is reputed to originate with the historical figure of the fifteenth-century Queen Isabella of Spain, who vowed not to change her undergarments until the Moors had been defeated.

Hooded Wheatear *Oenanthe monacha*

This species is found in Egypt and the Arabian Peninsula.

Hooded refers to the shape of the pied pattern of the male.

The specific name **monacha** alludes to the hooded cape of a monk (see also *Cinereous Vulture* and *Monk Parakeet*).

Desert Wheatear *Oenanthe deserti*

This is an occasional vagrant to Britain.

In spite of the names **Desert** and *deserti*, this species shuns full desert and prefers semi-desert habitats in North Africa and the Middle East.

Western Black-eared Wheatear *Oenanthe hispanica*

This species was formally split from the Eastern form in 2020.

Western in fact means 'Western Mediterranean'.

That is underwritten by the specific name *hispanica* (Spanish), though it is found more widely in the region.

Eastern Black-eared Wheatear *Oenanthe melanoleuca*

The IOC split this from the Western form in 2020.

Eastern means 'Eastern Mediterranean'.

The specific name *melanoleuca* means black and white: the buff coloration is much subdued in this form.

Cyprus Wheatear *Oenanthe cypriaca*

This species is endemic to **Cyprus** and was once treated as a race of Pied, which it resembles.

The specific name *cypriaca* also refers to Cyprus.

Pied Wheatear *Oenanthe pleschanka*

This is a Middle-Eastern species that is a scarce vagrant to Britain.

Pied is an obvious name for a male that is largely black and white.

The specific name *pleschanka* derives from Russian *plesch* for a bald spot, a reference created by the Russian Ivan Lepechin to the white head: like the Bald Eagle, the head is feathered in spite of the name.

Red-Rumped Wheatear *Oenanthe moesta*

This species is found widely across North Africa and into the Middle East.

Red-rumped is self-explanatory, though the colour is actually a tawny-rufous.

The specific name in this case is *moesta* (sad or mourning), which seems to be related to a very melancholy song.

Blackstart *Oenanthe melanura*

This species is found in Egypt, Israel, Jordan and the Arabian Peninsula.

The name **Blackstart** links neatly to the redstarts treated earlier. With an all-black tail and a specific name *melanura* to underline that, it was transferred in 2012, with other members of the former genus *Cercomela* (black-tail), into the genus *Oenanthe*.

Variable Wheatear *Oenanthe picata*

This Central Asian species has been recorded in Israel.

Variable is a justifiable name, since the bird has three morphs: some have white bellies, some black, and some have white caps.

The specific name *picata* occurs with other black and white species: it means 'daubed with tar'.

Black Wheatear *Oenanthe leucura*

This is a species found in Iberia, as well as in North Africa.

Black is true in all but the white rump, which is underlined by the specific name *leucura* (white tailed).

* On the face value, a name such as *leucura* is not of great use in a family of white rumps, but in a practical sense the white rump readily sets this bird apart from a Blue Rock Thrush in a distant sighting.

White-crowned (Black) Wheatear *Oenanthe leucopyga*

This is a species recorded as vagrant to Britain. It is normally breeds in North Africa.

White-crowned is self-evident.

Black has now been dropped by the IOC, though BOU still retains it.

The specific name *leucopyga* means 'white rump' (which, once again, is a little pointless in the wheatear family).

Hume's Wheatear *Oenanthe albonigra*

This species is found in Iran, Iraq and Kuwait and around the Arabian Gulf.

Allan **Hume** (1829–1912) was a British ornithologist based in India.

The specific name *albonigra* means 'black and white' (yet another unhelpful name in this family).

Finsch's Wheatear *Oenanthe finschii*

This species is found in Turkey and other parts of the Middle East, with some wintering in Cyprus and Egypt.

The German naturalist F.H.O. **Finsch** (1839–1917) is commemorated.

Maghreb Wheatear *Oenanthe halophila*

Maghreb is a region of north-west Africa where the bird is found.

The specific name *halophila* suggests that the bird is 'a lover of rock salt'. It certainly prefers rocky habitats.

Mourning Wheatear *Oenanthe lugens*

This is a species found in Egypt and neighbouring parts of the Middle East.

This is far from the blackest of wheatear species, which might be expected of a bird with that name. It is in fact distinctly pied, so that is not the reason. It seems probable that **Mourning** and its Latin equivalent, *lugens*, relate to its rather sad and monotonous song.

* It is unlikely that the name was earned by what was long considered to be an all-black 'morph', which was so little known that it was only designated as a separate species in 2021 (see next entry).

Basalt Wheatear *Oenanthe warriae*

This is a species of Jordan and South Syria, first identified as a taxon in 2011 and split in 2021 from the Mourning Wheatear.

Basalt comes from the basalt desert of its habitat.

The specific name *warriae* is the Latinised surname of Mrs F.E. Warr, a librarian at the British Museum of Natural History. It was a dedication of thanks by the researchers concerned, and is also one of the most recently created eponyms, of which there are few twenty-first-century examples.

Kurdish Wheatear *Oenanthe xanthoprymna*

Kurdish indicates the geographical region inhabited by the bird (parts of Turkey, Syria, Iran and Iraq).

The specific name *xanthoprymna* indicates that the bird has a 'yellow rear' (Greek *xanthos*, yellow, and *prumna*, rear or stern). It is actually a rich orange-yellow.

Red-tailed (Persian) Wheatear *Oenanthe chrysopygia*

The species was long treated as a subspecies of the neighbouring Kurdish Wheatear, but is now generally treated as a full species. The IOC uses the name **Red-tailed**, which seems obvious at face value. However it is confusingly similar to Red-rumped and somewhat contradicts the specific name *chrysopygia*, which means 'golden-rumped'.

The alternative name **Persian** Wheatear, the name used in the *Collins Bird Guide* (third edition, 2022) and by others, seems preferable for that reason. Persia is the historic name of modern Iran.

Dippers to Pipits

89. CINCLIDAE: DIPPERS

The family name is based on *Cinclus*.

(White-throated) Dipper *Cinclus cinclus*

This is a resident species in Britain.

White-throated is rarely used: there are no other relatives in the Western Palearctic and few others in the wider world.

Dipper is based in part on the bird's behaviour of bobbing, perhaps to triangulate its vision, perhaps to show off its fitness. The name is probably also influenced by its habit of dipping under water to feed.

The name *Cinclus* is based on a bobbing bird mentioned by Aristotle.

* Two subspecies – *gularis* (throated) and *hibernicus* (Irish) – are found in the British Isles. Both have rufous bellies, which makes the Continental form all the more obvious to birders when it makes a rare visit to Britain. That form has all-black undersides, so is known colloquially as a '**Black-bellied' Dipper**.

* *Water Ouzel* was once used widely in England for the Dipper (see thrushes) and that form is still used locally in North America for the all-grey American Dipper.

90. NECTARINIIDAE: Sunbirds

The family name is based on *Nectarinia*, one of the genera of a family that feeds on nectar.

These are mainly tropical birds of a group that evolved to fill a similar ecological niche in Africa and Asia to the New World hummingbirds and to the Australasian honeyeaters.

Pygmy Sunbird *Hedydipna platura*

This sub-Saharan species has been recorded in northern Chad and in northern Mauretania.

Pygmy suggests that this is a small species.

Sunbird describes birds that are attracted to sun-loving flowers in order to exploit their nectar.

Hedydipna in effect means 'dainty-dipping', for the nectar-eating habits.

The specific name *platura* suggests that the tail is 'oar-like': the male has two long spatulate tail plumes.

Nile Valley Sunbird *Hedydipna metallica*

Nile Valley is a simple description of this bird's range.

The specific name *metallica* emphasises the sheen of its upper plumage.

Palestine Sunbird *Cinnyris osea*

Palestine forms part of the bird's range.

Cinnyris is simply the name of an anonymous bird mentioned by Hesychius.

The specific name *osea* is derived from the Greek for 'sacred'. In the culture of Ancient Egypt, sunbirds were linked with death rituals.

Purple Sunbird *Cinnyris asiaticus*

This Asian species is well established in the Gulf region up to Kuwait and southern Iran.

Purple and *asiaticus* are self-explanatory.

91. PASSERIDAE: True Sparrows

The family name is based on *Passer*.

The word **Sparrow** has ancient roots in the Germanic languages. It first appeared in a recognisable English form in the Middle Ages as *sparewe*.

For several centuries the word 'sparrow' was used loosely to describe a number of small species: alongside House Sparrows and Tree Sparrows, there were Reed Sparrows, Hedge Sparrows, and so on. Hedge Sparrow is still sometimes used, though the name Dunnock is preferred in that case. In more formal, modern British usage, sparrow is restricted to the 'true' sparrows of the Passeridae. In American English, however, the word is also widely used for birds related to the buntings and known collectively as the 'American sparrows', the Passerellidae.

Pale Rockfinch *Carpospiza brachydactyla*

This Middle-Eastern species was formerly called the **Pale Rock Sparrow** (which is self-evident), but recent studies have raised unanswered questions about its genetics. The name **Rockfinch** may be temporary.

Carpospiza contains the Greek for 'wrist' and for 'finch', though the reason for that is unclear.

The specific name *brachydactyla* means short-toed (see also Treecreeper).

Rock Sparrow *Petronia petronia*

Surprisingly, since only the eastern part of the Mediterranean population is migratory, this species has been recorded in Britain.

For **Sparrow** see the introduction to this section.

Clearly this is a sparrow of **rocky** habitats.

Petronia was originally a Bolognese name for the bird, which derives from the Latin *petronius* (of the rocks).

(White-winged) Snowfinch *Montifringilla nivalis*

This is a montane species of Southern Europe.

White-winged is only used in a world context, since this is the only European species of Snowfinch.

Snowfinch is a traditional name for species living above the snowline. It is now considered to be a sparrow with finch-like characteristics.

Montifringilla describes the bird as a 'mountain finch (see *Fringilla*, Chaffinch).

The specific name *nivalis* returns to the 'snow' of the vernacular name (Latin *nix*, *nivis*, snow).

Yellow-throated Bush Sparrow *Gymnoris xanthocollis*

This is a migratory species that is found in summer in Turkey.

For **Sparrow** see the introduction to this section.

Yellow-throated is echoed in *xanthocollis* (yellow-necked).

Gymnoris refers to 'naked nostrils' (Greek *gumnos*, bare; *rhis, rhinos*, nostril).

(Eurasian) Tree Sparrow *Passer montanus*

This species now has a greatly reduced range in Britain and is absent from some parts altogether.

It is **Eurasian** because there is an unrelated species of 'Tree Sparrow' (*Spizella arborea*) in North America (where the introduced Eurasian form is sometimes colloquially called the *German Sparrow*).

The name **Tree** Sparrow was introduced by Pennant in 1770 to distinguish this more rural and previously unnamed form from the House Sparrow.

For **Sparrow** see the introduction to this section.

Passer is the Latin for 'sparrow'.

The inaccurate specific name *montanus* was sometimes anglicised as *Mountain Sparrow*.

Spanish Sparrow *Passer hispaniolensis*

Surprisingly, this relatively sedentary species has been recorded in Britain (probably ship-assisted, since most records are close to ports).

It is native to many parts of the Mediterranean, so **Spanish** is a slightly misleading name.

For **Sparrow** see the introduction to this section.

The specific name *hispaniolensis* reiterates the association with Spain.

Italian Sparrow *Passer italiae*

Italian Sparrow is a confusing concept and the subject of ongoing debate. The Italian population, at least, particularly in the northern half of the country, seems to be composed of 'stable hybrids', birds that originated as a cross between Spanish and House Sparrows and now regularly produce similar forms with stable intermediate characteristics when they breed together. Elsewhere, first-cross hybrids will generally resemble the stable form, but they vary in plumage detail. The 'stable hybrid' is now accepted by many authorities as a full species, hence *Passer italiae*, which is recognised by the IOC and by most other authorities.

House Sparrow *Passer domesticus*

This common sparrow associated itself with humans almost as soon as our ancestors started cultivating grain in the Middle East. It is still a relatively common species in Britain, in spite of a general reduction in numbers.

House reflects the bird's familiarity, as does the specific name *domesticus*.

Passer is simply the Latin for 'sparrow'.

* House and Tree Sparrows collectively boomed in Britain in the years when farming practices left waste grain in the fields, but declined as modern machinery led to tidier farms. For much of the eighteenth and nineteenth centuries they were seen as major agricultural pests and were persecuted by landowners, who encouraged local people to form 'sparrow clubs', which earned bounties for the number of birds killed. Today it is not surprising that there are far fewer of them in our much more sterile agricultural landscape, but not so well understood is the fact that they have also declined in urban habitats.

Iago Sparrow *Passer iagoensis*

This species is endemic to the Cape Verde Islands. In 2013, four ship-assisted vagrants were seen in the Netherlands.

Iago and *iagoensis* are derived from Sao Tiago Island.

Desert Sparrow *Passer simplex*

This is a sparrow of the North African sand **deserts**.

The specific name *simplex* tells of a bird with 'simple, plain' plumage.

Sudan Golden Sparrow *Passer luteus*

This sub-Saharan species has been recorded in the Cape Verde Islands, Western Sahara and Morocco. It was recorded in both France and the Canaries in 2023.

Sudan is only part of the bird's range, but distinguishes this from a similar Arabian form.

Golden and *luteus* both describe a bright yellow bird.

Dead Sea Sparrow *Passer moabiticus*

This species is found around the **Dead Sea**, but also in other parts of the Middle East.

The specific name *moabiticus* refers to the Moab region, which is adjacent to the Dead Sea

92. PLOCEIDAE: Weavers

The family name is based on the genus *Ploceus*.

A number of this family have established non-native free-flying colonies in various parts of Europe and the Middle East. These generally originate from the cage-bird trade, which is also true of the subsequent two families. Success in their new locations depends in part on climate compatibility.

Rüppell's Weaver *Ploceus galbula*

This species of the Horn of Africa and Arabia has been recorded in Kuwait.

Eduard **Rüppell** (1794–1828) was a German naturalist and explorer.

Weaver is a name given to this family of birds for their skills in building complex woven nests.

Ploceus means 'weaver' (from the Greek *plekō*, to plait).

The specific name *galbula* means 'small and yellowish'.

Village Weaver *Ploceus cucullatus*

This African bird has been recorded as a feral species in Spain and Portugal.

As the name **Village** implies, this species often lives close to humans in its native regions.

The specific name *cucullatus* means 'hooded' and refers to the male's black mask (see also Hooded Merganser *Lophodytes cucullatus*).

Black-headed Weaver *Ploceus melanocephalus*

This African species is well established as a feral species in Spain and Portugal.

Black-headed and *melanocephalus* mean the same.

Streaked Weaver *Ploceus manyar*

This South Asian species can be found living in Kuwait and Egypt.

Streaked is self-evident and serves to separate this in its native range from a plainer species.

The specific name *manyar* derives from a Tamil name for birds of this family

Yellow-crowned Bishop *Euplectes afer*

This African species has a strong feral population associated with rice cultivation in the Iberian Peninsula.

Yellow-crowned is self-evident, though the colour is only present when the male is in breeding plumage.

Bishop was a name given originally to the type species for this genus, the Southern Red Bishop, because its red and black were reminiscent of the robes of a Catholic bishop. (An alternative name for that species is *Grenadier* because of the soldier's black trousers and scarlet tunic.)

Euplectes prefaces *plekō*, to plait (see *Ploceus*) with *eu* (fine), another tribute to the bird's nesting skills.

The specific name *afer* simply means African.

The family name derives from *Estrilda*.

In this case, the use of 'finch' in the vernacular collective name is a tradition in which the name of the 'true' finches is applied to species that have evolved finch-like structures and habits. These are not at all closely related to the family that lends its name.

A number of this family have established non-native free-flying colonies in various parts of Europe and the Middle East. These generally originate from the cage-bird trade.

African Silverbill *Euodice cantans*

This mainly sub-Saharan species occurs naturally in southern Algeria.

African distinguishes this from the next species.

Silverbill is self-evident.

Euodice uses the prefix *eu* (fine) with *ōdikos* (singing).

The specific name *cantans* reinforces the generic name with the Latin for 'singing'.

Indian Silverbill *Euodice malabarica*

This species is found naturally in the **Indian** subcontinent but exists as a non-native feral species in Egypt and the Middle East.

The specific name *malabarica* refers to the Malabar Coast of India.

Scaly-breasted Munia *Lonchura punctulata*

This oriental species is widely established as a feral species in Portugal.

Munia is a name originating in Hindi.

Scaly-breasted is supported by the specific name *punctulata* (spotted).

Lonchura means 'spear-tailed'.

Orange-cheeked Waxbill *Estrilda melpoda*

This sub-Saharan species has a feral presence in Spain, with a Valencia population strong enough for the species to have been admitted to the Spanish list in 2019.

Orange-cheeked is self-evident.

Waxbill describes a bill that is sealing-wax red.

Estrilda (see next species).

The specific name *melpoda* means 'a singer of songs'.

Common Waxbill *Estrilda astrild*

This sub-Saharan species is found in a feral state in Spain, Portugal and Cape Verde among others.

Common is for the type species of the family.

Estrilda and *astrild* are based on Astrild, a Nordic name for Cupid, which was popularised by the seventeenth-century Swedish poet Georg Stiernhielm. The link appears to be between the flaming torch of the god and the flame-shaped scarlet facial mask of the species. The name was used by Linnaeus, though it may be linked with a name originally used in the cage-bird trade.

Black-rumped Waxbill *Estrilda troglodytes*

This sub-Saharan species is represented in Spain and Portugal by feral populations.

Black-rumped is self-evident.

The specific name *troglodytes* (cave dweller) recalls the name of the Wren, but that is only because it too makes a dome-shaped 'cave' nest.

Red Avadavat *Amandava amandava*

This is a native Asian species that has a number of implanted colonies in such places as Spain, Portugal, Italy and Egypt.

Red is self-evident.

Avadavat and the scientific name *Amandava* are corruptions that both derive from the name of the Indian city Ahmadabad.

Red-billed Firefinch *Lagonosticta senegala*

This sub-Saharan species is recorded in Morocco and Algeria.

Red-billed is self-evident.

Firefinch describes a scarlet-coloured bird.

Lagonosticta is a compound word describing the bird's spotted flanks.

The specific name *senegala* relates to Senegal, but the bird has a much wider range.

94. VIDUIDAE: WHYDAHS

The family name is based on *Vidua*.

This is yet another artificial introduction or escape.

Pin-tailed Whydah *Vidua macroura*

This sub-Saharan species has an established colony in Portugal.

Pin-tailed is a remarkably understated name for a tail that is almost twice as long as the body of a breeding male. Outside the breeding season the birds are finch-like in appearance.

Whydah is sometimes rendered as *Widow* or *Widowbird*, but Jobling quotes Edwards and Latham to point to the possibility that the former name is the original and that *Widow* was an Anglicised approximation. An old West African kingdom of that name existed until the eighteenth century and was an important trading gateway into that part of Africa.

Vidua is the Latin for 'widow': it seems that the English approximation influenced that choice.

95. PRUNELLIDAE: Accentors

The family name is based on *Prunella*.

Alpine Accentor *Prunella collaris*

This species has been recorded in Britain.

Alpine implies that it is found in mountainous areas (mainly in Southern Europe).

Accentor was a word devised by Bechstein to provide a collective name for this family to parallel the word 'warbler'. With so few Western species, there is little wonder that it was not a popular idea.

Prunella: see Dunnock below.

The specific name *collaris* refers to a small white 'necklace' which shows in breeding plumage.

Siberian Accentor *Prunella montanella*

This species has been recorded in Britain.

Siberian indicates that this is an Eastern species.

Prunella: see Dunnock below.

The specific name *montanella* implies that this is a mountain species, but the diminutive ending suggests that the bird (rather than the mountain) is a small one.

Radde's Accentor *Prunella ocularis*

This Middle-Eastern species is found in several countries of the Eastern Mediterranean.

Gustave **Radde** (1831–1903) was a German naturalist (see also *Radde's Warbler*).

Prunella: see Dunnock below.

The specific name *ocularis* (of the eye) seems intended to draw attention to the broad black eye-stripe that contrasts with a broad white supercilium to distinguish this species.

Black-throated Accentor *Prunella atrogularis*

This species has been recorded in Western Europe, but not in Britain.

Prunella: see Dunnock below.

Black-throated and *atrogularis* mean the same thing.

Dunnock *Prunella modularis*

This is a widespread and common species.

Dunnock is a well-preserved traditional name that was in written evidence in the fifteenth century, with local variants such as *dunek*. The word itself is based on *Dun* (brown) with the addition of the Anglo-Saxon diminutive suffix *-ock*. In short, it could be seen as the original version of the birder's enigma, the Little Brown Job. (*Dunlin* approximates to the same idea.)

Prunella: Gessner based the Latin form on the German *Braunelle* as early as 1555. That German folk name was an exact parallel of *Dunnock*, that is, *braun* with a diminutive ending.

The specific name *modularis* means melodious and reflects the bird's sweet warble.

* A common vernacular alternative, *Hedge Sparrow*, which is probably almost as old as Dunnock, still occurs in popular usage. An early twentieth-century push to promote Selby's 1825 confection *Hedge Accentor* (as used by Coward in 1926) was seen off in 1949, when the influential ornithologist Max Nicholson successfully championed the traditional name, Dunnock.

* The -*ock* suffix, which generally indicates smallness, also occurs in now-disused names such as *Pinnock* (also this species), *Ruddock* (Robin) and *Wrannock* (Wren), the former being in evidence as early as *c.*1000.

96. MOTACILLIDAE:
Wagtails and Pipits

The family name is based on *Motacilla*.

The English word **Wagtail** was first in evidence as *Wag tayle* in 1510 and slowly superseded an older form, *Wagstart* (compare *Redstart* and the later creation *Blackstart*), to become the standard name for the three British species, which were eventually distinguished by their own epithets.

The word *Motacilla* was found in the work of the Roman Varro and entered Modern Latin usage in 1555. However, it was misinterpreted as meaning something like a 'tail-shaker'. The result was that the *cilla* element started to mean tail.

Forest Wagtail *Dendronanthus indicus*

This Asian species has been recorded in Kuwait.

Forest underlines that this is more of a woodland species than other wagtails.

It has its unique genus *Dendronanthus*, which confirms its woodland preferences in *dendron-*, and its close relationship to the pipits in *anthus* (see below).

The specific name *indicus* relates to India, one of its main regions.

(Western) Yellow Wagtail *Motacilla flava*

This is a common summer migrant to Britain.

The word **Western** was added in 2017 with the split of the Eastern forms (see next species).

Yellow is apt for the dominant colour worn at the start of the breeding season.

The specific name *flava* means yellow, but that is only part of the picture: there are no fewer than ten subspecies throughout its breeding range and the males are sufficiently different as to be identifiable in breeding plumage. Caution is needed, since intergrades do occur too, but six subspecies have been accepted onto the British list:

– *flava* (yellow)	('Blue-headed Wagtail') in central Europe and western Russia
– *flavissima* (most yellow)	('British Yellow Wagtail') in Britain, north-west France, Belgium, Netherlands
– *thunbergi* (C.P. Thunberg)	('Grey-headed Wagtail') in Scandinavia and north European Russia
– *iberiae* (Iberian)	('Spanish Wagtail') in southern France, Iberia and north-west Africa
– *cinereocapilla* (ashy-headed)	('Ashy-headed Wagtail') in Sardinia, Italy and Sicily
– *feldegg* (C.F. von Feldegg)	('Black-headed Wagtail') in south-east Europe, Turkey, Caucasus, Iran, south Central Asia

The remaining four are recorded in the Western Palearctic:

– *pygmaea* (pygmy)	('Egyptian Wagtail') in Egypt

- *beema* (Beema River, India) ('Sykes's Wagtail') in northern Kazakhstan and southern Russia
- *leucocephala* (white-headed) ('White-headed Wagtail') in western Mongolia
- *lutea* (yellow) ('Yellow-headed Wagtail') in southern Russia

* In 1838, the British form, now *M. f. flavissima*, was described as a full species by Bonaparte as *Motacilla raii*, dedicated to the seventeenth-century naturalist John Ray. As a result it was known to authors such as Yarrell and Jenyns as Ray's Wagtail. Both Hartert (1912) and Coward (1926) listed it as *Motacilla flava rayi*. Unfortunately for Ray, Blyth's 1834 choice of *flavissima* eventually had precedence, so the use of the vernacular form soon faded.

Eastern Yellow Wagtail *Motacilla tschutschensis*

The split from the Western form was accepted by IOC in 2017. This form has been identified several times in Britain.

Eastern (as opposed to Western, above).

The specific name *tschutschensis* is a version of the Chukhotski Peninsula in Siberia.

Citrine Wagtail *Motacilla citreola*

This species is a scarce but regular vagrant to Britain.

Citrine and *citreola* both indicate a lemon-yellow colour.

Grey Wagtail *Motacilla cinerea*

This is a resident species, usually found near moving fresh water.

Grey and *cineria* say the same thing. Those names seem painfully inadequate for this elegant, lemon-bellied species, which deserves a much more colourful name. Ray was responsible in 1678 for the translation of Aldrovandi's earlier Latin form.

White Wagtail *Motacilla alba*

Pied Wagtail *Motacilla alba yarrellii*

White Wagtail is the species name generally reserved in Britain for the nominate form, which visits Britain annually from the Continent.

White is echoed by *alba*: this is in fact a grey-backed form, generally much paler than the British subspecies. Ray first used the term White Wagtail in 1678, probably as a translation of Gessner's *Motacilla alba*.

Pied Wagtail was coined by Latham in 1783 and has been considered the normal British name for the local subspecies since used by Selby in 1825. It is darker than the nominate form, the black and white contrasting to give a more striking appearance.

Its subspecific name *yarrellii* commemorates British naturalist, William **Yarrell** (1784–1856).

African Pied Wagtail *Motacilla aguimp*

This African species occurs in Egypt.

Pied should not be confused with the British form of the White Wagtail (see above).

The specific name *aguimp* derives from a local Namaqua name, *a-guimp* (shore-runner). That gem was preserved by Levaillant in 1805.

Pipit emerged as an echoic word in the late eighteenth and early nineteenth centuries for birds that had previously been known a 'titlarks' because they resembled small larks. The use of the name was finally established by Selby in 1833, when he distinguished between Meadow Pipit, Tree Pipit and Rock Pipit. This usage was confirmed by Yarrell in 1843.

The genus *Anthus* is derived from a Greek legend in which the young Anthus led his father's horses to a poor pasture and was killed by them. In the ensuing mayhem the gods turned him and his siblings into birds (a metaphor for death and an afterlife). Anthus became the pipit, Erodias a heron, Acanthis a finch and Schoenus a reed bird.

Richard's Pipit *Anthus richardi*

This is a regular vagrant to Britain from Siberia.

Charles **Richard** (1745–1835) was a French postal official in the north-eastern town of Lunéville, who was a collector of natural history specimens. His full details were lost for many years and were only rediscovered in about 2015.

Tawny Pipit *Anthus campestris*

This is a fairly regular vagrant to Britain from mainland Europe.

Tawny describes the colour, while *campestris* (of the fields) deals with the habitat.

Blyth's Pipit *Anthus godlewskii*

This is a rare Mongolian vagrant to Britain.

Edward **Blyth** (1810–1873) was an English zoologist based in India.

W.W. **Godlewski** (1831–1890) was a Polish naturalist who was exiled to Mongolia after an act of political dissent against the rule of Russia.

* This is a rare example of a name commemorating two individuals (see also Radde's Warbler, *Phylloscopus schwarzi*).

Long-billed Pipit *Anthus similis*

This is a species of Israel, Syria and Lebanon.

Long-billed is self-evident.

The specific name *similis* means 'similar', in this case seeming to relate to the Tawny Pipit.

Meadow Pipit *Anthus pratensis*

This is a common resident and passage species.

This was the original *Titlark* or *Meadow Lark*. The word **Meadow** was retained by Selby in 1833 to distinguish this from *Rock* and *Tree* pipits, and is supported by its Latin equivalent *pratensis*, to underline the sort of habitat preferred by the bird.

Tree Pipit *Anthus trivialis*

This is a summer migrant species.

Selby attached the **Tree** to this pipit because Montagu had noted that it 'rarely alights on the ground without previously perching on a tree'. It does in fact perch in tall trees during its displays.

The specific name *trivialis* was used by Linnaeus, who listed the species as a lark. The word means 'common' – though they are not so common today.

Olive-backed Pipit *Anthus hodgsoni*

This is a scarce vagrant from Siberia.

Olive-backed is factually correct.

Brian H. **Hodgson** (1800/1–1894) was an English naturalist working in India.

Pechora Pipit *Anthus gustavi*

This is a rare Siberian vagrant to Britain.

Pechora is a region of northern Russia.

The specific name *gustavi* is unusual for using the first name of the Dutch naturalist, Gustaaf Schlegel (1840–1903) (see also Boyd's Shearwater for a similar occurrence). The dedication was by Robert Swinhoe, who was a contemporary specialist in Chinese birds. The nature of the gesture suggests that they were comfortably acquainted.

Red-throated Pipit *Anthus cervinus*

This is a scarce vagrant to Britain.

Red-throated is an obvious name, but *cervinus* is much more subtle: it is the rufous colour of a stag (presumably a Red Deer).

Buff-bellied Pipit *Anthus rubescens*

This is a rare vagrant to Britain in its North American form, which was long treated as a subspecies of the Water Pipit/Rock Pipit complex. The Siberian form, *A. r. japonicus* ('from Japan') is recorded as a wintering species in Eastern parts of the region and has been recorded in France and the Netherlands.

Buff-bellied is a simple descriptive word, which is supported by the specific name *rubescens* (reddish).

However, a study published in 2023 by P. Alstrom et al. recommends that there should be a further split which, if accepted, would result in two different names:

American Pipit *Anthus rubescens*
Siberian Pipit *Anthus japonicus*

The name American Pipit has traditionally been used locally, so the potential changes will be more significant for British and other users.

Water Pipit *Anthus spinoletta*

This montane Continental species migrates to lowland areas in winter, including Britain.

This was formerly treated as the nominate subspecies of the Water Pipit/Rock Pipit complex.

The English name of this species emphasises its winter habitat, since this bird loves shallow water in areas such as estuaries, seashores, waterlogged fields, watercress beds and shallow scrapes. Even in its montane breeding habitats, it frequents 'wet meadows with pools, watercourses, or snowmelt' (Birds of the Western Palearctic Interactive, 2004–6).

The specific name *spinoletta* is an adaption of the Florentine name *spipoletta*.

(European) Rock Pipit *Anthus petrosus* (*littoralis*)

This is a resident species, which was long treated as a subspecies of the Water Pipit/Rock Pipit complex. The nominate form breeds in Britain and that is boosted by winter migrants of the Scandinavian race, *littoralis*.

The word **European** distinguishes this from an African Rock Pipit, but is generally not needed.

Rock is a habitat word and is supported by *petrosus*, which means much the same thing.

The subspecific name *littoralis* relates to its coastal preferences (which is true of the nominate form too).

Berthelot's Pipit *Anthus berthelotii*

This is a species endemic to the Canary Islands and Madeira.

Sabin **Berthelot** (1794–1880) was a French naturalist who studied the wildlife of the Canary Islands.

Finches to Cardinals

97. FRINGILLIDAE: Finches

The family name is based on *Fringilla*.

Finch probably originated as an echoic variant of the 'pink' call of a Chaffinch: it has parallels in the German *Fink* and the Dutch *Vink*. The word evolved with a number of qualifying prefixes as single-word names for the most common species, *Chaffinch* and *Goldfinch* being apparently the oldest usages and others – *Bullfinch*, *Greenfinch* and *Hawfinch* – being adopted later. More exotic forms, such as Citril Finch, are not compressed. Today the word serves as a collective name for all members of the Fringillidae, and thus embraces Brambling, Linnet, Redpoll, Crossbill and others, which do not have finch in the name. The word is used elsewhere to describe unrelated birds with finch-like characteristics, such as the Estrildid finches.

Common (Eurasian) Chaffinch *Fringilla coelebs* (*gengleri*)

This is a common resident species.

Chaffinch is a name rooted in the Old English *ceaffinc*. It was so-named for its habit of feeding on the waste grain left in chaff (the separated husks of the grain after threshing).

Fringilla is the Latin word for 'finch'.

The specific name *coelebs* means 'celibate': this was used by Linnaeus when he noted that the females wintered further south than the males.

* The Chaffinch is now less common in Britain, where it was long the most familiar finch species. The resident British form is the subspecies *gengleri* (named for German ornithologist J. **Gengler** (1863–1931), but is joined in winter by migrants of the nominate race originating from the Continent.

* In the traditional, non-mechanised farming methods used when the name Chaffinch was coined, the whole ears of cereal crops were placed on a threshing floor and beaten with a flail to separate the husks from the seed. The lighter husks (chaff) were then wafted clear with a winnowing fan, but some grain might be mixed with the chaff and this would attract the birds. See also Lapwing, where the scientific name *Vanellus* derives from the Latin for a 'winnowing fan' (referring to the broad wings).

Until recently, a total of seventeen subspecies of what was formerly known as Common Chaffinch were recognised throughout the range, though these were often divided into three distinct groups. Recuerda et al. (2021) recommend a wholesale repositioning of nine of the more southerly subspecies, with a resultant five-species split 'based on genomic and morphological differences'. That was accepted by the IOC in early 2023. The provisional recommendation suggests that the seven northern forms will be known as the Eurasian Chaffinch (above) and the remainder as outlined below.

African Chaffinch *Fringilla spodiogenys*

This new species groups the three North African subspecies under the specific name *spodiogenys*, which describes their 'ash-grey cheeks'. The males are entirely grey headed and lack the pink face of the other forms.

Azores Chaffinch *Fringilla moreletti*

This upgrades a single subspecies that is named after the French artist and naturalist Pierre **Morelet** (1809–1892), who collected in the **Azores**.

Madeira Chaffinch *Fringilla madeirensis*

This again upgrades a single subspecies that is limited to **Madeira**.

Canary Islands Chaffinch *Fringilla canariensis*

This new species incorporates four former **Canary Islands** subspecies under one umbrella name.

> The original Blue Chaffinch is less-closely related to the other chaffinches. As the result of a separate study it too has now been formally split into two new species, following a 2015 recommendation.

Tenerife Blue Chaffinch *Fringilla teydea*

This is a species of Tenerife in the Canary Islands.

Blue is literal, though the male is in fact slaty-blue and the female more drab.

The specific name *teydea* is taken from Mount Teyde on Tenerife.

Gran Canaria Blue Chaffinch *Fringilla polatzeki*

This endangered species is endemic to the island of **Gran Canaria** and was split from the Tenerife form in 2016.

The blue-grey in this form shades to white on the undertail and to black on the face.

Johann **Polatzek** (1838/9–1927) was an Austrian naturalist.

Brambling *Fringilla montifringilla*

This is a moderately common winter visitor and a close relative of the chaffinches.

Brambling is first in evidence (as Bramlyng) in Turner in 1544. The implication is that this is a word of even older origins. In spite of its apparently transparent meaning, Lockwood felt that it has nothing to do with brambles: the bird is, after all, more associated with woodlands at all seasons. He explained the anomaly by pointing out that the blackness of the male bird strongly suggested that this should be a *Brandling*, a rendering of the blackness as *burnt*, as in Brant/Brent Goose. His recommendation of a permanent name-change was not taken up.

The specific name *montifringilla* suggests a link to that of the Snow Finch, and in doing so reminds us that the Brambling breeds in upland birch forests as a *mountain-finch*.

Evening Grosbeak *Hesperiphona vespertina*

This American species has been recorded in Britain.

The only part of this bird's name that is reliable is **Grosbeak**, a word related to its large bill, via the French *grosbec*. That name was adapted from Belon and introduced into English by Ray in 1678. It was used initially for the Hawfinch and later embraced other species with similarly large bills.

The concept of **Evening** is contentious: when the first one was collected in Michigan in 1820, the bird was mistakenly reported as 'singing in the evening'. The specimen was sent to William Cooper, a New York scientist, who built that story into the names. The facts were better understood by 1836, but what remains is the vernacular name and the fact that *Hesperiphona* means 'an evening singer', while *vespertina* (of the evening) underlines the idea.

It does seem, though, that the scientific names are more subtle than first appears; Choate points out that both elements can be interpreted as relating to the west (via a link to the setting sun), and that was certainly true of Cooper's New York perspective on a bird that was not known in the Eastern States at that time. A current post by Kenn Kaufmann, on the Audubon website, usefully explains that the bird was then only known in the western part of North America and has only spread eastwards in modern times.

Hawfinch *Coccothraustes coccothraustes*

This is a resident species in Britain.

The 'haw' element of **Hawfinch** comes from the berry of the hawthorn tree. The name was first seen in Ray in 1674. He also introduced the alternative name *Common Gros-beak*, which was used (without hyphen) by Pennant and White, among others. Hawfinch was eventually preferred by Selby, Gould and Yarrell during the first half of the nineteenth century and dominated thereafter.

The name *Coccothraustes* is based on a Greek word composed of *kokkos* (seed) and *thrauō* (break).

* The 'seed-breaking' of the Hawfinch is enabled by a stout bill, a hard upper palate and strong jaw muscles, which enable it to extract larger tree seeds from husks and shells, including wild cherry stones.

Pine Grosbeak *Pinicola enucleator*

This is a rare vagrant to Britain from Scandinavia and Russia.

Pine and *Pinicola* tell us that the bird frequents pine trees.

The specific name *enucleator* recounts the fact that the bird 'removes the kernels' from the pine cone.

(Eurasian) Bullfinch *Pyrrhula pyrrhula*

This is a resident species that is boosted by winter visitors of the nominate race.

Eurasian is needed in a world context.

Pyrrhula means 'flame-coloured', a reference to the male's spectacular breast.

* The British Isles have their own subspecies, *pileata* (capped), which is smaller than the nominate race. The continental form is sometimes recorded in winter.

Azores Bullfinch *Pyrrhula murina*

This species was split from the Eurasian form in 1993.

It is endemic to the **Azores**.

This is a less bright form, so its greyish tones earn it the name *murina* (mouse-grey).

Asian Crimson-winged Finch *Rhodopechys sanguineus*

This is a high-altitude species of Turkey and Iran and elsewhere in **Asia**.

Crimson-winged perhaps exaggerates the intensity of the colour, which is more accurately reflected in *Rhodopechys*, which literally means 'pink-armed'.

However, the specific name *sangineus* swings the meaning towards 'blood-red'.

African Crimson-winged Finch *Rhodopechys alienus*

The **African** form, found primarily in Morocco, is well separated geographically from the Middle-Eastern form. It was granted separate status in 2011.

The specific name *alienus* refers to this as a 'stranger', possibly as a reference to its large separation from the very similar Asian form. It was described and named almost 60 years after its congener.

Trumpeter Finch *Bucanetes githagineus*

This mainly North African species is found in Spain and is a very rare vagrant to Britain.

The name **Trumpeter** derives from the 'toy-trumpet' call.

That meaning is repeated in *Bucanetes*, which means much the same in Greek.

The specific name *githagineus* relates to the name of the plant, *Agrostemma githago*, corncockle. The bird eats a range of seeds.

Mongolian Finch *Bucanetes mongolicus*

This is the Middle Eastern equivalent of the Trumpeter Finch.

Mongolian and *mongolicus* suggest that the bird's range stretches right across Asia.

Common Rosefinch *Carpodacus erythrinus*

This species of the Eurasian mainland is seen fairly regularly in Britain and has bred.

Common separates it from other related species.

Rosefinch is for the distinctive rose-red of the adult male.

Carpodacus means 'fruit biter'.

The specific name *erythrinus* means 'red'.

Great Rosefinch *Carpodacus rubicilla*

This species is found at altitude in the Caucasus.

Great is no exaggeration: it is *much* larger than the Common Rosefinch.

The specific name *rubicilla* tells of a 'red tail', though 'red rump' would be more accurate.

Siberian Long-tailed Rosefinch *Carpodacus sibiricus*

This Siberian species has been recorded in Russia and Denmark.

All aspects of this bird's name are self-evident.

Sinai Rosefinch *Carpodacus synoicus*

This is an Arabian species.

The names **Sinai** and *synoicus* refer to the Sinai Peninsula.

Pallas's Rosefinch *Carpodacus roseus*

This Far-Eastern species has been recorded in Eastern Europe. The odd British sightings have probably originated from cage-bird escapes.

Peter **Pallas** (1741–1811) was a Prussian zoologist and botanist who worked in Russia.

The specific name *roseus* emphasises the pinkness of this species.

(European) Greenfinch *Chloris chloris*

This is a common resident species.

It is **European** in a world context: the other species are all Asian.

Greenfinch first appeared in Palsgrave in *c.*1532 as *grene fynche*, but the single-word version we use now was the preference of Pennant in 1768.

Chloris derives from the Greek *khlōris* (greenfinch) referring to *khlōros* (green).

* An alternative local name *Green Linnet* was first recorded by Ray.

Desert Finch *Rhodospiza obsoleta*

The **deserts** of the Middle East are home to this species.

The name *Rhodospiza* means 'rosy finch'.

The specific name *obsoleta* means 'plain', which may be a little unjust.

Twite *Linaria flavirostris*

This is a fairly scarce and local resident in northern Britain.

Twite is one of the simplest of bird names, rooted in the sound, 'tweet', made by a small bird. That traditional name survived a period when *Mountain Linnet* was the favoured name, used as a parallel to Common Linnet (below).

Linaria relates more to the habits of the Common Linnet than to the Twite.

The specific name *flavirostris* means 'yellow-billed'. However, the bill is only yellow in winter, when it is a useful field mark to distinguish from Linnet: it takes on a darker grey-brown in summer.

(Common) Linnet *Linaria cannabina*

This is a common and widespread species.

There are two other linnet species (in Africa and Asia), but **Common** is rarely used.

Both **Linnet** and the genus *Linaria* derive from *Linum*, the name of the flax plant, which produces stem fibres used to make linen and also seeds (processed commercially as linseed) that are attractive to the bird.

In addition, the seeds of hemp, *cannabina*, are also eaten by the Linnet.

REDPOLLS

The current position is that IOC and BOU accept the separation of the species as below. This is an area where speciation has long been unresolved. However, there are differences that are diagnostic in the field.

Common Redpoll *Acanthis flammea*

This is a fairly scarce winter visitor from the Continent.

Common is used internationally, but the term **Mealy** Redpoll is popularly used by British birders to describe a whiter bird that looks 'dipped in meal', meaning flour. That was, in fact, the former formal name and appears on Coward's 1926 list.

Redpoll uses an archaic word for the head, to describe the red cap that is a feature of these birds. (In Modern English the term 'going to the polls' to vote derives from the idea of a head count.)

Acanthis returns to the legend of Anthus and the horses (see Pipits). As one of the siblings caught up in the disaster, his sister Acanthis was transmuted into a finch.

The specific name *flammea* refers to the 'flame-red' of the breeding male.

Lesser Redpoll *Acanthis cabaret*

This is the form that breeds in Britain.

Lesser indicates that this is a smaller species than Common Redpoll.

The specific name *cabaret* was used by Brisson and came from the French for 'a kind of finch' ('*Espèce de gros-bec*'). That is probably explained by the fact that one of the French names of the teasel plant is *cabaret des oiseaux*. In that usage *cabaret* equates to an eating place: the seeds of the plant are favoured by finches.

Arctic Redpoll *Acanthis hornemanni*

Hornemann's Arctic Redpoll *Acanthis hornemanni hornemanni*
Coues's Arctic Redpoll *Acanthis hornemanni exilipes*

Both forms of this species are rare winter vagrants to Britain.

Arctic is the formal international name, which reveals its northerly habits. In America the preferred name is **Hoary** Redpoll, which emphasises the 'frosty white' of the plumage ground-colour. That was the name used on Coward's 1926 British list for the Coues's form (below).

The specific name *hornemanni* is after Jens W. Hornemann (1770–1841), a Danish botanist.

Birders in Britain recognise two subspecific forms: the nominate (and rarer) form, **Hornemann's**, which comes from the west, and **Coues's**, which is the Eurasian form. Elliott **Coues** (1842–1899) was a founder-member of the American Ornithologists Union. It is unlikely that the 2023 AOS decision to abolish eponyms will affect this informal usage.

Coues's form is *A. h. **exilipes***, which means 'slender-footed'.

* The relationship of the Eurasian form to the American ornithologist and of the American form to the Norwegian is just another confusing fact in the complications of redpoll names

Crossbills are also a taxonomically difficult subset of the finch family, which have long caused confusion, debate and disagreement on matters of speciation. The exact status of the Scottish Crossbill is challenged by some authorities, while the huge spectrum of the Red Crossbill complex, which formally covers the whole of Eurasia and North America, seems to be in need of re-examination.

The fact and extent of the crossing of the bill in this family is related to the evolution of specialist feeding habits: the entire family depends on the cones of various species of tree, with different populations exploiting different tree species. As a consequence, the shape of the bill relates to the principal diet.

Parrot Crossbill *Loxia pytyopsittacus*

While this is a resident in the north of Scotland, it is a rare vagrant elsewhere in Britain following occasional irruptions from the Continent.

The larger, stouter bill of this species earns it the name **Parrot**. The bill barely crosses at the tip.

For **Crossbill** and *Loxia* see Red Crossbill below.

The specific name *pytyopsittacus* joins two elements meaning 'pine' and 'parrot'. This name hints at the bird's evolutionary adaptation to the heavier cones and stouter seeds of the Scots pine.

* As well as the stouter bill, it has a more robust head and neck structure, which shows in the field.

Scottish Crossbill *Loxia scotica*

This is recognised as a full species by IOC and in that respect is currently Britain's only endemic species, so that status is jealously guarded against doubts and challenges (though it seems possible that Red Grouse will join it after a period of separation).

This is a very local endemic species in northern Scotland, hence the specific **Scottish** and *scotica*.

For **Crossbill** and *Loxia* see Red/Common Crossbill below.

* This species also has a stout bill and is intermediate in habits and structure between Parrot and Red Crossbill. Studies have shown that is separated from both by a difference of song, which appears to prevent interbreeding.

(Red/Common) Crossbill *Loxia curvirostra*

This is a moderately common species in appropriate habitats in Britain.

What is generally known as Common Crossbill in Britain now carries the American vernacular name **Red** Crossbill. That does not seem to be a helpful name in a family where males of all species are more or less red and the females of each population green.

In this species, the bill crosses very prominently at the tip. In spite of such a visible oddity, there appears to have been no widespread vernacular English name for this species until the name **Crossbill** was adapted from Gessner's version (below). That name was established formally by Pennant in 1768 and first used for this species.

In 1555 Gessner had created the Modern Latin form *Loxia*, which was derived from the Greek *loxos*, crosswise. That in turn had been based on his native Swiss-German *Krützvogel*.

The specific name, *curvirostra*, uses Latin more directly to describe a 'curved bill'.

* While the European form is the nominate subspecies, there are some 19 subspecies over a huge geographical world range. The bill structures of individual subspecies appear to vary according to the structure of the cones in the relative diets. It seems almost inevitable that there will be a number of splits and wide renaming in the future.

Two-barred Crossbill *Loxia leucoptera*

This species breeds in northern Russia, but is a vagrant to Britain.

Two-barred refers to the double white wing-bar, which is also the subject of the specific name *leucoptera* (white-winged).

* The bill is crossed, but is longer and thinner than in the previous species. Larch seeds and rowan berries are its main foods.

(European) Goldfinch *Carduelis carduelis*

This is a common species.

The word **European** is only needed in an international context.

Goldfinch is entirely because of the broad gold wing-bar, which also distinguishes juveniles with an undeveloped head-pattern.

Carduelis refers to the thistle, *Carduus*, whose seeds provide a large part of the bird's diet.

* In spite of the name, the American Goldfinch is more closely related to the Siskins.

Citril Finch *Carduelis citrinella*

This Southern European species has been noted only very rarely in Britain.

Both **Citril** and *citrinella* are based on the yellowish colour (citrine) of the bird.

Corsican Finch *Carduelis corsicana*

The names explain clearly that this is a species found in **Corsica** – but it is also found in Sardinia.

Red-fronted Serin *Serinus pusillus*

This is a species inhabiting montane areas from Greece eastwards.

Red-fronted describes a bold red forehead patch (from the Latin *frons*).

For **Serin** and *Serinus* see next species.

The specific name *pusillus* means very small.

(European) Serin *Serinus serinus*

This is a scarce but regular vagrant to Britain.

European separates this from other similar species.

The name **Serin** is a French form that probably evolved from *citrinus* for its yellow colour.

The Modern Latin form, *Serinus*, is therefore very contrived and artificial, a concoction of the German, C.L. Koch in 1816.

Syrian Serin *Serinus syriacus*

The names **Syrian** and *syriacus* are indicative of the breeding area of this species.

Canary *Serinus canaria*

The name of the original wild species of **Canary** comes from the Canary Islands where it still found, as well as on Madeira.

The islands gave their name to the bird, but probably gained their own collective name *Canaria* from the many large dogs (Latin *canis*) that lived there in ancient times. This phenomenon was mentioned by Pliny the Elder during the first century.

(Eurasian) Siskin *Spinus spinus*

This is a widespread species, but is unevenly distributed in Britain.

Eurasian again distinguishes this species in a world context, but is not needed generally.

Siskin is first in evidence in Turner in 1544. He created it from a German form, possibly with Slavonic roots that had travelled with the cage-bird trade (Lockwood).

The word *Spinus* started life as the Greek *spinos*, an unidentified bird mentioned in Aristophanes.

BUNTINGS

The word *bunting* began its life as a local English name for a small, dumpy, rural bird, now the Corn Bunting, a rather featureless species that harms nothing and disturbs nobody. It was a term of affection, such as might be used for a chubby child: the word features in the nursery rhyme 'Bye baby bunting'. Like so many simple names, such as Robin, Wren and Finch, it was eventually taken up by ornithologists and became formally attached to other related or similar birds at home and then abroad, together with the qualifying words that catalogued them more exactly.

The German word Embritz (Bunting) is at the root of the generic name *Emberiza*, which now encompasses a range of Eurasian buntings.

Prior to 2008, the two Eurasian species that are now classed in the Calcariidae were treated part of that family, and so retain some links in their vernacular names.

98. CALCARIIDAE: Longspurs

The family name is based on *Calcarius*.

Lapland Longspur *Calcarius lapponicus*
(Lapland Bunting)

This is a scarce winter visitor to Britain.

Lapland and *lapponicus* indicate that it is a species of northern Scandinavia and Russia.

The use of **Bunting** is traditional in Britain, but the IOC now prefers **Longspur**, with reference to the rear-facing claw.

Calcarius relates to the spur (Latin *calcar*).

Snow Bunting *Plectrophenax nivalis*

This species breeds in the Scottish mountains and is a scarce winter visitor to the rest of Britain.

In spite of being classed in this new family, the IOC has left it with its traditional name.

Snow and *nivalis* mean the same, indicating that it is a species of high latitude and/or high latitude.

Plectrophenax refers first to the spur (Greek *plēktron*) and then, in *phenax*, to the fact that it is an imposter (imitating a fighting cock perhaps).

99. EMBERIZIDAE: Buntings

The family name is based on *Emberiza*.

Only four bunting species breed regularly in Britain, but a surprisingly large number of other buntings occur occasionally as winter vagrants from various parts of Eastern Europe and Asia. Some of those are recorded only very rarely, others more frequently.

Further species occur elsewhere within the Western Palearctic.

Corn Bunting *Emberiza calandra*

This species is resident in Britain.

Corn represents the bird's habitat preference as a grassland and arable species.

For **Bunting** see the introductory comment.

Emberiza is rooted in an Old German word *Embritz* (bunting).

The specific name *calandra* (lark) relates to an ancient tradition that treated this bird as 'the poor man's lark', often trapped and eaten. For part of the twentieth century and until about 2010, the Corn Bunting was placed in the unique genus *Miliaria*, which was based on the millet used to trap and fatten the birds for the table.

* Historically, this species has been named and renamed many times. The move to *Emberiza* is in fact a restoration, since that was how it appeared in Hartert (1912) and Coward (1926).

Yellowhammer *Emberiza citrinella*

This is another British resident. The name of this common farmland species has followed a bumpy trail.

Yellowhammer started life as a '*yellow ammer*', *ammer* being an Old English word of West Germanic origins, meaning 'bunting'. It shares a much older Germanic root, *amarzo*, with both *embritz* (above) and with the Modern German *Ammer*. The later addition of the 'h' obscured and confused its original sense. During the mid-nineteenth century the formal name was updated by Yarrell to *Yellow Bunting*, a name that underlined its family status. That was still favoured by Hartert in 1912 and by Coward in 1926, in spite of an 1883 BOU move to reinstate Yellowhammer, which eventually prevailed in general usage.

The specific name *citrinella* is the diminutive form of *citrinus* (a shade of yellow).

Pine Bunting *Emberiza leucocephalos*

This species breeds east of the Urals, and is vagrant to Britain.

Pine suggests that it is associated more with woodland than is the closely related Yellowhammer.

The specific name *leucocephalos* (white-headed) refers to the male's clean white crown.

Rock Bunting *Emberiza cia*

This is a widespread species in Southern Europe and a rare vagrant to Britain.

Rock represents its favoured habitats, which are generally above the tree-line.

The specific name *cia* is from a local Italian form, probably based on the call.

Grey-necked Bunting *Emberiza buchanani*

This is the equivalent of Ortolan and Cretzschmar's, but is found in the Caucasus, eastwards, and winters in India.

Grey-necked is self-evident.

Dr F. **Buchanan-Hamilton** (1762–1829) is commemorated in the specific name *buchanani*. He was a physician, zoologist and explorer, initially known as Buchanan and later as Hamilton, but generally recorded under the double name.

Cinereous Bunting *Emberiza cineracea*

Breeding partly in western Turkey, this species winters around the Red Sea.

Both **Cinereous** and *cineracea* refer to the ashy-greyness of the bird.

Ortolan Bunting *Emberiza hortulana*

This is a widespread European species, sometimes recorded on passage through Britain.

The French-rooted form **Ortolan** is echoed in the Italian-based specific name *hortulana*, both of which have their roots in the Latin *hortus* (garden).

* It is a large bunting and is partly migratory. In spite of universal EU protection, a pretentious culinary tradition involving the consumption of these birds still persists in some strata of French society. Unfortunately, little seems to be done to enforce the law.

Cretzschmar's Bunting *Emberiza caesia*

This is an Eastern Mediterranean species that has been recorded in Britain.

Phillip **Cretzschmar** (1786–1845) was a German naturalist.

The specific name *caesia* describes the blue-grey colour of the bird's head.

Cirl Bunting *Emberiza cirlus*

This is one of Britain's rarest breeding species, limited to parts of Devon and Cornwall, where it is at the northern extremity of its range. It was once more widespread.

Cirl and *cirlus* both derive from local Italian names for the species and are very probably based on the song of the bird.

Striolated Bunting *Emberiza striolata*

This species is found in Sinai and Arabia.

Striolate and *striolata* mean streaked.

House Bunting *Emberiza sahari*

This is a North African species.

House reflects the fact that this confiding species is found close to human habitations.

The specific name *sahari* refers to the Sahara.

Chestnut-eared Bunting *Emberiza fucata*

This is a Far-Eastern species, found from Pakistan to Japan. It has been recorded in Britain and Sweden.

Chestnut-eared refers to the rufous ear patches, while *fucata* means 'painted red'.

Little Bunting *Emberiza pusilla*

This is a species of the Northern taiga forests, which is a regular but scarce visitor to Britain.

Little and *pusilla* say the same thing.

Yellow-browed Bunting *Emberiza chrysophrys*

This is a species from eastern Siberia, which is a rare vagrant to Britain.

Yellow-browed is exaggerated further by the specific name *chrysophrys*, which means 'golden-browed'.

Rustic Bunting *Emberiza rustica*

This is another Northern species that is a scarce visitor to Britain.

Rustic and *rustica* say the same thing: that this is a simple, rural bird. It is a happy coincidence that the brown is of a distinctly rusty shade.

Yellow-breasted Bunting *Emberiza aureola*

This is a species of the boreal forests that is seen occasionally in Britain.

Yellow-breasted is self-evident, and it is underlined by *aureola* (golden).

Chestnut Bunting *Emberiza rutila*

This is a species of the Siberian taiga that has been recorded in Western Europe, though British sightings may have been escaped cage-birds.

The male is a **chestnut**-rufous colour; this is further described by the specific name *rutila* (golden-red).

Black-headed Bunting *Emberiza melanocephala*

This is a species of south-east Europe and the Middle East that is sometimes seen in Britain.

Black-headed and *melanocephala* mean the same thing.

Red-headed Bunting *Emberiza bruniceps*

All historical British records of this species are suspect, because it was a widely imported cage-bird. Modern Western European records may include some genuine vagrants.

Red-headed is slightly contradicted by the specific name ***bruniceps*** (brown-headed).

Black-faced Bunting *Emberiza spodocephala*

This is an East Asian species. It is a very rare vagrant to Western Europe, including Britain.

While it is **Black-faced** in the vernacular, it is 'ashy-headed' according to the specific name ***spodocephala***. That is not contradictory: the grey head shades to black around the eyes and bill.

Pallas's Reed Bunting *Emberiza pallasi*

This is a species from northern Russia that has been recorded rarely in Britain.

Peter **Pallas** (1741–1811) was a Prussian zoologist and botanist who worked in Russia, and whose name appears in a number of eastern species.

(Common) Reed Bunting *Emberiza schoeniclus*

This is the commonest of the British buntings.

Common is only used in a world context

While the bird does frequent **reeds**, it is also found in drier habitats and, nowadays especially, in rape crops.

The specific name ***schoeniclus*** reinforces its link to the reeds. In this case Gessner modified Aristotle's name *skhoiniklos* for a small waterside bird. Schoenus (also a reed) was one of the unfortunate siblings of Anthus and Acanthis (see pipits and redpolls).

VAGRANT AMERICAN SPECIES

A number of American passerines turn up in the British Isles, or in other parts of Europe, often storm-blown and sometimes ship-assisted. Some of these are included in previous sections, examples being wildfowl, waders, gulls and vireos. The phenomenon is not that uncommon because the seasonal migration of many species makes them vulnerable to the prevailing westerly winds, and in particular to Atlantic storms. In the autumn a great many birds use a short-cut, known as the Atlantic flyway. This crosses the open ocean from Canada and New England towards Florida and South America. On that crossing birds may encounter autumn storms, which build in the tropics and then veer east across the open Atlantic. Such storms may pick up and carry migrating birds towards European shores. Nobody can ever calculate the number of birds that perish during such events, but a small number of survivors may make it across to Europe to be seen as 'rare vagrants'. Ship assistance, where birds have landed on a vessel to seek refuge, may be a more significant factor with smaller species, as evidenced by the number of finds that are made close to ports.

100. PASSERELLIDAE:
AMERICAN SPARROWS

The family name is based on *Passerella*.

The American Sparrows of the Passerellidae were long treated as a subset of the bunting family, but now are treated separately. Both the vernacular and scientific terms are misleading, since these birds are more closely related to Old World buntings than they are to the 'true' sparrows.

Lark Sparrow *Chondestes grammacus*

Lark is for its resemblance to members of that family.

Chondestes is a compression of two words (Greek *khondros* and *odestes*), which together mean 'grain-eating'.

The specific name *grammacus* is an erroneous version of *grammicus* (streaked).

Red Fox Sparrow *Passerella iliaca*

This species has been recorded in Ireland, Iceland, Estonia and Finland.

In 2015 this was spilt from a number of variants of **Fox** Sparrow, so named for their warm brown colour, and is distinguished from them by the word **Red** because it is the richest in shade.

Passerella is likened to a small *Passer* (true sparrow).

The specific name *iliaca* refers to the bird's flanks (see also Redwing).

American Tree Sparrow *Spizelloides arborea*

This species has been recorded in Sweden.

The addition of **American** is necessitated by the presence in North America of the introduced *Eurasian* Tree Sparrow.

Tree Sparrow reflects its arboreal habits, which are underlined by the specific name *arborea*.

Spizelloides is a curious concoction, based on *Spizella*, itself a diminutive of the Greek *spiza* (sparrow). The addition of *-oides* indicates that it resembles *Spizella*, which is another genus of this family.

Dark-eyed Junco *Junco hyemalis*

Dark-eyed separates this species from a 'Yellow-eyed' relative.

Junco originally refers to the rush plant, though none of this genus is attracted to such habitat. When he named it, Wagler was working on a museum specimen, with little knowledge of the bird's actual behaviour. Today the word is almost synonymous with 'sparrow'.

The specific name *hyemalis* (winter) was given because the Northern-breeding birds appeared in the United States in that season. They are also colloquially known as 'snow-birds' for the same reason.

White-crowned Sparrow *Zonotrichia leucophrys*

White-crowned is self-evident.

Zonotrichia means 'banded hair', and applies to a genus in which breeding males have a heavily zoned and generally black-and-white head.

The specific name *leucophrys* refers to the prominent white brow.

* The most famous vagrant individual of this species is probably the one seen at Cley, Norfolk, in 2008. So many birders contributed to a local collection intended to repair verges after the 'twitch' that the surplus enabled the local church to install a new stained glass window that depicts the bird.

White-throated Sparrow *Zonotrichia albicollis*

White-throated and *albicollis* say much the same thing.

Savannah Sparrow *Passerculus sandwichensis*

Savannah is not about habitat, but about the location, Savannah, Georgia, where the first specimen was obtained by Wilson.

Passerculus was a creation of Bonaparte and means 'little sparrow'.

The specific name *sandwichensis* is for another location, this one some 4,500 miles or more from Savannah, Georgia. The type specimen for this species was obtained at Sandwich Bay in the Aleutian Islands off Alaska. In short, the species is found over a massive range, so little wonder that there are 17 named subspecies.

* The spelling here is different from the form used for the Sandwich Tern (*sandwichensis/ sandvicensis*).

Song Sparrow *Melospiza melodia*

Song Sparrow does indeed have a sweet song, delivered from prominent perches.

Melospiza literally means a 'song finch', and that is reinforced by the name *melodia* (melodious).

Lincoln's Sparrow *Melospiza lincolnii*

This species was recorded in Iceland.

Thomas **Lincoln** (1812–1871) was an American explorer and naturalist who travelled with Audubon and discovered this sparrow. However, the AOS announced in 2023 that all eponyms in vernacular names of American species will eventually be replaced.

Eastern Towhee *Pipilo erythrophthalmus*

Eastern is used because there are several species of towhee.

Towhee originated as an echoic name, based on the bird's call.

Pipilo is also based on the bird's chirp.

The specific name *erythrophthalmus* tells us that the bird is red-eyed.

101. ICTERIDAE: Icterids

The family name is based on *Icterus*.

The Icterids are another American family, members of which are found throughout the Americas. They take their name from a Greek word, *ikteros* (jaundice yellow), which also occurs in the unrelated Icterine Warbler of Europe. Many of this family have yellow in their plumage, though there is a lot of black too.

Yellow-headed Blackbird *Xanthocephalus xanthocephalus*

This species has been recorded in Iceland and the Netherlands.

Yellow-headed is self-evident and that is also the meaning of *Xanthocephalus*.

Bobolink *Dolichonyx oryzivorus*

The name **Bobolink** probably derives from the bird's bubbling song.

Dolichonyx states that the bird is 'long clawed'.

The specific name *oryzivorus* means 'rice-eating', but in truth it is both granivorous and insectivorous. The latter habit is attested by an alternative popular name, *Armyworm bird*, because it consumes large numbers of plant-eating moth larvae.

Baltimore Oriole *Icterus galbula*

Baltimore is only indirectly connected to the well-known city. Catesby recorded the species in 1731 as the *Baltimore-bird* because 'its colours were the same as those of Baltimores, the colonial proprietors of Maryland' (Choate). However the AOS announced in 2023 that all eponyms in vernacular names of American species will eventually be replaced.

Oriole borrows the name of the Eurasian Golden Oriole and applies it to a number of icterid species that resemble it a general way.

The specific name *galbula* (Latin *galbus*, yellow) was originally associated with the Golden Oriole.

Red-winged Blackbird *Agelaius phoeniceus*

Red-winged describes the scarlet 'epaulettes' that are used in display.

In America, **Blackbird**, like *oriole*, was borrowed for birds that superficially resembled the European bird of that name.

Agelaius is from the Greek meaning 'gregarious'. The birds form huge winter flocks that are often seen as crop pests.

The specific name *phoeniceus* is from the Greek for 'scarlet', a further reference to the shoulder flash.

Brown-headed Cowbird *Molothrus ater*

Brown-headed describes a male that is otherwise glossy purple-black, a fact that is recorded in the specific name *ater*.

Cowbird is another species that owes its name to Catesby, though it started life as *Cowpen-bird*, because he saw it near cattle pens. It was also sometimes called the *Buffalo Bird* because it followed the herds across the prairies. In both cases it was exploiting the creatures disturbed by the animals.

Molothrus is apparently an erroneous usage that, according to Coues, Swainson repeated so often that it became the accepted form. It has to do with a 'struggle to impregnate' and is nonsense. The intention was *molobros* (greedy).

* One curious fact about this bird is that it is a nest parasite. Like the Common Cuckoo it lays its eggs in the nests of other species and is not fussy: it is on record as having attempted to parasitise 140 species, including the odd hawk, several waders and a rail. Because it is still expanding its range, it is very experimental and its eggs are frequently rejected.

Common Grackle *Quiscalus quiscula*

There is a single record from the Netherlands that seems feasible, but the bird is kept as a cage-bird, so occasional records elsewhere in Europe are unaccepted.

Common indicates that there other several other species and that this one is fairly widespread.

Grackle is an anglicised version of the Latin *graculus*, once used for 'Jackdaw'. Larger grackles are almost crow-like.

Quiscalus and **quiscula**: the actual route to the modern name was via *Gracula quiscula*, with the current generic name being created from the original specific.

The roots of this word puzzled Choate, but Jobling suggests that it originates in the Carib word *Quisqueya* (the mother of all lands).

102. PARULIDAE:
AMERICAN WOOD-WARBLERS

The family name is based on *Parula*, but, most unusually, that genus no longer exists, since it is has been subsumed into *Setophaga* during recent revisions.

In the vernacular form, these are the American wood-warblers, a phrase that was concocted to distinguish them from the Old World warblers, which were once thought of as a more cohesive entity than they are today. There has been some considerable taxonomic revision in this area, but mainly of genera – though the revision of Yellow Warbler was very radical.

Ovenbird *Seiurus aurocapilla*

This species has been recorded in Britain.

This species earns the name **Ovenbird** by building a ground-level, oven-shaped nest of grass. It otherwise has no connection with the 'ovenbirds' (Furnariidae) of Central and South America, some of which build a nest of mud shaped like a clay ground-oven.

Seiurus means 'wagtail'.

The specific name *aurocapilla* means 'golden headed', a reference to a Goldcrest-like orange crown-stripe.

Northern Waterthrush *Parkesia noveboracensis*

This species has been recorded in Britain.

Northern distinguishes this bird from the *Louisiana* Waterthrush.

The name **Waterthrush** indicates that the birds frequent water margins. The reference to 'thrush' is misleading: it is merely to do with their heavy breast spotting, which makes them look like a miniature thrush.

Parkesia commemorates Dr K.C. **Parkes** (1922–2007), a US ornithologist.

The specific name *noveboracensis* is an intriguing concoction: the Roman name of the city of York was *Eboracum*, which led to the adjective *boracensis*. With *nove* added, this word refers to New York.

Golden-winged Warbler *Vermivora chrysoptera*

This species has been recorded in Britain.

Golden-winged and *chrysoptera* mean the same thing.

Vermivora means worm-eating (though in the broader American sense of grubs and caterpillars).

Blue-winged Warbler *Vermivora cyanoptera*

This species has been recorded in Ireland.

Blue-winged and *cyanoptera* mean the same thing.

Black-and-white Warbler *Mniotilta varia*

This species has been recorded in Britain.

Black-and-white is accurate, but minimal: the black is a streaking on a white ground. It is a rather mundane name for a jazzy little warbler with unusual nuthatch-like agility.

Mniotilta is another name that misfires. This time it was Vieillot's mistake: he tried to describe the bird as 'moss-plucking' and chose 'seaweed' by mistake.

The specific name *varia* is used in this case to mean 'variegated', which fails to describe accurately black-and-white streaking.

Prothonotary Warbler *Protonotaria citrea*

This species has been recorded in Iceland.

Prothonotary and *Protonotaria* both relate to the plumage of the bird, which recalls the golden-yellow robes of a Byzantine lawyer. Although they are rather pedantic-looking names, invented in a French form by the Comte de Buffon, they do at least give more interest than some of the descriptive 'book-names' used for this family.

The specific name *citrea* means yellow.

Tennessee Warbler *Leiothlypis peregrina*

This species has been recorded in Britain.

The name **Tennessee** is another of Wilson's geographical links to the finding of a first specimen, in this case in the US state of that name.

Leiothlypis suggests that these are plain (*leios*) warblers. *Thlypis* starts life in Greek as a small bird of some sort and later comes to denote parulid warblers in American ornithology.

The specific name *peregrina* describes this long-distance migrant as 'wandering' (see also the origins of Peregrine Falcon).

Connecticut Warbler *Oporornis agilis*

This species was recorded in the Azores.

As so often, Wilson named **Connecticut** Warbler after the place where it was first noted.

Oporornis is from the Greek *opōra* (autumn) and *ornis* (bird).

The specific name *agilis* speaks of an active bird.

Mourning Warbler *Geothlypis philadelphia*

This species was recorded in the Azores in 2023.

Mourning reflects the dark grey of the upperparts, head and breast this bird, which is rich yellow below.

Geothlypis see below.

The specific name *philadelphia* records the city near which Wilson first found it.

Common Yellowthroat *Geothlypis trichas*

This species has been recorded in Britain.

Common distinguishes this from several other yellowthroats.

Yellowthroat is self-evident.

Geothlypis describes a 'ground warbler'. This species feeds in low vegetation.

The specific name *trichas* (based on the Greek for 'thrush') reinforces the image of the generic name as a ground-hugging bird.

Hooded Warbler *Setophaga citrina*

This species has been recorded in Britain.

Hooded describes the black cowl of the male bird.

Setophaga means 'moth-eater'.

The specific name *citrina* means yellow, which applies to the undersides of the male and to the female more generally.

American Redstart *Setophaga ruticilla*

This species has been recorded in Britain.

American distinguishes this from its European namesakes.

Redstart was another name borrowed from the Old World. This is a wholly unrelated species.

The specific name *ruticilla* also means red-tailed.

* In this case, only adult males fit the name: females and juveniles have lemon yellow rather than orange-red.

Cape May Warbler *Setophaga tigrina*

This species has been recorded in Britain.

Cape May is yet another Wilson name based on a geographical location, this time one where his friend George Ord collected the specimen bird.

The specific name *tigrina* reflects the bold stripes of its undersides.

Northern Parula *Setophaga americana*

This species has been recorded in Britain.

Northern distinguishes this from, for example, the *Tropical* Parula.

Parula recalls the genus *Parus*, because the bird is 'tit-like'.

The specific name *americana* is self-evident.

Magnolia Warbler *Setophaga magnolia*

This species has been recorded in Britain.

This is yet another of Wilson's names: he shot his first specimen in a **Magnolia** tree.

Bay-breasted Warbler *Setophaga castanea*

This species has been recorded in Britain.

Bay-breasted is for the rich brown patches displayed by both sexes.

The specific name *castanea* means chestnut – which is a little confusing, given the vernacular name of another close relative (see *Chestnut-sided Warbler* below).

Blackburnian Warbler *Setophaga fusca*

This species has been recorded in Britain.

Blackburnian is an unusual name because of its connection to a rare phenomenon of the eighteenth century, a British female collector and naturalist, Mrs Anna Blackburne (1726–1793). In the course of her work, she corresponded with figures such as Linnaeus, Pallas, Forster and Pennant. Even so, she used the title Mrs to give more weight to her status an unmarried woman. Her connection with the bird began when her brother sent her a specimen from America. However, the AOS announced in 2023 that all eponyms in vernacular names of American species will eventually be replaced.

The specific name, *fusca*, means 'dark' or 'dusky', but rather understates the attractiveness of a male, which exhibits orange, yellow and white as well as black.

American Yellow Warbler *Setophaga aestiva*

This species has been recorded in Britain.

A major upheaval in the taxonomy of the widespread Yellow Warbler complex saw this form separated by the IOC in 2012 from its former classification and redesignated as a species in its own right, hence the recent addition of the name **American** and a new specific name.

The **Yellow** of the breeding male is bright (enhanced by scarlet stripes on the breast – see *petechia* below).

The new specific name is *aestiva*, because it appears in North America in summer.

* The former specific name *petechia* (from an Italian word describing red spots on skin) still appears in older references, but is now owned by its close cousin, the Mangrove Warbler.

Chestnut-sided Warbler *Setophaga pensylvanica*

This species has been recorded in Britain.

Chestnut-sided tells us that the flanks of this species are chestnut-brown.

The specific name *pensylvanica* relates to the state of Pennsylvania, where the first specimen was collected. Note the modification of the spelling of William Penn's contribution to that name.

Blackpoll Warbler *Setophaga striata*

This species has been recorded in Britain.

The **Blackpoll** warbler gets its name for its black cap (*poll* means head, as in the European redpolls).

The specific name *striata* means 'striated' or 'streaked'.

Black-throated Blue Warbler *Setophaga caerulescens*

This species has been recorded in Iceland.

Black-throated Blue Warbler is more of a description than a name, and provides an example of the worst sort of invented 'book-name'.

The specific name *caerulescens* means blue. The male is strikingly blue, black and white, though the female has neither of the features mentioned in the name.

Palm Warbler *Setophaga palmarum*

This species has been recorded in Iceland.

Palm and *palmarum* may suit this bird in its warmer wintering climates, but in summer it breeds in Northern coniferous forests, where the name seems slightly incongruous.

Cerulean Warbler *Setophaga cerulea*

This species has been recorded in Iceland.

Cerulean and *cerulea* both mean blue, though this bird is less intensely coloured than *S. caerulescens* above.

Myrtle Warbler *Setophaga coronata*

This species has been recorded in Britain.

The bird was originally considered a full species and was given the name **Myrtle** Warbler for its love of myrtle berries. Then, for many years, it was lumped with Audubon's Warbler (its Western equivalent) as a subspecies of **Yellow-rumped** Warbler, a name based upon the butter-yellow rump shared by both species. Both are once more considered to be full species. However, the vernacular name *Butterbutt* is still a birder's favourite.

The specific name of the Yellow-rumped, *coronata* (for its yellow crown), stayed with the Myrtle Warbler after the split.

Yellow-throated Warbler *Setophaga dominica*

This species has been recorded in the Azores.

Yellow-throated is self-evident.

The specific name *dominica* in this case refers to the Caribbean island of San Domingo (Hispaniola), where the species was first found.

Prairie Warbler *Setophaga discolor*

This is another record from the Azores.

Prairie is an inaccurate name, as this is more a species of open woodlands.

The specific name *discolor* means 'of different colours': it is a dominantly grey and yellow bird with black streaks and a white under-tail. The male also has a patch of red-brown on the back.

Black-throated Green Warbler *Setophaga virens*

This species has been recorded in Germany and Iceland.

Black-throated Green Warbler is another descriptive book-name, though the female lacks the black throat.

The specific name *virens* tells us that the bird is green, though that only applies to its upper-parts – but to both sexes in this case.

Canada Warbler *Cardellina canadensis*

This species has been recorded in Iceland and, for the first time in Britain in 2023.

Canada and *canadensis* are self-evident, though part of the population breeds south of the Canada/United States border.

Cardellina comes from an Italian dialect word for a (European) Goldfinch (see *Carduelis*) and works here because the bird is mainly yellow. This genus has now replaced *Wilsonia*.

Wilson's Warbler *Cardellina pusilla*

This species has been recorded in Britain.

The aforementioned **Wilson** has his own warbler. He travelled widely in North America, recording and painting a large number of species, and publishing a nine-volume American Ornithology. However the AOS announced in 2023 that all eponyms in vernacular names of American species will eventually be replaced.

The specific name *pusilla* tells us that it is 'very small'.

103. CARDINALIDAE:
CARDINALS AND ALLIES

This family takes its name from *Cardinalis*, the genus of the iconic Red Cardinal, *Cardinalis cardinalis*, so-called for its 'red robe' plumage, and well known as a feature on North American Christmas cards.

A few of this family, all of them migratory, have been recorded in Britain.

Summer Tanager *Piranga rubra*

Summer indicates that this is a summer migrant to North America.

The name **Tanager** is of South American origins. It starts life in the Tupi language as *Tangara*, a dancer. The 'true' tanagers are classed in the family Thraupidae, but a few tanager-like members of the Cardinalidae have acquired the vernacular name.

Piranga originates in another Tupi word for a small bird.

The specific name, *rubra* (red) is for the male's all-red summer plumage.

Scarlet Tanager *Piranga olivacea*

Scarlet Tanager has a darker red plumage than the Summer Tanager.

Curiously, the specific name *olivacea* (olive green) applies to all females but only to winter-plumaged males.

Rose-breasted Grosbeak *Pheucticus ludovicianus*

The breeding male **Rose-breasted** Grosbeak is largely black and white but has a crimson breast that is much darker than suggested by the word **rose**.

Grosbeak originated as a name for certain European finches (see Hawfinch) with large bills (from the French *Gros-bec*) and there are two of that family in North America. Here it is used for a bird of similar structure but of a different family.

Pheucticus is rooted in Greek and means 'shy'.

The specific name *ludovicianus* relates to Louisiana (*Ludovicus* is the Latin rendering of the name of Louis XIV of France, after whom the first French colony was named).

Dickcissel *Spiza americana*

This species has been recorded in Norway.

Dickcissel is thought to be an echoic name based on the bird's sibilant song. The bird was originally known by Wilson and Audubon as the *Black-throated Bunting*, but the rather odd popular name seems to have come into general usage in the late nineteenth century.

Spiza is a Greek word for a bird that was probably the Chaffinch.

The specific name *americana* is self-evident.

Blue Grosbeak *Passerina caerulea*

This is yet another Azores record.

Blue describes an ink-blue male and that is echoed by *caerulea*.

Passerina means 'sparrow-like'

Indigo Bunting *Passerina cyanea*

This species has been recorded occasionally in Britain.

Indigo indicates the ink-blue of the breeding male.

Bunting is used here in a borrowed sense, as the bird does not belong to the Emberizidae.

The genus *Passerina* means 'sparrow-like'.

The specific name *cyanea* means 'cyan blue'.

GLOSSARY AND EXPLANATORY NOTES

BIOLOGICAL TERMS

GENETICS

The modern understanding of genes and of DNA dates from a massive breakthrough in the mid-twentieth century. The mapping of the genomes of living creatures has added a potent tool to the analytical powers of modern scientists. DNA sequencing can now be used to identify species and their relationships from a different perspective, adding greatly to the toolbox used to define speciation – and provoking what the IOC website calls 'the taxonomic industry'.

TAXONOMY

This concerns the systematic organisation and presentation of the relationships and hierarchies within the world of birds (and of other organisms too, of course), which presents that data in some sort of logical order. In the case of this reference, the framework is first the sequencing of birds by *Family*, following IOC usages and, within that, the sequencing by *Genera*.

PASSERINES AND NON-PASSERINES

In the Linnaean hierarchy, the highest level within the *Class* Birds is the *Order*. The largest Order is that of the *Passerines*, the perching birds, which include over 50% of the world's species. These are sometimes called the songbirds, which is a little unhelpful considering the vocal ineptitude of corvids, for instance. For the purposes of this work, the division is largely ignored, since the main framework used is the Family. However, the first half contains the *Non-Passerines* and the latter half the *Passerines*, with the division coming after the Psittaculidae (Old World parrots) and before the Tyrannidae (tyrant flycatchers). In species terms, this represents a 50/50 split.

FAMILY

A group of birds in different genera that have close common attributes is referred to as a family. An obvious example would be ducks, geese and swans, which are all waterbirds with a common ancestry and a generally similar biological structure.

GENUS (PLURAL GENERA)

Birds that are closely related and exhibit common characteristics are grouped into a Genus. Thus, a number of large, long-necked species of web-footed waterbirds fit into the Genus *Cygnus* (swan). Within a genus, characteristic differences, such as plumage, song, habits and so on, may separate members into species: *Cygnus olor* (Mute Swan), *Cygnus atratus* (Black Swan) and so on. A **congener** is a species belonging to the same genus as another.

SPECIES

The exact definition of a species can be arbitrary, which is one reason why not all changes proposed are accepted by all authorities in the scientific world. The plumage differences between Common Teal and Green-winged Teal are quite obvious in the field, but there is still hesitation in some quarters about whether these are truly separate species. Italian Sparrow and Hooded Crow offer particularly interesting examples, because in both cases the definition of a species enters a grey and disputed area. The Bean Goose/Pink-footed Goose complex is almost constantly under review. Such matters are part of ongoing investigations, which are typical of the modern preoccupation with genetic studies. **Conspecific** means belonging to the same species.

SUBSPECIES

While some species are considered to be monotypic (of one form) throughout their range, others have two or more subspecies. Some subspecies are fairly obvious, as with Brent Geese, which can usually be distinguished visually. In some species, the currently listed subspecies may be of arbitrary definition. In other cases, recent genetic research has confirmed the fact that former subspecies should become full species. This area has not been treated exhaustively in the species entries, though some more obvious cases, such as the Mew Gull of the title, are considered.

The word **race** may sometimes replace subspecies in informal usage. The term **nominate subspecies** refers to the form bearing the species name, as in *Branta bernicla* **bernicla**.

'SPLITS' AND 'LUMPS'

These terms of popular convenience describe in crisp, if inelegant, style the results of decisions to divide one species into two or more new species (see Chaffinch), or conversely to unite two or more species as one (see Osprey). Although such usage may appear to be quite modern, the terms were in fact employed in the same way in Hartert's 1912 Checklist.

HYBRIDS

Hybridisation (the interbreeding of two species) is not seen widely in normal wild situations. It tends to be more obvious in some families than in others. It is perhaps most often noted among wildfowl, but it has been recorded in species as diverse as herons, waders and finches. A case of hybridisation between White and Black Storks was reported in 2023. Perhaps the most famous hybrid in Europe is the Italian Sparrow, now given species status as a 'stable hybrid', that is, one that can replicate itself. There have been a few occasions when 'new species' were named (e.g. Cox's Sandpiper), only for the bird to be later revealed as a hybrid. In general, wild hybrids are not named. The *Collins Bird Guide* (third edition, 2022) offers some telling illustrations of *Aythya* duck hybrids.

EVOLUTION

Darwin and Wallace articulated the concept of the evolution of species. Some staggering facts are revealed by taking note of a bird's evolutionary path. By a freak of history, all wrens in the world take their collective English name from the one species that crossed the Aleutian Bridge from North America. Fossil studies have revealed that hummingbirds once existed in Europe and that they have a close common ancestry with the swifts. DNA sequencing in particular has contributed greatly to the understanding of the relationships between bird species and of speciation itself.

CONVERGENT EVOLUTION

Some unrelated creatures evolve physical similarities owing to the development of similar lifestyles. For instance, swallows and swifts, which are widely separated in evolutionary terms, independently developed a torpedo-shaped body, a broad gape and long curved wings suited to their habit of hunting insects in the air. The Great Crested Grebe and the Red-breasted Merganser show great similarities in structure and aquatic feeding styles, yet they are not closely related.

BROOD PARASITISM

In its best-known form, this refers to the habits of species that place their eggs in the nests of other species and leave them to be hatched and raised by the host parents. This is true of many members of the cuckoo family, but also of the North American species Brown Cowbird, which has been recorded in the region. The young of the cowbirds behave differently from those of the cuckoos, but the net result is that foster parents raise the young of another species. Some species may deposit eggs in the nest of another female of the same species to gain similar benefits.

KLEPTOPARASITISM

This term is used to describe the habits of some species that rob others of their food. The habit is particularly rife among skuas, gulls and frigatebirds, which force terns and auks to drop or regurgitate their food and then steal it. The distinctions are sometimes blurred, because larger gulls and skuas are also predators and may kill and consume the food-carriers.

SEXUAL DIMORPHISM

In many species the two sexes are unalike. Such differences may be slight, as in many geese, or very marked, as in Capercaillie. Among our garden birds, the Blackbird and House Sparrow are good examples. While many males advertise themselves with showy plumage, dull, drab or cryptic plumages help to camouflage the nesting females from potential predators. In many species of songbird, the male is the sole or better songster. Such words as *duck* and *drake* are based on visible differences, while the fact that the female Ruff was traditionally known as the Reeve may relate to its lesser value as the smaller bird in late Tudor markets.

REVERSED SEXUAL DIMORPHISM

In some species dimorphism is reversed. In many raptor species, the female is the larger. That relates in part to the fact that males and females do quite different jobs and may even specialise in different prey. The falconer's name, *tiercel*, for the male Peregrine Falcon, derives from the fact that it is one-third smaller than the female. In the case of phalaropes and painted-snipes, the female is the more colourful and seeks out one or several duller mates to raise her young.

LEK AND LEKKING

Used as a noun, the word lek designates the neutral display ground on which some species of male birds, such as Black Grouse or Ruff, meet to display and to win the right to mate with a number of females. As a verb, to lek means to meet for such display. The action is known as lekking. The unusual-looking word appears to have older Germanic roots, but was introduced into English from Swedish as recently as 1867.

LEGS, KNEES AND FEET

The anatomy of birds' legs appears to be reversed compared with a human's. The true 'knee' is usually hidden by plumage. What appears to be a reversed 'knee' in many species is the equivalent of the human 'heel'. The name *Thick-knee* compounds a general misunderstanding of that joint. Strictly speaking, that makes names like Red-legged Partridge and Greenshank somewhat inaccurate. The apparent 'lower leg' derives from a fusion of foot bones to which are joined the toes, the latter generally seen as the 'feet' of the bird.

PLUMAGE

The words used to describe plumage are sometimes complicated (though the fiddliest terms have been avoided in this book). Terms like *crown*, *breast*, *belly* and *under-tail* are generally straightforward. Some are less obvious:

- *Rump* refers to the area of the lower back immediately above the base of the tail.
- *Mantle*: in a strict sense this refers to the upper-back of the bird. It is often used more loosely, as in gulls at rest, to refer to the upper-parts, including the folded wings.
- *Bars* are lines formed by the different colours or shades of feathers in contrast to the general colour (Two-barred Crossbill, Barred Warbler, 'ringtail').
- *Spots* may be distinct and separate marks (Spotted Sandpiper), but the word is often used loosely in names such as in Spotted Flycatcher and Lesser Spotted Woodpecker.
- *Speculum*: in certain species, some ducks in particular, the speculum is a colourful and usually iridescent wing-bar that often aids in identification.
- *Moult*: worn plumage needs to be renewed, so feathers are shed periodically and new ones are grown. Some birds change their appearance by age or season.
- *Eclipse*: in the likes of colourful male ducks, moult involves a period of drabness that helps to camouflage them while flight is impaired. This state is known as the *eclipse*.
- *Fingers*: the spread primary feathers at the wingtip of a soaring raptor are sometimes referred to as the fingers.

FACIAL FEATURES

- *Bill* and *beak* are effectively synonymous. The former has roots in Old English and is used in a number of bird names, such as *Razorbill*, *Crossbill* and *Spoonbill*. The latter derives from the Old French element of English and only occurs in *Grosbeak*.
- *Frons* is the Latin term for the 'forehead' of a bird and relates to the names of the 'white-fronted' geese and Red-fronted Serin. In each case the front of the head carries the relevant colour.
- The *frontal shield* is a plate of mainly keratinous material found on that part of the 'forehead' of some species, such as coot and moorhen.
- *Supercilium* is the eyebrow stripe (Eyebrowed Thrush).
- *Eye-stripe* is in line with the eye.
- *Lores* refer to the areas between the bill and the eyes of the bird.
- *Cere* is an area of waxy skin found over the base of the upper bill of a bird, most noticeably in such species as pigeons and parrots.
- *Moustachial stripe* runs from the corner of the mouth (Moustached Warbler, Bearded Tit). A contrasting white facial patch accounts for Whiskered Tern.
- *Wattles* are patches or extensions of 'warty' skin which serve for temperature regulation and display. These feature in such names as Red-wattled Lapwing, Lappet-faced Vulture and Bald Ibis.

- *Crests* and *'Ears'* are feathered display features found on the head of a number of species and appear in various ways in names such as Hooded Merganser, Goldcrest and Long-eared Owl, and also in several scientific names.

TERMS RELATED TO MOVEMENTS OF BIRDS

VAGRANCY AND VAGRANTS

A century ago, Thomas Coward referred to occasional birds as 'wanderers' but today the preferred term is 'vagrants'. Such birds, of a wide range of species, occur infrequently outside their normal range, particularly due to a navigational malfunction during migration or as the result of extreme weather events, so odd arrivals must be considered seriously. The degree of vagrancy is variable: while some species may occur annually and in moderate numbers, others may have been recorded only once within the area covered. A very high number of species covered in these texts qualify only because they have wandered well outside their normal range. The British list contains species that breed right across the Northern Hemisphere, with some from Africa too and others from the southern oceans. Conversely, of all the species treated here, there is only one example of an exclusively Latin American land bird being recorded in mainland Europe. The topic is thoroughly addressed by Lees and Gilroy (2021).

MIGRATION

The advantages of moving to a different area to breed, or to avoid a seasonally hostile climate, support the maximisation both of breeding opportunities and of survival. For example, birds such as the Whimbrel migrate out of cold areas when food becomes locked away by ice. Many others, such as larks, pipits, warblers and swallows, leave arid areas when seed or insect production is due to reduce seasonally. Moving into another zone where there is enough food for survival or for raising young generally outweighs the risks of the journey. The costs of such movements are eventually borne by individuals for the benefit of the species as a whole. The timing and direction of migration varies between species and even between populations of the same species.

'TRANSMUTATION' AND HIBERNATION

The phenomenon of migration was observed long before it was understood. Aristotle noted that Wheatears arrived with the vine blossoms. He also linked the arrival of winter Robins with the disappearance of Redstarts. His explanation of that, and of other similar exchanges, was that the birds changed their appearance by 'transmutation'. This was not unreasonable, given that domestic and ornamental birds changed their appearance during moult.

Another ancient belief was that birds, like mammals or insects, might hibernate. Hirundines, in particular, were thought to hibernate in the mud of the reedbeds where they were seen to roost while on passage.

Aristotle's views remained the norm, perpetuated by the likes of Turner, until at least the seventeenth century, but cracks began to appear in that dogged belief as people like Willughby and Ray began to study nature in the field. Even so, as late as 1789, Gilbert White, one of the most critical of observers, had not wholly discarded the possibility that swallows hibernate.

IRRUPTION

This term refers to an erratic form of short-distance migration which occurs when there is a population boom and/or a shortage of food supplies. In Britain, Waxwings are the best-known irruptive species. These occur very irregularly, whenever competitive pressure within the population forces some individuals to seek out resources in regions that are peripheral to their normal winter territory. Some owls and certain seed-eaters such as Parrot Crossbills and Nutcrackers are also subject to such movements.

ALTITUDINAL MIGRATION

Some upland or montane breeding species, such as Merlin, Wallcreeper and Snowfinch, are found at lower altitudes in the same region in winter. They retreat from cold and inhospitable altitudes into levels where their food sources are more secure. In this case, a change of altitude serves a similar purpose to a change of latitude.

SEDENTARY SPECIES

While some birds migrate, some remain all year-round in the same general area. Both terms, migratory and sedentary, are not absolute: some species abandon summer haunts, such as suburban gardens, to feed in open countryside a few miles away; some parts of the population of, for instance, Song Thrush move out of colder zones to mingle with sedentary cousins in milder areas.

OVERSHOOTING

Describes the process of accidentally travelling past the intended breeding or wintering grounds. Many species move by night or get caught in an over-favourable wind and simply travel too far. The consequences might be fatal, though some recover their bearings. Most spring sightings of Hoopoe in Britain are of 'overshoots'.

'CARRYING'

A displaced bird may join a flock of another species that is following its normal pattern of movement and then find itself in an alien location. The phenomenon is known as 'carrying'. This is most obvious with, for example, a Cackling Goose travelling with Barnacle Geese, but a winter flock of Dunlin may readily hide a vagrant 'peep', such as a Baird's Sandpiper, while a Pallas's Warbler might be found in company with a flock of migrant Goldcrests.

SHIP ASSISTANCE

Land birds that cross seas and oceans during migrations or storms sometimes find resting places onboard ship and may be transported to an alien landfall. In that way, unlikely vagrants found within a few miles of a port are often considered to be 'ship assisted'. One Snowy Owl, which arrived at the port of Felixstowe in Suffolk in 2001 had taken refuge from a storm on a ship travelling down the St Lawrence River in Canada and remained onboard during the Atlantic crossing, presumably sustained by eating smaller refugees. Similar assisted travel accounts for the odd House Crows in Europe, for a good proportion of small passerines that survive the Atlantic crossing and for the odd 'sedentary' sparrow that takes an accidental boat trip.

WEATHER AND CLIMATE

WEATHER EVENTS

Birds on the move are very vulnerable to shifting weather patterns that affect the timing, direction, safety and success of migratory movement. The Western Palearctic would have a much smaller list of species but for the numbers of birds that get caught up and carried by Atlantic storms in particular. The migration of many species down the Atlantic Flyway off North America makes it likely that some will be trapped in autumn storms that brew up in tropical waters to move northwards before veering east towards Europe.

An easterly wind during migration to or from Eurasia may throw rare vagrants onto British shores. One-off weather events, such as major storms, flooding, drought or harsh winters can affect breeding success or survival. Populations may fluctuate as a result, but might be expected to recover in subsequent years. One of today's problems is that extreme weather is becoming more frequent and recovery time is at risk.

CLIMATE CHANGE

Weather is temporary but climate change is more permanent, and the changes are marked by the habits of birds. Britain now has breeding species, such as the Little Egret, that were formerly considered to belong in Mediterranean countries. At the same time, wintering wildfowl such as Bewick's Swan, White-fronted Goose and Smew no longer have to travel so far to find haven from the frozen Arctic, so fewer reach British shores. Since the pace of climate change seems now to be faster than the potential for most species to evolve a response to it, the future for some species looks very bleak.

HUMAN INTERVENTIONS

NON-NATIVE SPECIES/EXOTICS

Natural range expansion by, for example, the Cattle Egret or the Collared Dove does not involve the direct intervention of humans, and such species soon become accepted as 'native'. On the other hand, human action is normally responsible for the introduction of non-native birds to an adopted area. Human history is such that birds have long been collected, moved and traded, for one reason or another, into regions that are not their natural ones. This applies to domesticated birds, of course, but here the focus is on birds that live freely in the wild: Egyptian Geese, for instance, were imported into Britain during the seventeenth century as an ornamental species, but escaped and bred in sufficient numbers to now have a place on the British list.

INTRODUCTIONS

Birds are introduced by human intervention in a number of ways into non-native environments, usually deliberately, for commercial or ornamental purposes. The shooting industry is responsible for a range of exotics, including a massive annual restocking of Common Pheasants, whose heritage is generally a meaningless jumble of exotic subspecies. Red-legged Partridges are also massively restocked but this is at least a European species. Others, such as the Golden Pheasant and Peacock, have been released for ornamental purposes or simply to fulfil a whim, as was the case for the Little Owl in Britain.

REINTRODUCTIONS

Some native species become extinct or endangered. Over the past half-century or so there have been many initiatives that have brought back to Britain, for example, the White-tailed Eagles and Great Bustards, and have boosted weak populations of the Red Kite and Common Crane. In Spain and other parts of Europe, the Northern Bald Ibis has been a major focus. On the other hand, some species may reoccur as the result of illegal release: this might account for some of the Eagle Owls and Goshawks in Britain, though natural repopulation by migrants could be considered possible.

ESCAPES

Birds are often kept as cage-birds or in wildfowl collections. Inevitably, some non-native birds do escape to live freely among wild birds, although many do not survive for long. Mass escapes sometimes result from storm damage to aviaries or pens, and that seems to be the source of Black Swans originating from the Continent. Ornamental gamebirds, parrots and caged songbirds are likely to escape, or even to be released deliberately. They may wander, survive and even proliferate if numbers are sufficient and conditions right. Red-headed Bunting sightings have long been treated with official disdain, because so many were imported by the cage-bird trade, but now that Europe has banned their importation, such birds that do turn up might merit closer consideration.

'FERAL' BIRDS

When enough escaped or released non-native species establish a sustained breeding population, they may be accepted as 'feral' birds, which are self-sustaining, naturalised wild populations of exotic species (e.g. Canada Geese or Ring-necked Parakeets in Britain). Such species may or may not impose a threat to native species: when escaped Ruddy Ducks reached that status, they were extirpated from Britain for fear of compromising the genetic purity of the related White-tailed Duck in Europe. A breeding colony of exotic flamingos in Germany has already produced hybrid forms, and conservationists are concerned for the future purity of the Greater Flamingo. Some species are more readily tolerated than others.

COLLECTING AND COLLECTORS

Historically, scientists gathered data on birds: this was often only possible when dead birds were available for analysis. Birds were collected by trapping or by shooting, often skinned and preserved for transport, and retained in museums as dried skins, as skeletons or as stuffed exhibits. Today, such tools as high-resolution photography of live birds can be linked to data from mist-net captures and even to DNA from faecal matter, so there is less need for such collection. In fact, only a decade or so ago, the world of bird-lovers was scandalised when an American ornithologist shot and collected a specimen of a redis-covered 'lost' kingfisher species that already seemed to be on the verge of extinction. During the nineteenth century, collection reached horrendous proportions as wealthy private collectors amassed dead birds and other creatures in their mansions. This gave a commercial value to rarities, with the result that greed drove the final extinction of several species, including the Great Auk, when rarity value was matched by cash value. Rothschild's huge private collection eventually became the core of the British Natural History Museum's array of animals and bird specimens. Large sums were often paid to the field collectors who physically gathered the specimens. Such work was often unglamorous: several explorers and collectors, such as Kennicott, Wahlberg, Hasselqvist and Fischer, died as a direct or indirect result of their work. Wallace's own account of his time as a professional collector is not pleasant reading. Nonetheless, the Verreaux family built their fortune on such activity.

OTHER COMMERCIAL EXPLOITATION

During the nineteenth century in particular, wild creatures were treated as commercial resources and many species were exploited mercilessly for fashion or for food. Grebes, egrets, birds of paradise and even some stork species were exploited to near extinction for their feathers. The collection of wild birds for food was widespread: this was the motor that drove the extinction of the once very numerous Passenger Pigeons in America. The commercial exploitation of wildfowl by professional hunters using punt guns and nets continued in Britain until the early part of the twentieth century: one inheritor of that tradition, Kenzie Thorpe, was still around to help Peter Scott set up the WWT in 1946 and to work as a wildlife pundit on Anglia TV into the 1960s. Meanwhile, Victorian game-shooting interests and their inheritors did their best to extirpate birds of prey and corvids in particular, and that heritage still prejudices the future of some species.

SOME LANGUAGE ISSUES

'ONOMATOPOEIC' AND 'ECHOIC'

These two words mean much the same (though the second is a lot easier to spell). Both relate to a name that is based on the sound a bird makes. *Cuckoo* and *Curlew* are obvious examples; *Chough* is another. *Crex crex* works beautifully in the Latinate form. *Whimbrel* and *Gadwall* have evolved just enough to appear less obvious examples.

EAST, WEST, NORTH AND SOUTH

In bird names, these terms have different meanings in different contexts: their values are not absolute and each usage depends on a geographical viewpoint. When used in the context of the Western Palearctic, the term 'Eastern species' refers to a bird from the Orient. The same is relevant for such names as Levant Sparrowhawk and for Latin-based forms such as *sinensis* and *orientalis*, which have a Eurocentric bias. That is perfectly reasonable for the simple historical reasons that both the English language and the Linnaean system are the products of European history.

In many bird names, however, such terms have a different geographical viewpoint:

- In *Western* Sandpiper or *Eastern* Phoebe, the words relate specifically to the boundaries of North America.

- In *Eastern* and *Western* Subalpine Warblers, the focus is solely on a Mediterranean context.

- *Northern* Grey Shrike distinguishes a species in North America from a southern relative (*Loggerhead* Shrike) that is not named as its polar opposite. The now-abandoned *Southern* Grey Shrike (replaced by Iberian) served only in a European context.

- *Northern* Bald Ibis separates a species of North Africa and the Middle East from *Southern* Bald Ibis, a South African species.

In short, names that include such terms need some interpretation, and it seems regrettable that they are used as a fashionable 'quick fix' in many modern splits.

HYPHENS

Beware of hyphens in English bird names: they have been used widely in the past and conventions have now changed. Coward's list has a great many that now appear to be totally pointless. The IOC has a simple code that explains when and where hyphens should and should not be used in bird names. Some of these rules may seem arbitrary, but the safe approach might be to follow the usage on the current list.

LATINISED COLLECTIVES

Terms such as *corvid*, *hirundine* and *larid* are sometimes used as convenient collective words. These are generally anglicised versions of the scientific family names and are useful where there are no convenient alternatives such as 'wildfowl', 'gamebirds' or 'waders', but there is no strict convention in this.

CLASSICAL LEGENDS

With all the good bird legends and stories around the world, it seems at first odd that so many bird names are rooted in the legends of Ancient Greece and Rome. This bias is entirely in keeping with the fact that the Linnaean system is built upon Latin and Greek. A Classical education was based on the study of Greek and Roman writers, who included politicians, military men and agronomists on the one hand, and playwrights, philosophers, poets and historians on the other. It was therefore inevitable that names such as *Nisus*, *Pandion*, *Diomedes*, *Picus*, *Philomela*, *Procne* and *Penelope* would be woven into the textures of scientific ornithology. A good many of those involved in the study and naming of American wildlife during the nineteenth century had a similar educational bias. If we were to start again today, it is highly likely that few such names would make it to the starting line, because modern culture is so much less aware of the Classics.

TRANSFORMATION AS PUNISHMENT

Classical legends are frequently morality tales, arising as guidelines or warnings about acceptable behaviour. They often underline the fact that rule-breakers will be in trouble. Consequently, figures committing blasphemy; those betraying their nation; those spurning more powerful figures; those committing rape or murder; or even the gormless lad, Anthus, who puts the horses in the wrong field: none can escape retribution. What is more, their sin or folly often involves the innocents around them. A common punishment is the transformation of those concerned into some sort of bird. The image is a thinly veiled metaphor of death, their newly acquired wings representing an afterlife.

NATIVE AMERICAN LANGUAGES

There is very little trace today of Native North American languages in bird names: the only two words that seem to show some evidence of such origins are *Sora* and *Dowitcher*, and both appear on the British list. The domination of European culture was so complete that the Turkey, a native American species, is known in America by its late medieval English name.

Conversely, the Tupi-Guarani native languages of modern Brazil, among others of South and Central America, gave rise to a number of bird names that are fairly widely used in American English. Among those, words such as *Tanager* and *Anhinga* appear on the Western Palearctic list.

VERNACULAR AND FOLK NAMES

The term *vernacular*, in the context of this work, refers to the current English-language name of the species.

Folk name refers to those forms that evolved locally and were traditionally used in regional areas of Britain (or in some cases, America). Some of these forms remain as standard vernacular names, but others have died out and have mostly been forgotten, or have faded into the background. Many such names are to be found in literature, as well as in the history of ornithology. Lockwood's *Oxford Book of British Bird Names* (1984) covers such matters well, while Choate's *Dictionary of American Bird Names* (1973) is also a treasure-trove.

ANTHROPOMORPHISM

Humans have long applied human attributes to other entities, whether it be in the depiction of deities in human form; in the portrayal of birds and animals in legend and fable (the 'wise' owl and the 'cunning' corvid); in the tradition of giving 'pet' names to birds and animals (Robin Redbreast and Brock the Badger); in the favouring of upright species like puffins and owls; or even in the judgement of the 'behaviour' of certain species (the persecution of raptors and corvids). There is a strong thread of anthropomorphism running through our relationships with birds.

CONFIDING

Some creatures adapt more readily than others to the presence of humans. The term 'confiding' describes the behaviour of wild creatures that adapt readily to our presence, often lured by food. The Robin is probably the most familiar example, and almost every family will know that 'feeding the ducks' will bring birds close to. It was that habit that allowed Mute Swans to be semi-domesticated by such as the priory at Abbotsbury in Dorset, or Greylag Geese and Mallards to become the root-stock of farm birds.

JIZZ

The roots of this word are somewhat disputed: a popular view is that it relates to the Second World War air-spotter's acronym GISS (a general impression of size and shape) and that idea is not unhelpful. However, *jizz* was used in print in a way well understood by birdwatchers as early as 1922. It was used then by Thomas Coward, whose name crops up frequently in these texts. In short, it is a term widely employed by birdwatchers to describe the combination of shape, movement, habits, size and colour that together form the impression that an individual bird is a given species. The skill is instinctive, though it develops with experience: a small child can recognise the distinct shape of a Blue Tit feeding upside down on the end of a frail twig, just as an inexperienced adult can soon learn that a stiff-winged 'gull' might be a Fulmar.

Little Brown Jobs (LBJs) are the unidentified small brown birds that frustrate even experienced birdwatchers. Such birds as *Phylloscopus* warblers need to be seen well and are often elusive as well as confusing. The term consigns many such birds to the bin of frustration. One cure for that frustration is to learn the jizz of a species, as well as its calls and songs. If the LBJ gets away unidentified, there is always tomorrow. In terms of bird names, both Dunnock and its scientific name *Prunella* derive from words meaning 'brown and small', which is one reason why the bird was also long called a Hedge Sparrow, since 'sparrow' once served as a catch-all name for small brown birds. The Dunlin too has a similar meaning and one only has to look at all the 'small' names in that genus to realise that LBJs are an essential element of birding.

BIRDWATCHER JARGON

The jargon begins with the labels that are applied to people interested in birds and birdwatching. *Birdwatchers* are British, *birders* are American, but we tend to blur those terms today. *Ornithologists* are clearly academic. Francis Willughby, with his private income, was an '*amateur ornithologist*' but his collaborator, John Ray, was a '*professional ornithologist*' because he needed a wage in order to pursue the activity.

Twitcher is a very modern term, used originally in an ironical sense to convey the nervous anxiety of a curious subset of birders addicted to rarity-chasing. For some, there is no greater insult than to be called a *twitcher*, a word that is much loved by the press and the general public alike because anyone with binoculars is a bit odd anyway. Some birders might admit publicly to being a '*lister*' (though that term can be applied to both record keeping and to number-chasing) but few would willingly wear the label '*twitcher*'. If the vocabulary includes words such as '*dipped*' (missed), '*mega*' (a major rarity) or '*gripping*' (very exciting) the twitcher is incurable.

Laridophiles (gull specialists) and *laridophobes* (gull-haters) are birder subspecies.

Unsurprisingly, the birds themselves are also subject to a clannish jargon, which includes words such as '*barwit*' and '*blackwit*' for the two godwits, and '*mipit*' for Meadow Pipit. Use of such words insures the user against the label '*dude*', which will surely be attached to anyone using the name *Bearded Reedling* because that came from a book.

As for the photographer who refuses to kill the macho shutter-noise in a quiet bird hide or the dog-walker who strolls in front of a row of telescopes, the language is pure Anglo-Saxon.

ORGANISATIONS AND THEIR ACRONYMS

AOS: The American Ornithological Society (2016) (www.americanornithology.org)
The former **AOU**, the American Ornithologists' Union (1883), amalgamated with the **COS**, the Cooper Ornithological Society (1892).

BOU: The British Ornithologists' Union (1858) (www.bou.org.uk)

BOURC: The BOU Records Committee has published the British list since 1883.

BTO: The British Trust for Ornithology (1933) (www.bto.org)
Study projects, surveys and the *Atlas of British Birds*.

ICZN: The International Commission on Zoological Nomenclature (1895)
The organisation oversees the International Code of Zoological Nomenclature (www.iczn.org).

IOC: The International Ornithological Congress (first convened 1884)
This is now under the organisational umbrella of the **International Ornithologists Union (IOU)**. The checklist is maintained as the **IOC Bird List** (www.worldbirdnames.org).

RSPB: The Royal Society for the Protection of Birds (1889) (www.rspb.org.uk)
Protection of birds and their habitats.

WWT: The Wildfowl and Wetlands Trust (1946) (www.wwt.org.uk)
Conservation of wetland birds and habitats.

ALTERNATIVE ORGANISATIONS INVOLVED IN TAXONOMY AND NOMENCLATURE

The Clements Checklist of the Birds of the World (www.birds.cornell.edu/clementschecklist)

HBW Alive/Birdlife International (datazone.birdlife.org/species/taxonomy)

The Howard & Moore Complete Checklist of the Birds of the World, 4th Edition (www.howardandmoore.org)

The IOC has formed two subcommittees with representatives of those bodies, to coordinate, where possible, the decision-making processes for taxonomy and nomenclature in particular. Eventually there will be an IOU checklist based on the current IOC list. This will become the new world standard.

WHO'S WHO?
THE NAMES BEHIND THE NAMES

The names of many people appear in the species notes. Most of those would be classed as 'ornithologists' in one way or another because at some point they contributed to our knowledge of birds, though a great many would have described themselves otherwise; as naturalists; as botanists; as philosophers; as civil servants; as military or naval officers; as priests; as mariners; as explorers; as academics; and even as poets, dramatists and artists. The story of bird names is just about as broad as it could be. Such names are generally given some context that will enable the user to find out more if need be: some of their stories are worth pursuing: Pliny, for example, may have transmitted some useful snippets about larks and bustards, but the story of his final hours during the eruption of Vesuvius will not fit in here, any more than would the full story of Peter Pallas's epic explorations.

The names of many people are recorded in the bird-names themselves, either in the vernacular forms or in the scientific names. Such names (eponyms) are dealt with in the species accounts and are listed in the biographical index that follows. These are usually dedicated to people significant in the discovery or description of a bird: they may be at the cutting edge of ornithology or they may be sponsors, though not all are of those types: there are a few surprises. The vast majority of such names come from the eighteenth and nineteenth centuries, which were the peaks of exploration and discovery, but it is apparently not too late for the current generation: the most recent eponym included here is hardly a decade old, though new eponyms are very rare, particularly in the heavily studied Western Palearctic. A close search of the accounts will find a few examples of late twentieth-century eponyms and just two twenty-first-century examples. New dedications are more likely to be found among subspecies, but these are not treated exhaustively in this work.

There was, and remains, a certain ethic in the use of such names: it was considered bad form to dedicate a name to oneself, though some did it. Most frequently, the name was a tribute to a friend, a patron or to someone admired.

With the benefit of historical hindsight, some of those accolades are now deemed to be inappropriate. That is not a problem in itself, since such names have come and gone historically, as evidenced by Ray's Wagtail in Britain. However, in November 2023, the AOS announced the decision to remove all eponyms from the vernacular names of American birds because some of those commemorated were linked historically to aspects of slavery and colonialism. Whether or not such blanket erasure is justifiable or even vaguely fair to all concerned, the result cannot be seen as a purely American issue, firstly for the fact that it involves a number of birds on the British and Western Palaearctic lists and secondly because it sets a precedent by overtly politicising bird names. A secondary aspect of the American announcement is the intention to replace eponyms with descriptive or habitat names. Some past efforts in that genre do not inspire confidence. Significantly, there is no mention in the announcement that this might be a good opportunity to rediscover some lost Native American names.

This matter will inevitably open debate on the use of eponyms in the wider world of birds. It is to be hoped that future discussion will, at the very least, take each case on its merits and will also acknowledge that eponyms are part of the texture, colour and history of bird names. On the face value of the American announcement, we are due to lose, for instance, the one memorial to Anna Blackburne, a rare example of an eighteenth-century female collector and naturalist. Eponyms are so often significant markers in the very story of ornithology and indeed in the story of plants, mammals, insects and fungus too. In any case, the device is almost as old as language itself and is a perfectly legitimate tool.

It is manifestly obvious that history happened without our permission, but to eradicate it entirely from the story of bird names greatly impoverishes an important aspect of ornithology.

	EPONYM	PAGE
Abdim, Bey El-Arnaut (1780–1827) Turkish governor of Wadi Halfa in Sudan. The dedication was by Martin Lichtenstein (q.v.) who collected in that area.	Abdim's Stork, *Ciconia abdimii*	34
Adalbert, Prince of Bavaria (1828–1875) Patron and possibly friend of Reinhold Brehm (below). Adalbert married into the Spanish nobility. (This dedication is sometimes misattributed to his contemporary, Adalbert of Prussia.)	Spanish Imperial Eagle, *Aquila adalberti*	159
Adams, Edward (1824–1856) English Royal Navy surgeon, naturalist, Arctic explorer. Sailed with Ross in 1848 and Collinson in 1852 to search for Franklin's lost expedition.	White-billed Diver, *Gavia adamsii*	119
Aeschylus (525/524–456/455 BCE) Greek playwright reputedly killed by a tortoise dropped by a bird of prey.		155
Albertus Magnus (1200–1280) German Dominican scholar who studied birds of prey and first named the Peregrine Falcon *Falco peregrinus*.		188, 196
Albin, Eleazar (1690–1742) English naturalist, artist, author of *A Natural History of Birds* (1731–1738).		136
Aldrovandi, Ulisse (Aldrovandus) (1522–1605) Important pioneer Italian naturalist, author of books on ornithology and other topics of natural history.		160, 187, 237, 287
Alexander, Lieutenant Boyd (1873–1910) British collector and explorer. Unusually, his forename and his surname are used in two different species.	Cape Verde Swift, *Apus alexandri* Boyd's Shearwater, *Puffinus boydi*	39, 131
Allsop, Kenneth (1920–1973) Author and broadcaster. *Adventure Lit Their Star* (1949) (Little Ringed Plover).		72
Amherst, Countess Sarah (1762–1838) A British naturalist and collector who lived in India. She was commemorated by naturalist/taxidermist Benjamin Leadbeater.	Lady Amherst's Pheasant, *Chrysolophus amherstiae*	31
Anouilh, Jean (1910–1987) French dramatist, author of *Antigone*, 1944, a philosophical critique of power and integrity.		58

	EPONYM	PAGE
Aristophanes (*c*.450–388 BCE) Greek dramatist, known in particular for *The Birds*, which places the Hoopoe in the role of the king.		176, 205, 300
Aristotle (384–322 BCE) Greek philosopher and natural scientist, whose works on natural history were long treated with unquestioning reverence.		12, 18, 30, 58, 81, 89, 140, 155, 186, 198, 222, 231, 247, 249, 261, 270, 276, 305, 323
Arminjon, Admiral Vittorio (1830–1897) Italian explorer from whose ship, *Magenta*, the first specimen was collected by E. Giglioli in 1865.	Trindade Petrel, *Puffinus arminjoniana*	128
Audouin, Jean V. (1797–1841) French entomologist and ornithologist. The gull was dedicated to him by fellow zoologist Payraudeau in 1826.	Audouin's Gull, *Larus audouinii*	100
Audubon, John James (1785–1851) French-born naturalist and artist working in America, known for *The Birds of America* (1827–38). The Audubon Society is named after him.	Audubon's Shearwater, *Puffinus lherminieri*	90, 92, 99, 130, 183, 257, 294, 308, 315, 317
Baillon, Louis-Antoine (1778–1855) French naturalist and collector.	Baillon's Crake, *Zapornia pusilla* Tropical Shearwater, *Puffinus bailloni*	57, 130
Baird, Spencer F. (1823–1887) US ornithologist, founder of the US National Museum.	Baird's Sandpiper, *Calidris bairdii*	83, 324
Baldenstein, Conrad von (1784–1878) Swiss naturalist. First to describe Willow Tit.		231
Baldner, Leonhard (1612–1694) Strasbourg naturalist known to Willughby and Ray who provided them with illustrations. Probable source of name *Roller*.		177
Bannerman Dr David A. (1886–1979) Scottish ornithologist, collector, author and curator.	Cape Verde Buzzard, *Buteo bannermani*	167
Barolo, Marchese Carlo T.F. di (1782–1838) Italian nobleman at court of Napoleon I, later philanthropist in Turin. Dedication by Charles-Lucien Bonaparte in 1857.	Barolo Shearwater, *Puffinus baroli*	131
Barrow, Sir John (1764–1848) British explorer-naturalist, founder Royal Geographical Society.	Barrow's Goldeneye, *Bucephala islandica*	22, 23
Bartram, William (1739–1823) Called 'the grandfather of American ornithology' because of his influence on Alexander Wilson.	Upland Sandpiper, *Bartramia longicauda*	78
Bechstein, Johann (1757–1822) German naturalist and pioneer conservationist, dubbed the 'father of German ornithology'.		183, 284

	EPONYM	PAGE
Belon, Pierre (1517–1564) French naturalist. *Histoire de la nature des oyseaux* (1555).		67, 293
Bent, Arthur C. (1866–1954) American ornithologist and prolific author. *Life Histories of North American Birds* (1961–8).		22
Bergius, Karl. H. (1790–1818) German naturalist, collecting in South Africa, where he died penniless.	Greater Crested Tern, *Thalasseus bergii*	107
Berners, Abbess Juliana (1388–?) English author (presumed) of *The Book of St Albans* (earliest evidence of Starling murmuration).		255
Berthelot, Sabin (1794–1880) French naturalist in the Canaries. *L'Histoire Naturelle des Îles Canaries* (1835–50) with Philip Barker Webb (q.v.).	Berthelot's Pipit, *Anthus berthelotii*	290
Bewick, Thomas (1753–1828) Wood-engraver and natural history author. *A History of British Birds* (1797/1804).	Bewick's Swan, *Cygnus columbianus bewickii*	8, 9, 236, 325
Blackburne, Mrs Anna (1726–1793) British collector and naturalist. Unmarried – used the title to give gravitas to her status as a member of the scientific community. The warbler was sent by her brother and dedicated by Müller.	Blackburnian Warbler, *Setophaga fusca*	31, 314, 333
Blyth, Edward (1810–1873) English zoologist, curator of the museum of the Royal Asiatic Society of Bengal, India.	Blyth's Pipit, *Anthus godlewskii* Blyth's Reed Warbler *Acrocephalus dumetorum*	229, 287, 288
Boddaert, Pieter (1730–1795) Dutch naturalist accredited with naming no fewer than 190 taxa in the early stages of the Linnaean system.		236
Boie, Friedrich (1789–1870) German naturalist. Author of a number of bird genera and species. *Auszüge aus dem System der Ornithologie* (1844). Genus *Spatula*.		13, 169, 220
Bolle, Carl August (1821–1909) German naturalist, visited Cape Verde and Canaries.	Bolle's Pigeon, *Columba bollii*	49
Bonaparte, Charles-Lucien (1803–1857) Nephew of Napoleon I. French ornithologist working in America and later Italy. This dedication was the work of Swainson and Richardson (both q.v.).	Bonaparte's Gull, *Chroicocephalus philadelphia*	21, 52, 58, 84, 98, 99, 127, 131, 160, 192, 200, 217, 256, 257, 287, 307
Bonaparte, Zenaïde (1801–1854) Daughter of Joseph Bonaparte, cousin/wife of Charles-Lucien.	Mourning Dove, *Zenaida macroura*	52

	EPONYM	PAGE
Bonelli, Franco (1784–1830) Italian ornithologist and entomologist, Professor of zoology at the University of Turin and curator. Discovered the warbler and eagle, which were named by Vieillot.	Bonelli's Eagle, *Aquila fasciata* Eastern Bonelli's Warbler, *Phylloscopus orientalis* Western Bonelli's Warbler, *Phylloscopus bonelli*	xii, 160, 223
Brehm, Alfred (1829–1884) German zoologist, son of Christian. Collected in Spain and North Africa. Directed the Hamburg Zoological Garden.		212
Brehm, Christian (1787–1864) German pastor, ornithologist, and prodigious collector based at Rentendorf, Germany. He classified and named species collected by his sons, among others.		49, 159, 258, 264
Brehm, Reinhold (1830–1891) German ornithologist, son of Christian. Settled in Spain and studied local Natural History. See Adalbert (q.v.) re Spanish Imperial Eagle.		159
Brehm, Thekla (1833–1857) Daughter of Christian Brehm. The dedication marked her early death.	Thekla's Lark, *Galerida theklae*	212
Brisson, Mathurin (1723–1806) French zoologist. *Ornithologie* (1760).		69, 71, 119, 137, 170, 216, 269, 297
Bruce, Rev Henry James (1835–1909) American missionary working in India.	Pallid Scops Owl, *Otus brucei*	170
Bruce, James (1730–1794), Scottish explorer who located the source of the Blue Nile.	Bruce's Green Pigeon, *Treron waalia*	52
Brünnich, Morten T. (1737–1827) Danish ornithologist and curator. *Ornithologia borealis*, 1764.	Brünnich's Guillemot, *Uria lomvia*	113, 114
Buchanan-Hamilton, Dr Francis (1762–1829) Scottish physician, zoologist and explorer.	Grey-necked Bunting, *Emberiza buchanani*	303
Buffon, Georges-L. Leclerc, Compte de (1707–1788) French naturalist, Director of the Jardin du Roi. His 36-volume *Histoire naturelle, générale et particulière* (1749–67) focused heavily on birds.		44, 68, 112, 153, 182, 193, 207, 211, 217, 261, 312
Bulwer, Rev. James (1794–1879) Artist and naturalist living on Madeira.	Bulwer's Petrel, *Bulweria bulwerii* Jouanin's Petrel, *Bulweria fallax*	131
Butler, Lt Col. Edward A. (1843–1916) English ornithologist during service with British Army.	Omani (Hume's) Owl, *Strix butleri*	172, 173
Cabanis, Jean (1816–1906) German ornithologist and curator. He collected in America and founded the *Journal für Ornithologie* in 1853. (*Tryngites*)		84

	EPONYM	PAGE
Cabot, Dr Samuel (1815–1885) American surgeon-physician and ornithologist. Described the tern, which was taken in the Yucatan, in 1847.	Cabot's Tern, *Thalasseus acuflavidus*	107
Caligula (12–41 CE) Infamous Emperor of Rome as Gaius Julius Caesar Augustus.		230
Canute (Cnut) the Great (*c.*990/95–1035) King of Denmark, Norway and England	Knot, *Calidris canutus*	81
Cassin, John (1813–1869) American naturalist, traveller and author based at Philadelphia. Described 198 species, including Ross's Goose.		7
Catesby, Mark (1683–1749) English explorer-naturalist who spent seven years in America. He provided the earliest account of American flora and fauna and imported the name *Oystercatcher*.		14, 68, 181, 185, 309, 310
Cetti, Francesco (1726–1778) Italian Jesuit educator, mathematician and zoologist in Sicily. *Natural History of Sardinia* (1774–77).	Cetti's Warbler, *Cettia cetti*	221
Charles I (1600–1649)/**Charles II** (1630–1685), Kings of England Both kings (Latin *Carolus*) issued charters for the establishment of the Carolina States in America. That later transmitted into the scientific names of several species found there.		4, 16, 33, 54, 252
Charleton, Walter (*c.*1619–1707) English natural philosopher. In *Onomasticon Zoicon* (1668) he was mainly concerned with recording birds and their names in several languages.		21, 176, 217
Chaucer, Geoffrey (*c.*1342–1400) English poet. In the *Parliament of Fowles* he provided many examples of bird names in Medieval English.		32
Clarke, William E. (1853–1938) English ornithologist, collector and curator, President of the BOU in 1918.	Song Thrush ssp., *Turdus philomelos clarkei*	258
Clot, Antoine B. (Clot Bey) (1793–1868) French doctor working for many years in Egypt for the Ottoman viceroy to establish better medical care. He was awarded the honorific *Bey* for his services. The dedication was by Bonaparte.	Thick-billed Lark, *Ramphocoris clotbey*	210
Columella, Lucius (*c.*4 CE–*c.*70 CE) Roman writer on agriculture, including bird names.		14
Cook, James (1728–1779) British navigator-explorer, whose ships carried first Banks and then Forster as expedition naturalists.		68, 93, 110, 173
Cook, Samuel (1787–1856) British collector and naturalist in Spain.	Iberian Magpie, *Cyanopica cooki*	201

	EPONYM	PAGE
Cooper, William (1798–1864) New York naturalist (Evening Grosbeak, Cooper's Hawk) and father of James Cooper, who founded the Cooper Ornithological Society.		294
Cory, Charles B. (1857–1921) American ornithologist, who first collected this species off Massachusetts.	Cory's Shearwater, *Calonectris borealis*	128, 129
Coues, Elliott (1842–1899) Founder-member of the American Ornithologists Union. *Key to North American Birds* (1872).	Coues's Arctic Redpoll (ssp.), *Acanthis hornemanni exilipes*	297
Coward, Thomas (1867–1933) English ornithologist and writer (see Bibliography).		xxiii, 16, 23, 29, 46, 49, 50, 60, 72, 78, 82, 84, 85, 89, 90, 92, 100, 112, 147, 148, 162, 163, 168, 169, 171, 188, 206, 244, 249, 254, 258, 261, 263, 264, 285, 287, 297, 302, 323, 328, 329
Cretzschmar Phillip (1786–1845) German anatomist and naturalist who named several species.	Cretzschmar's Bunting, *Emberiza caesia*	42, 303
Dabbene, Roberto (1864–1938) Italian-born ornithologist who settled in Argentina to study the birds there.	Tristan Albatross, *Diomeda dabbenena*	122
Darwin, Charles (1809–1882) English naturalist, geologist. His study of pigeons contributed to his formulation of the Theory of Evolution.		48, 320
Degland, Côme-Damien (1787–1856) French naturalist. *Ornithologie européenne* (1849).	White-winged Scoter, *Melanitta deglandi*	21
Denham, Major Dixon (1786–1828) English naturalist and explorer of West Africa, who led a colourful and sometimes controversial life.	Denham's Bustard, *Otis denhami*	42
Dunn, Captain Henry N. (1864–1952) British Army surgeon, later hunter and collector.	Dunn's Lark, *Eremalauda dunni*	214
Dupont, Léonard P. (1795–1828) French naturalist and collector in Africa. The bird was so-named by Vieillot.	Dupont's Lark, *Chersophilus duponti*	214
Edwards, George (1693–1773) English artist, naturalist, who coined the names Horned Grebe and Eared Grebe.		22, 24, 61, 118, 126

	EPONYM	PAGE
Eleonora of Arborea (1347–1404) Sardinian warrior princess and national heroine. She reformed local law, which included protection of hawk and falcon nests. Giuseppe Gené (q.v.) was responsible for the dedication.	Eleonora's Falcon, *Falco eleonorae*	186
Erckel, Theodor (1811–1897) Assistant to E. Rüppell, who described this species.	Erckel's Spurfowl, *Pternistis erckelii*	34
Eversmann, Alexander E.F. (1794–1860) Prussian doctor and biologist who collected in Russia and Central Asia.	Yellow-eyed Dove, *Columba eversmanni* Eversmann's Redstart, *Phoenicurus erythronotus*	50, 267
Eyton, Thomas C. (1809–1880) English ornithologist who created genus *Chroicocephalus*.		98, 127
Fea, Leonardo (1852–1903) Italian zoologist, explorer of Burma, West Africa and Cape Verde, where he collected the petrel, later named by Salvadori in 1899.	Fea's Petrel, *Pterodroma feae*	127
Feldegg, C.F. von (1779–1845) Austrian army officer.	Yellow Wagtail, *Motacilla flava feldegg*	286
Finsch, F.H. Otto (1839–1917) German naturalist, author of *Die Papageien* (1867) (parrots)	Finsch's Wheatear, *Oenanthe finschii*	272
Fischer, Gustav (1848–1886) German physician and explorer in Africa. The bird was named by Reichenow in 1887, the year after Fischer's death.	Fischer's Lovebird, *Agapornis fischeri*	190, 326
Fischer von Waldheim, Johann G. (1771–1853) German naturalist working in Russia, where his academic prowess earned him an ennoblement.	Spectacled Eider, *Somateria fischeri*	20
Fleming, John (1785–1857) Scottish church minister and naturalist, whose *Philosophy of Zoology* (1822) influenced Darwin (Velvet Scoter, Arctic Skua etc.).		21, 112
Forster, Johann R. (1729–1798) German naturalist of Scottish descent. Joined Cook's Second Pacific voyage and later wrote widely. In later life took academic posts in Germany.	Forster's Tern, *Sterna forsteri*	93, 110, 173, 314
Franklin, Benjamin (1706–1790) American polymath, US Founding Father.		152, 165, 193
Franklin, Sir John (1786–1847) Royal Navy officer and British explorer. Two voyages of exploration in Canadian Arctic and Governor of Van Diemen's Land, before the fatal attempt to find the North-West Passage.	Franklin's Gull, *Leucophaeus pipixcan*	4, 100, 102
Frivaldsky von Frivald, Dr Emerich (1799–1870) Hungarian naturalist who described the Eurasian Collared Dove.		51

	EPONYM	PAGE
Gené, Giuseppe (1800–1847) Italian naturalist. Succeeded Bonelli (q.v.) at Turin.	Slender-billed Gull, *Chroicocephalus genei*	98
Gengler, Josef (1863–1931) German ornithologist commemorated in British subspecies of Chaffinch.		292
Georgi, Johann G. (1729–1802) German botanist, naturalist and geographer associated with Peter Pallas in Siberian explorations. Baikal Teal.		13
Gessner, Conrad (1516–1565) Swiss polymath, prolific author and influential naturalist. *Historiae animalium* (1551–58).		89, 137, 177, 185, 201, 212, 237, 285, 287, 298, 305
Giraldus (Gerald of Wales) (*c*.1146–*c*.1223) Cambro-Norman Archdeacon of Brecon. Legend of Barnacle Geese.		4
Gmelin, Johann F. (1748–1804) German naturalist. After Linnaeus's death he edited and published the 13th edition of *Systema Naturae* (1793). He is considered the authority in the naming of 290 species of bird.		23, 25, 236, 256, 267
Godlewski, Wictor (1831–1900) Polish naturalist who served 12 years in the Siberian mines for revolt against Russia. After that time he settled there to study the natural history of the region.	Blyth's Pipit, *Anthus godlewskii*	288
Godman, Frederick DuCane (1844–1894) English ornithologist, founder-member of the BOU.		49
Graells, Mariano de la Paz y de la Agüerra (1809–1898) Spanish zoologist.	Lesser Black-backed Gull ssp., *Larus fuscus graellsii*	105
Gray, George R. (1808–1872) English zoologist at the British Museum. Brother of John E. Gray.	Gray's Grasshopper Warbler, *Helopsaltes fasciolatus*	63, 169, 216, 232
Gray, John E. (1800–1875) English ornithologist at the British Museum. Brother of George R. Gray.	Indian Pond Heron, *Ardeola grayii*	146
Güldenstädt, Johann A. (1745–1781) Latvian-German naturalist/explorer in Russia.	Güldenstädt's Redstart, *Phoenicurus erythrogastrus*	267
Hartert, Ernst (1859–1933) German ornithologist who was for much of his life in England working closely with Jourdain, Ticehurst and Witherby on the influential *Hand-list of British Birds* (1912).		xx, 49, 85, 112, 169, 188, 258, 287, 302, 320
Hartlaub, K.J. Gustav (1814–1900) German physician and ornithologist specialising in African species.		228

	EPONYM	PAGE
Hasselqvist, Fredrik (1722–1752) Swedish traveller, naturalist. Undertook to collect for Linnaeus, but died of an illness on his way back from Africa. Cattle Egret.		74, 146, 179, 326
Heine, Ferdinand (1809–1894) German ornithologist.	Turkestan Short-toed Lark, *Alaudala heinei*	214
Hemprich, Wilhelm F. (1796–1825) German naturalist and explorer.	Sooty Gull, *Ichthyaetus hemprichii*	101
Hernandez de Toledo, Francisco (1514–1587) Spanish court physician, naturalist, explorer.		100
Herodotus (*c.*484–*c.*425 BCE) Greek historian and geographer who wrote of the crocodile bird.		75
Hesychius of Alexandria (fifth or sixth century BCE), Greek grammarian, lexicographer whose work records many Greek bird names.		18, 155, 160, 198, 213, 244, 277
Heuglin, M.T. (1824–1876) German explorer and ornithologist who made one expedition to Spitsbergen and Novaya Zemlya.	Heuglin's Gull, *Larus fuscus heuglini*	105, 106
Hey, Michael (1798–1832) German collector.	Sand Partridge, *Ammoperdix heyi*	32
Hill, John (1714–1775) British naturalist and author, and one of the most controversial figures of his age. Squacco Heron.		145
Hodgson, Brian. H. (1800/1–1894) English naturalist working in India.	Olive-backed Pipit, *Anthus hodgsoni*	289
Hopkins, Gerard Manley (1844–1889) English poet: 'The Windhover' (1877).		185
Hornemann, Jens W. (1770–1841) Danish botanist.	Arctic Redpoll, *Acanthis hornemanni*	297
Hume, Allan (1829–1912) British civil servant in India, and passionate ornithologist. His collection and accounts of local birds earned him the accolade of the 'father of Indian ornithology'.	Hume's Short-tailed Lark, *Calandrella acutirostris* Hume's Whitethroat, *Curruca althaea* Hume's Wheatear, *Oenanthe albonigra* Hume's Leaf Warbler, *Phylloscopus humei*	xii, 108, 172, 173, 213, 214, 223, 237, 238, 272
Hutchins, Thomas (1730–1790) British surgeon, naturalist, Hudson Bay Company.	Cackling Goose, *Branta hutchinsii*	4
Hutton, Captain T. (1807–1874) Served with the British Army in Afghanistan.	Afghan Babbler, *Argya huttoni*	243
Ingoldsby, Thomas, pen name of R. Barham (1788–1845) English writer, humourist: 'The Jackdaw of Rheims'.		202
Isabella I of Castile (1451–1504) Queen of Spain with King Ferdinand. Isabelline.	Isabelline Wheatear, *Oenanthe isabellina*	xxii, 198, 199, 231, 270

	EPONYM	PAGE
Levaillant (Vaillant), François (1753–1824) An eminent explorer and naturalist.	Levaillant's Woodpecker, *Picus vaillantii*	158, 184, 288
L'Herminier, Félix Louis (1779–1833) French naturalist working in America and Guadeloupe.	Audubon's Shearwater, *Puffinus lherminieri*	130
Lichtenstein, Martin H.C. (1780–1857) German zoologist.	Lichtenstein's Sandgrouse, *Pterocles lichtensteinii*	42, 47
Lincoln, Thomas (1812–1883) American explorer and naturalist who travelled with Audubon.	Lincoln's Sparrow, *Melospiza lincolnii*	308
Linnaeus, Carl (1707–1778) Swedish botanist, zoologist, 'father of modern taxonomy' after whom the Linnaean System is named.		xx, 8, 15, 38, 46, 58, 60, 61, 68, 74, 80, 89, 100, 113, 114, 116, 119, 122, 128, 135, 137, 146, 186, 187, 200, 205, 206, 209, 216, 228, 237, 248, 249, 252, 258, 264, 283, 289, 292, 314
Lorenz, Theodore K. (1842–1909) German zoologist collecting in Russia.	Mountain Chiffchaff, *Phylloscopus lorenzii*	225
MacGillivray, William (1796–1852) Scottish Naturalist (named Ross's Gull and linked to Audubon).		99
MacQueen, General Thomas R. (1792–1840) Collector: presented specimen to the British Museum c.1832.	MacQueen's Bustard, *Chlamydotis macqueenii*	41
Marmora, Gen. Alberto F. La (1789–1863) Italian general and naturalist.	Marmora's Warbler, *Curruca sarda*	240
Mauri, Ernesto (1791–1836) Italian botanist.	Western Sandpiper, *Calidris mauri*	85
McCormick, Robert (1800–1890) British naval surgeon.	South Polar Skua, *Stercorarius maccormicki*	111
McDougall, Dr Peter (1771–1814) Scottish physician and collector.	Roseate Tern, *Sterna dougallii*	109
Ménétries, Edouard (1802–1861) French zoologist.	Menetries's Warbler, *Curruca mystacea*	239
Merrett, Christopher (1614/15–1695) English physician, scientist. Compiled a useful list of birds.		xxi, 67, 202
Michahelles, Karl W. (1807–1834) German zoologist.	Yellow-legged Gull, *Larus michahellis*	105
Milne-Edwards, Alphonse (1835–1900) French zoologist; son of Henri (q.v.).	Ashy-throated Parrotbill, *Suthora alphonsiana*	242

	EPONYM	PAGE
Milne-Edwards, Henri (1800–1885) French naturalist.	Cape Verde Shearwater, *Calonectris edwardsii*	129
Mlokosiewicz, L.F. (1831–1909) Polish naturalist specialising in the Caucasus region.	Caucasian Grouse, *Lyrurus mlokosiewiczi*	30
Molina, Juan (1740–1829) Chilean Jesuit priest, botanist who envisaged evolution (quoted by Darwin). Snowy Egret name *thula*.		149
Moltoni, Edgardo (1896–1980) Italian naturalist.	Moltoni's Warbler, *Curruca subalpina*	239
Montagu, George (1753–1815) British ornithologist.	Montagu's Harrier, *Circus pygargus*	xxi, 54, 62, 109, 140, 164, 210, 289
Monteiro, Luis (1962–1999) Portuguese ornithologist.	Monteiro's Petrel, *Hydrobates monteiroi*	124, 125
Morelet, Pierre (1809–1892) French artist and naturalist collecting in the Azores.	Azores Chaffinch, *Fringilla moreletti*	293
Moussier, Jean (1795–1850) French surgeon-naturalist.	Moussier's Redstart, *Phoenicurus moussieri*	267
Naumann, Johann Andreas (1744–1826) German naturalist, father of Johann Friedrich (q.v.).	Naumann's Thrush, *Turdus naumanni*	260
Naumann, Johann Friedrich (1780–1857) German artist and naturalist.	Lesser Kestrel, *Falco naumanni*	185
Neumayer, Franz (1791–1842) Austrian botanist and collector.	Western Rock Nuthatch, *Sitta neumayer*	247
Newton, Alfred (1829–1907) English zoologist, ornithologist leading founder of BOU. *A Dictionary of Birds* (1893–96).		xxi, 89, 263
Nicholson, Max (1904–2003) Irish-born English ornithologist, co-founder WWF. Dunnock.		285
Nordmann, Alexander von (1803–1866) Finnish naturalist.	Black-winged Pratincole, *Glareola nordmanni* Nordmann's Greenshank, *Tringa guttifer*	91, 93
Nuttall, Thomas (1786–1859) English botanist, zoologist, collector in North America. (Named Forster's Tern.)		110, 162, 257
Oddie, Bill (b.1941) English comedian, ornithologist, broadcaster, writer who did much to popularise birds on TV, including the word *Bonxie*.		111
Ord, George (1781–1866) American ornithologist. Tundra Swan, Bonaparte's Gull, etc.		9, 99, 313

	EPONYM	PAGE
Oustalet, Jean-Frédéric F.E. (1844–1905) French zoologist and ornithologist. Cape Verde Shearwater.		129
Ovid (43 BCE–17/18 CE) Roman poet. *Metamorphoses*.		176, 264
Pagnol, Marcel (1895–1974) French writer, film-maker, *La Gloire de mon Père*, etc.		33
Pallas, Peter (1741–1811) Prussian zoologist and botanist who established his reputation at The Hague and was invited by Catherine the Great of Russia to be professor at the St Petersburg Academy of Sciences. He led a major six-year expedition into Siberia and a later one into southern Russia, and eventually retired back in Germany.	Pallas's Sandgrouse, *Syrrhaptes paradoxus* Pallas's Gull, *Ichthyaetus ichthyaetus* Pallas's Fish Eagle, *Haliaeetus leucoryphus* Pallas's Leaf Warbler, *Phylloscopus proregulus* Pallas's Grasshopper Warbler, *Helopsaltes certhiola* Pallas's Rosefinch, *Carpodacus roseus* Pallas's Reed Bunting, *Emberiza pallasi*	46, 100, 101, 106, 115, 166, 224, 232, 245, 296, 305, 314, 324, 332
Palsgrave, John (1485–1554) English priest, tutor at court of Henry VIII. Woodpecker.		181, 296
Parkes, Dr Kenneth C. (1922–2007) US ornithologist.	Northern Waterthrush, *Parkesia noveboracensis*	311
Pennant, Thomas (1726–1798) Welsh naturalist and writer, who standardised many English names via the reputation of his *British Zoology. Vol. II*, Birds (1776).		xxi, 3, 6, 8, 9, 18, 19, 20, 24, 49, 54, 55, 60, 62, 67, 69, 71, 90, 91, 102, 106, 109, 111, 112, 114, 118, 119, 146, 165, 172, 178, 181, 182, 187, 197, 198, 221, 223, 236, 240, 256, 267, 279, 294, 296, 298, 314
Phipps, Constantine (Lord Mulgrave) (1744–1792) British explorer, later politician. Ivory Gull.		97
Pliny the Elder (Gaius Plinius Secundus) (*c.*23–79 CE) Roman writer, naturalist. Lark and Bustard names.		24, 41, 58, 212, 300, 332
Polatzek, Johann (1839–1927) Austrian ornithologist working in Canaries.	Gran Canaria Blue Chaffinch, *Fringilla. polatzeki*	293
Preuss, Paul (1861–1926) German botanist and collector in West Africa and elsewhere.	Preuss's Cliff Swallow, *Petrochelidon preussi*	221

	EPONYM	PAGE
Radde, Gustav (1831–1903) German naturalist and finder of warbler, which bears both his name and that of his expedition leader.	Radde's Warbler, *Phylloscopus schwarzi* Radde's Accentor, *Prunella ocularis*	224, 245, 284, 288
Ray, John (1627–1705) English parson-naturalist and writer who worked with Willughby on compiling a comprehensive *Ornithology* (1678).		xxi, 3, 8, 12, 13, 19, 21, 67, 90, 106, 111, 113, 114, 142, 160, 163, 172, 177, 198, 212, 237, 267, 268, 287, 293, 294, 296, 323, 330, 332
Reeves, Rev. John (1774–1856) Inspector of tea, British East India Company.	Reeves's Pheasant, *Syrmaticus reevesii*	31
Richard, Charles (1745–1835) French postal official at Lunéville, collector of natural history.	Richard's Pipit, *Anthus richardi*	288
Richardson, John (1787–1865) British explorer of Canada with Franklin.	Richardson's Goose, *Branta hutchinsii*	xii, 4, 99, 101, 102, 112
Ross, Bernard (1827–1874) Irish agent of the Hudson Bay Company.	Ross's Goose, *Anser rossii*	7
Ross, Rear Admiral Sir James (1800–1862) British polar explorer.	Ross's Gull, *Rhodostethia rosea*	99
Rüppell, Eduard (also **Rueppell**) (1794–1884) German naturalist and explorer.	Rüppell's Warbler, *Curucca ruppeli* Rüppell's Weaver, *Ploceus galbula*	34, 73, 111, 156, 239, 281
Sabine, General Sir Edward (1783–1883) British Arctic explorer. President of the Royal Society.	Sabine's Gull, *Xema sabini*	97, 98
Savi, Paolo (1798–1871) Italian ornithologist.	Savi's Warbler, *Locustella luscinioides*	232
Savigny, Marie (1777–1818) French zoologist. Osprey genus *Pandion*.		153
Say, Thomas (1787–1834) American naturalist.	The genus *Sayornis*	192
Schintz, Heinrich R. (1777–1861/62) Swiss naturalist.	Dunlin (subspecies), *Calidris alpina schintzii*	83, 84
Schlegel, Gustaaf (1840–1903) Dutch naturalist. An unusual case of the use of forename (see also Boyd Alexander).	Pechora Pipit, *Anthus gustavi*	289
Schrenck, Leopold Von (1826–1894) Russian zoologist and geographer.	Von Schrenck's Bittern, *Ixobrychus eurhythmus*	144
Schwarz, Ludwig (1822–1894) German astronomer and leader of Radde's (q.v.) expedition.	Radde's Warbler, *Phylloscopus schwarzi*	224, 288

	EPONYM	PAGE
Scopoli, Giovanni (1723–1788) Italian physician and naturalist who formally described a dozen species of bird, as well as insects, etc.	Scopoli's Shearwater, *Calonectris diomedea*	128, 129
Scott, Sir Peter (1909–1989) English ornithologist, founder of WWT, co-founder of WWF.		24, 327
Seebohm, Henry (1832–1895) English ornithologist.	Atlas Wheatear, *Oenanthe seebohmi*	270
Selby, Prideaux (1788–1867) English ornithologist, wildlife artist. Several definitive English names.		69, 164, 285, 287, 288, 289, 294
Shakespeare, William (1564–1616) 'The Bard'. English playwright, poet. Several uses of interesting historical forms.		xix, xxii, 50, 136, 147, 201, 259
Sharpe, Richard B. (1847–1909) British zoologist who first described this woodpecker.	Iberian Green Woodpecker, *Picus sharpei*	184
Shaw, George (1751–1813) English botanist, zoologist, co-founder of Linnaean Society.		54
Shirihai, Hadoram (b.1962) Israeli naturalist, *Birds of Israel* (1996).	Desert Owl, *Strix hadorami*	172
Sibbald, Sir Robert (1641–1722) Scottish physician, antiquarian. Ptarmigan.		29
Smithson, James (1765–1829) English Chemist etc, founder of Smithsonian Museum.	American Herring Gull, *Larus smithsonianus*	104
Sophocles (497/6–406/5 BCE) Greek tragedian. *Antigone*.		58
St Cuthbert (*c*.634–687 CE) Scottish Bishop of Lindisfarne, who established the first 'nature reserve' on Inner Farne Island.		20
Steinbüchel, Mrs Marion (b.1977) German biologist in team working in Canary Islands.	Gran Canaria Robin, *Erithacus marionae*	263
Stiernhielm, Georg (1598–1672) Swedish civil servant and poet. Estrildid finches.		282
Stejneger, L.H. (1851–1943) Norwegian-American biologist.	Amur Stonechat, *Saxicola stejnegeri* Stejneger's Scoter, *Melanitta stejnegeri*	22, 269
Steller, G.W. (1709–1746) German naturalist and explorer.	Stellers's Eider, *Polysticta stelleri*	19
Stephens, James F. (1772–1852) English entomologist and zoologist. Genus *Mareca*.		14
Sturm, J.H.C.F. (1805–1862) German bird artist and collector.	Dwarf Bittern, *Ixobrychus sturmii*	144

	EPONYM	PAGE
Wagler, Johann (1800–1832) German herpetologist, ornithologist. Franklin's Gull.		100, 148, 307
Wahlberg, Johan A. (1810–1856) Swedish naturalist and collector who was killed by an elephant.	Wahlberg's Eagle, *Hieraaetus wahlbergi*	158, 326
Wallace, Alfred Russell (1823–1913) British collector, zoologist and author working in Southeast Asia. The Wallace Line, Wallacea, Theory of Evolution.		232, 320, 327
Warr, Mrs F.E. (b.1938) Librarian at the British Natural History Museum.	Basalt Wheatear, *Oenanthe warriae* (2020)	273
Webb, Mary (1881–1927) English romantic novelist and poet who popularised the use of 'murmuration'.		255
Webb, Philip Barker (1793–1854) British botanist working mainly on the Canary Islands.	Vinous-throated Parrotbill, *Suthora webbiana*	242
White, Gilbert (1720–1793) Parson-naturalist. Regular observations. *The Natural History of Selborne* (1789). Warblers, House Martin, etc.	White's Thrush, *Zoothera aurea*	165, 220, 223, 224, 225, 237, 257, 323
Whitehead, John (1860–1899) English naturalist.	Corsican Nuthatch, *Sitta whiteheadi*	247
Willughby, Francis (1635–1672) English ornithologist, worked with John Ray on systematics etc. Died prematurely and work completed by Ray.		xxi, 145, 153, 155–6, 177, 184, 268, 323, 330
Wilson, Alexander (1766–1813) Scottish-American 'father of American ornithology'.	Wilson's Petrel, *Oceanites oceanicus* Wilson's Snipe, *Gallinago delicata* Wilson's Phalarope, *Phalaropus tricolor* Wilson's Warbler, *Cardellina pusilla*	78, 87, 88, 121, 162, 307, 312, 313, 316, 317
Worm, Ole (Wormius) (1588–1654) Danish naturalist. Eider.		20
Yarrell, William (1784–1856) English ornithologist. *A History of British Birds* (1843). Bewick's Swan and others.	Pied Wagtail, *Motacilla alba yarrellii*	xxi, 8, 9, 36, 60, 164, 169, 197, 236, 244, 287, 288, 294, 302
Zino, Paul Alec (1916–2004) Madeiran businessman and conservationist. Rediscovered 'Madeiran' Petrel (now split).	Zino's Petrel, *Pterodroma madeira*	124, 127

REFERENCES

Attenborough, D. (1998) *The Life of Birds*. London: BBC Books.

Audubon, J.J. (1827–38) *Birds of America*. Available at https://www.audubon.org/birds-of-america.

Avery, M. (2014) *A Message from Martha: The extinction of the Passenger Pigeon and its relevance today*. London: Bloomsbury.

Balmer, D.E., Gillings, S., Caffrey, B.J., Swann, R.L., Downie, I.S. and Fuller, R.J. (2013) *Bird Atlas 2007–11: The Breeding and Wintering Birds of Britain and Ireland*. Thetford: BTO Books.

Bircham, P. (2007) *A History of Ornithology*. London: Collins.

BirdGuides (2004–6) Birds of the Western Palearctic Interactive. http://www.birdguides.com.bwp

Birdguides (July 2023) *Valencia's Category C Birds*. Birdguides newsletter by email.

Birkhead, T., Wimpenny, J. and Montgomerie, B. (2014) *Ten Thousand Birds: Ornithology since Darwin*. Princeton, NJ, and Darwin: Princeton University Press.

Birkhead, T. (2016) *The Most Perfect Thing: Inside (and outside) a bird's egg*. London: Bloomsbury

Birkhead, T. (2018) *The Wonderful Mr Willughby: The First True Ornithologist*. London: Bloomsbury.

Boelens, B. and Watkins, M. (2003) *Whose Bird? Men and women commemorated in the common names of birds*. London: Christopher Helm.

Bosworth Smith, R. (1913) *Bird Life and Bird Lore*. Cambridge: Cambridge University Press.

Brown, L. (1982) *British Birds of Prey: A Study of Britain's 24 Diurnal Raptors*. London: Collins.

Bucknell, N., Clews, B., Righelato, R. and Robinson, C. (2013) *The Birds of Berkshire – Atlas and Avifauna*. BBAG, Reading

Burnham, K. and Newton, I. (2011) 'Seasonal movements of Gyrfalcons'. *Ibis* 153, July: 468–84.

Campbell, B. (1974) *The Dictionary of Birds in Colour* London: BCA/Michael Joseph.

Centre National des Ressources Textuelles et Lexicales [online etymological French dictionary]. Available at: http://www.cnrtl.fr/etymologie/.

Choate, E.A. (1973) *The Dictionary of American Bird Names*. Boston: Gambit.

Cocker, M. and Mabey, R. (2005) *Birds Britannica*. London: Chatto and Windus.

Collar N. (2013) 'A species is whatever I say it is'. *British Birds* 106, March (pp. 130–142)

Coward, T.A. (1920) *The Birds of the British Isles and Their Eggs (First and Second Series)*. London: Frederick Warne.

Coward T.A. (1926) *The Birds of the British Isles: Migration and habits (Third Series)*. London: Frederick Warne.

Cumming, W.P., Hillier, S.E., Quinn, D.B. and Williams, G. (1974) *The Exploration of North America 1630–1776*. London: Paul Elek Ltd.

Darwin, C. (1979) *The Origin of Species (Abridged R.E. Leakey)*. London: BCA/Faber and Faber.

Dennis, R. (2021) *Restoring the Wild: Sixty years of re-wilding our skies, woods and waterways*. London: William Collins.

Dictionnaire Littré en Ligne, Reverso (n.d.). Available at littre.reverso.net/dictionnaire-francais.

Elkins, N. (1999) *Weather and Bird Behaviour* (second edition). London: Poyser.

Elphick, C., Dunning, J.B. and Sibley, D. (2001) *The Sibley Guide to Bird Life and Behaviour*. London: Christopher Helm.

Ferguson-Lees, J. and Christie, D.A. (2001) *Raptors of the World*. London: Christopher Helm.

Fisher, J. (1966) *The Shell Bird Book*. London: Ebury Press and Michael Joseph.

Game and Wildlife Conservation Trust (n.d.). Available at http://www.gwct.org.uk/game/.

Gill F, Donsker, D. and Rasmussen, P. (eds) (2023) IOC World Bird List (v13.1). https://doi.org/10.14344/IOC.ML.13.1.

Hartert, E., Jourdain, F.C.R, Ticehurst, N.F. and Witherby, H.F. (1912) *A Hand-List of British Birds*. London: Witherby and Co.

Heisman, Rebecca (2017) *What's In A Name? How genome mapping can make it harder to tell species apart*. Available at https://www.allaboutbirds.org/news/whats-in-a-name-how-genome-mapping-can-make-it-harder-to-tell-species-apart/.

Holloway, S. (1996) *The Historical Atlas of Breeding Birds in Britain and Ireland: 1875–1900*. London: Poyser.

Jameson, C.M. (2013) *Looking for the Goshawk*. London: Bloomsbury.

Jenkins, Alan C. (1978) *The Naturalists, Pioneers of Natural History*. London: Hamish Hamilton.

Jobling, J.A. (2010) *Helm Dictionary of Scientific Bird Names from Aalge to Zusii*. London: Christopher Helm.

Knepp (n.d.) 'White Storks'. Available at https://knepp.co.uk/rewilding/reintroductions/white-stork/.

Lees, A. and Gilroy, J. (2021), *Vagrancy in Birds*. London: Helm.

Linnaeus, C. (1796), *Systema Naturae per regna tria naturae* (tenth edition). openlibrary.org, Internet Archive. Available at https://archive.org/details/carolialinnsyst01gmelgoog

Lockwood, W.B. (1984) *The Oxford Book of Bird Names*. Oxford: Oxford University Press.

Löfgren, L. (1984) *Ocean Birds: Their Breeding Biology and Behaviour*. London and Canberra: Croom Helm.

Madge, S. and McGowan, P. (2002) *Pheasants Partridges and Grouse*. London: Christopher Helm.

Moorhead, A. (1966) *The Fatal Impact: An account of the invasion of the South Pacific, 1767–1840*. London: Hamish Hamilton.

Morris, F.O., ed. T. Soper (1850/1985) *British Birds*. London: Peerage Books.

Mynott, J. (2018) *Birds in the Ancient World*. Oxford: Oxford University Press.

Newton, I. (2010) *Bird Migration*. London: Collins (New Naturalist Library).

Perrins, C. (ed.) (2003) *The New Encyclopaedia of Birds*. Oxford: Oxford University Press.

Recuerda, M., Illera, J.C., Blanco, G., Zardoya, R. and Milá, B. (2021) Sequential colonization of oceanic archipelagos led to a species-level radiation in the common chaffinch complex (Aves: *Fringilla coelebs*). *Molecular Phylogenetics and Evolution*. https://doi.org/10.1016/j.ympev.2021.107291.

Reedman, R. (2013) 'The origins of the vernacular name of the Common Scoter'. *British Birds* 106, July, p. 417

Reedman, R. (2016) *Lapwings, Loons and Lousy Jacks: The How and Why of Bird Names*. Exeter: Pelagic.

Reilly, J. (2018) *The Ascent of Birds: How Modern Science is Revealing Their Story*. Exeter: Pelagic.

Sedinger, J.S. (ed.) (1996) *The Emperor Goose: An Annotated Bibliography*. University of Alaska Institute of Arctic Biology.

Sibley, D.A. (2014) *The Sibley Guide to Birds* (second edition). New York: Alfred A. Knopf.

Sims, E. (1985) *British Warblers*. London: Collins (New Naturalist Library).

Stoddart, A. and McInerny, C.J. (2017) 'The "Azorean Yellowlegged Gull" in Britain'. *British Birds* 110, November: 666–74.

Stoll, N.R. et al. (eds) (1961) *The International Code of Zoological Nomenclature Adopted by the XV International Congress of Zoology*. London: The International Trust for Zoological Nomenclature.

Svensson, L., Mullarney, K. and Zetterström, D. (2022) *Collins Bird Guide* (third edition). London: HarperCollins.

Toms, M. (2014) *Owls*. London: Collins (New Naturalist Library).

Tree, I. (2018) *Wilding*. London: Picador, Pan Macmillan.

Wallace, A.R. (1969/2014) *The Malay Archipelago*. London: Penguin Classics.

Walters, M. (2003) *A Concise History of Ornithology*. London: Helm.

Warham, J. (1990) *The Petrels: Their ecology and breeding systems*. London: Academic Press.

Watson, D. (1977) *The Hen Harrier*. London: Poyser.

Werkman, E. (2021) *The Amazing Story of the Montagu's Harrier*. Available at https://harrierconservationinternational.com.

Wernham, C.V., Toms, M.P., Marchant, J.H., Clark, J.A., Siriwardena, G.M. and Baillie, S.R. (eds) (2002) *The Migration Atlas: movements of the birds of Britain and Ireland*. London: T. and A.D. Poyser.

White, G. (1789 Revised1993) *The Natural History of Selborne*. London: Thames and Hudson.

Yalden, D.W. and Alborella, U. (2009) *The History of British Birds*. Oxford: Oxford University Press.

THE FOLLOWING WEBSITES ALSO PROVIDED MUCH USEFUL DATA

Audubon Society. Available at https://www.audubon.org

Avibase. The World Bird database Available at https://avibase.bsc-eoc.org

Birdguides. Available at https://www.birdguides.com

Birdlife International. Available at https://www.birdlife.org

The British Trust for Ornithology. Available at https://www.bto.org

Ebird. Available at https://ebird.org

IOC World Bird List, v 13.2. Available at https://www.worldbirdnames.org

Oiseaux.net (in English and French). Available at: https://www.oiseaux.net/birds

Wikipedia. Available at https://en.wikipedia.org/wiki

Photographic captions and credits

The pictures were all taken by the author, with the exception of the Hoopoe, which was generously contributed by my former colleague and regular birding companion, Adrian Brown.

Mute Swan *Cygnus olor* at Whiteknights Park, Reading.

Turtle Dove *Streptopelia turtur* at RSPB Otmoor, Oxfordshire.

Great Crested Grebe *Podiceps cristatus* at Moor Green Lakes Nature Reserve, Berkshire.

Pied Avocet *Recurvirostra avosetta* at Titchfield Haven Nature Reserve, Hampshire.

Common Greenshank *Tringa nebularia* at Moor Green Lakes Nature Reserve, Berks.

Common Gull *Larus canus* on the beach at RSPB Titchwell, Norfolk.

Great Northern Diver *Gavia immer* at Farmoor Reservoir, Oxfordshire.

Little Egret *Egretta garzetta* at Langstone Harbour, Hampshire.

Red Kite *Milvus milvus* over Earley, Berkshire.

Eurasian Hoopoe *Upupa epops* photographed in Sri Lanka by Adrian Brown.

Western Jackdaw *Coloeus monedula* at All Saints' Church at Swallowfield, Berkshire.

Common Chiffchaff *Phylloscopus collybita* at RSPB Otmoor, Oxfordshire.

Eurasian Wren *Troglodytes troglodytes* at Moor Green Lakes Nature Reserve, Berkshire.

Fieldfare *Turdus pilarus* a rare hard-weather visitor to the author's Berkshire garden.

Grey Wagtail *Motacilla cinerea* at Swallowfield Park, Berkshire.

Corn Bunting *Emberiza calandra* on the Oxfordshire Downs.

INDEX OF VERNACULAR NAMES

INDEX OF SCIENTIFIC NAMES

Printed in the USA
CPSIA information can be obtained
at www.ICGtesting.com
CBHW021939220324
5728CB00001B/4